T0300963

STOCHASTIC RESONANCE

From Suprathreshold Stochastic Resonance to Stochastic Signal Quantization

Stochastic resonance occurs when random noise provides a signal processing benefit, and has been observed in many physical and biological systems. Many aspects have been hotly debated by scientists for nearly 30 years, with one of the main questions being whether biological neurons utilise stochastic resonance. This book addresses in detail various theoretical aspects of stochastic resonance with a focus on its extension to suprathreshold stochastic resonance, in the context of stochastic signal quantization theory. Models in which suprathreshold stochastic resonance occur support significantly enhanced "noise benefits", and exploitation of the effect may prove extremely useful in the design of future engineering systems such as distributed sensor networks, nano-electronics, and biomedical prosthetics.

For the first time, this book reviews and systemizes the topic in a way that brings clarity to a range of researchers from computational neuroscientists through to electronic engineers. To set the scene, the initial chapters review stochastic resonance and outline some of the controversies and debates that have surrounded it. The book then discusses the suprathreshold stochastic resonance effect as a form of stochastic quantization. Next, it considers various extensions, such as optimization and tradeoffs for stochastic quantizers. On the topic of biomedical prosthetics, the book culminates in a chapter on the application of suprathreshold stochastic resonance to the design of cochlear implants. Each chapter ends with a review summarizing the main points, and open questions to guide researchers into finding new research directions.

STOCHASTIC RESONANCE

From Suprathreshold Stochastic Resonance to Stochastic Signal Quantization

MARK D. McDONNELL

Research Fellow
University of South Australia
and
The University of Adelaide

NIGEL G. STOCKS

Professor of Engineering
University of Warwick

CHARLES E.M. PEARCE

Elder Professor of Mathematics
The University of Adelaide

DEREK ABBOTT

Professor of Electrical and Electronic Engineering
The University of Adelaide

CAMBRIDGE
UNIVERSITY PRESS

CAMBRIDGE
UNIVERSITY PRESS

Shaftesbury Road, Cambridge CB2 8EA, United Kingdom

One Liberty Plaza, 20th Floor, New York, NY 10006, USA

477 Williamstown Road, Port Melbourne, VIC 3207, Australia

314–321, 3rd Floor, Plot 3, Splendor Forum, Jasola District Centre, New Delhi – 110025, India

103 Penang Road, #05–06/07, Visioncrest Commercial, Singapore 238467

Cambridge University Press is part of Cambridge University Press & Assessment, a department of the University of Cambridge.

We share the University's mission to contribute to society through the pursuit of education, learning and research at the highest international levels of excellence.

www.cambridge.org
Information on this title: www.cambridge.org/9780521882620

First published 2008
First paperback edition 2012

A catalogue record for this publication is available from the British Library

Library of Congress Cataloging-in-Publication data
Stochastic resonance : from suprathreshold stochastic resonance to stochastic signal quantization / Mark D. McDonnell . . . [et al.].
p.   cm.
Includes bibliographical references and index.
ISBN 978-0-521-88262-0
1. Stochastic processes.   2. Signal processing.   3. Mathematical physics.
I. McDonnell, Mark D., 1975–   II. Title.
QC20.7.S8S78   2008
621.382′2–dc22
2008024827

ISBN   978-0-521-88262-0   Hardback
ISBN   978-1-107-41132-6   Paperback

In memory of Charles Edward Miller Pearce (1940 -2012)

Contents

Figures

List of figures

Tables

Preface

Quantization of a signal or data source refers to the division or classification of that source into a discrete number of categories or states. It occurs, for example, when analogue electronic signals are converted into digital signals, or when data are binned into histograms. By definition, quantization is a lossy process, which compresses data into a more compact representation, so that the number of states in a quantizer's output is usually far fewer than the number of possible input values.

Most existing theory on the performance and design of quantization schemes specifies only deterministic rules governing how data are quantized. By contrast, *stochastic quantization* is a term intended to pertain to quantization where the rules governing the assignment of input values to output states are stochastic rather than deterministic. One form of stochastic quantization that has already been widely studied is a signal processing technique called *dithering*. However, the stochastic aspect of dithering is usually restricted so that it is equivalent to adding random noise to a signal *prior* to quantization. The term *stochastic quantization* is intended to be far more general, and applies to the situation where the rules of the quantization process are stochastic.

The inspiration for this study comes from a phenomenon known as *stochastic resonance*, which is said to occur when the presence of noise in a system provides a better performance than the absence of noise. Specifically, this book discusses a particular form of stochastic resonance – discovered by Stocks – known as *suprathreshold stochastic resonance*, and demonstrates how and why this effect is a form of stochastic quantization.

The motivation is two-fold. First, stochastic resonance has been observed in many forms of neuron and neural systems, both in models and in real physiological experiments. The model in which suprathreshold stochastic resonance occurs – sometimes called a *pooling network* – was designed to model a population of neurons, rather than a single neuron. Unlike single neurons, the suprathreshold stochastic resonance model supports stochastic resonance for input signals that are not entirely or predominantly subthreshold. Hence, it has been conjectured that the suprathreshold stochastic resonance effect is utilized by populations of neurons to encode noisy sensory information, for example in the cochlear nerve.

Second, although stochastic resonance has been observed in many different systems, in a wide variety of scientific fields, to date very few applications inspired by stochastic resonance have been proposed. One of the reasons for this is that in many circumstances utilizing stochastic resonance to improve a system is dismissed as suboptimal when compared with optimizing that system to operate without requiring stochastic resonance. This is because, given a system or device that is *a priori* nonlinear, a designer has the choice of (i) trying to operate it in a quasi-linear regime by locking operation to a near-linear part of the input–output transfer function, or (ii) allowing the system to be operated throughout its full nonlinear characteristic. The first alternative has historically been preferred, because linear systems are far easier to understand and analyse. However, if the system is allowed to run freely, then it is possible to utilize stochastic resonance, in a very carefully controlled manner, to enhance performance. Although adjusting certain parameters in the nonlinear system other than noise may be the preferred option, sometimes this cannot be achieved, and utilizing noise via stochastic resonance can provide a benefit. Given that stochastic resonance is so widespread in nature, and that many new technologies have been inspired by natural systems – particularly biological systems – new applications incorporating aspects of stochastic resonance may yet prove revolutionary in fields such as distributed sensor networks, nano-electronics, and electronic biomedical prosthetics.

As a necessary step towards confirming the above two hypotheses, this book addresses in detail for the first time various theoretical aspects of stochastic quantization, in the context of the suprathreshold stochastic resonance effect. The original work on suprathreshold stochastic resonance considers the effect from the point of view of an information channel. This book comprehensively reviews all such previous work. It then makes several extensions: first, it considers the suprathreshold stochastic resonance effect as a form of stochastic quantization; second, it considers stochastic quantization in a model where all threshold devices are not necessarily identical, but are still independently noisy; and, third, it considers various constraints and tradeoffs in the performance of stochastic quantizers. To set the scene, the initial chapters review stochastic resonance and outline some of the controversies and debates that have surrounded it.

Foreword

Due to the multidisciplinary nature of stochastic resonance the Foreword begins with a commentary from Bart Kosko representing the engineering field and ends with comments from Sergey M. Bezrukov representing the biophysics field. Both are distinguished researchers in the area of stochastic resonance and together they bring in a wider perspective that is demanded by the nature of the topic.

The authors have produced a breakthrough treatise with their new book *Stochastic Resonance*. The work synthesizes and extends several threads of noise-benefit research that have appeared in recent years in the growing literature on stochastic resonance. It carefully explores how a wide variety of noise types can often improve several types of nonlinear signal processing and communication. Readers from diverse backgrounds will find the book accessible because the authors have patiently argued their case for nonlinear noise benefits using only basic tools from probability and matrix algebra.

Stochastic Resonance also offers a much-needed treatment of the topic from an engineering perspective. The historical roots of stochastic resonance lie in physics and neural modelling. The authors reflect this history in their extensive discussion of stochastic resonance in neural networks. But they have gone further and now present the exposition in terms of modern information theory and statistical signal processing. This common technical language should help promote a wide range of stochastic resonance applications across engineering and scientific disciplines. The result is an important scholarly work that substantially advances the state of the art.

Professor Bart Kosko
Department of Electrical Engineering, Signal and Image Processing Institute, University of Southern California, December 2007

A book on stochastic resonance (SR) that covers the field from suprathreshold stochastic resonance (SSR) to stochastic signal quantization is a long-anticipated major event in the world of signal processing.

Written by leading experts in the field, it starts with a didactic introduction to the counterintuitive phenomenon of stochastic resonance – the noise-induced increase of order – complete with a historical review and list of controversies and debates. The book then quickly advances to the hot areas of signal quantization, decoding, and optimal reconstruction.

The book will be indispensable for both students and established researchers who need to navigate through the modern sea of stochastic resonance literature. With the significance of the subject growing as we advance in the direction of nanotechnologies, wherein ambient fluctuations play an ever-increasing role, this book is bound to become an influential reference for many years to come.

Sergey M. Bezrukov
National Institutes of Health (NIH)
Bethesda, Washington DC, USA
July 2007

Acknowledgments

Putting this book together was a lengthy process and there are many people to thank. The field of stochastic resonance (SR) is one that has been rapidly evolving in recent years and has often been immersed in hot debate. Therefore we must thank all our colleagues in the SR community for providing the springboard for this book.

Many discussions have been influential in crystallizing various matters, and we especially would like to acknowledge Andrew G. Allison, Pierre-Olivier (Bidou) Amblard, David Allingham, Said F. Al-Sarawi, Matthew Bennett, Sergey M. Bezrukov, Robert E. Bogner, A. N. (Tony) Burkitt, Aruneema Das, Paul C. W. Davies, Bruce R. Davis, Simon Durrant, Alex Grant, Doug Gray, David Grayden, Leonard T. Hall, Priscilla (Cindy) Greenwood, Greg P. Harmer, David Haley, Laszlo B. Kish, Bart Kosko, Riccardo Mannella, Ferran Martorell, Robert P. Morse, Alexander Nikitin, Thinh Nguyen, David O'Carroll, Al Parker, Juan M. R. Parrondo, Antonio Rubio, Aditya Saha, John Tsimbinos, Lawrence M. Ward, and Steeve Zozor.

Special thanks must go to Adi R. Bulsara and Matthew J. Berryman who read early drafts and provided comments that greatly improved clarity. Also a special thanks is due to Withawat Withayachumnankul for his wonderful graphics skills and assistance with some of the figures.

Finally, we would like to thank Simon Capelin, our editor at Cambridge University Press, for his patience and overseeing this project through to completion, and assistant editor Lindsay C. Barnes for taking care of the manuscript.

1

Introduction and motivation

1.1 Background and motivation

We begin by briefly outlining the background and motivation for this book, before giving an overview of each chapter, and pointing out the most significant questions addressed.

Although the methodology used is firmly within the fields of signal processing and mathematical physics, the motivation is interdisciplinary in nature.

The initial open questions that inspired this direction were:

(i) How might neurons make use of a phenomenon known as *stochastic resonance*?
(ii) How might a path towards engineering applications inspired by these studies be initiated?

Stochastic resonance and sensory neural coding

Stochastic resonance (SR) is a counter-intuitive phenomenon where the presence of *noise* in a nonlinear system is essential for optimal system performance. It is not a technique. Instead, it is an effect that might be observed and potentially exploited or induced. It has been observed to occur in many systems, including in both neurons and electronic circuits.

A motivating idea is that since we know the brain is far better at many tasks compared to electronic and computing devices, then maybe we can learn something from the brain. If we can ultimately better understand the possible exploitation of SR in the brain and nervous system, we may also be able to improve aspects of electronic systems.

Although it is important to have an overall vision, in practical terms it is necessary to consider a concrete starting point. This book is particularly focused on an exciting new development in the field of SR, known as *suprathreshold stochas-tic resonance* (SSR) (Stocks 2000c). Suprathreshold stochastic resonance occurs in a parallel array of simple threshold devices. Each individual threshold device receives the same signal, but is subject to *independent* additive random noise. The output of each device is a binary signal, which is unity when the input is greater than the threshold value, and zero otherwise. The overall output of the *SSR model* is the *sum* of the individual binary signals. Originally, all threshold devices were considered to have the same threshold value.

Early studies into SSR considered the effect from the point of view of infor-mation transmission and the model in which SSR occurs as a communication channel. Furthermore, the model in which SSR occurs was originally inspired by questions of sensory neural coding in the presence of noise. Unlike previously studied forms of SR, either in neurons or simple threshold-based systems, SR can occur in such a model for signals that are not entirely or predominantly subthresh-old, and for both very small and quite large amounts of noise. Although each threshold device in the SSR model is very simple in comparison with more bio-logically realistic neural models, such devices have actually been used to model neurons, under the name of the McCulloch–Pitts neural model (McCulloch and Pitts 1943).

The aim of this book is to comprehensively outline all known theoretical and numerical results on SSR and to extend this theory. It is anticipated that the land-scape presented here will form a launching pad for future research into the specific role that SSR may play in real neural coding. Note that the goal is not to prove that living systems actively exploit SR or SSR – these are ongoing research areas in the domain of neurophysiology and biophysics. Rather, our starting point is the theoretical mathematical underpinning of SR-type effects in the very simple McCulloch–Pitts model.

This is in line with the time-honoured approach in physics and engineering; namely, to begin with the simplest possible model. As this book unfolds, it will be seen that the analysis of SR in arrays of such simplified neural models gives rise to rich complex phenomena and also to a number of surprises. Explanation, or indeed discovery, of these effects would not have been tractable if the starting point were an intricate neural model. Analysis of the simple model lays the foundation for adding further complexities in the future.

The intention is that the mathematical foundation provided by this book will assist future neurophysiologists in asking the right questions and in performing the right experiments when establishing if real neurons actively exploit SSR. In the meantime, this book also contributes to the application of SSR in artificial neu-ral and electronic systems. To this end, the book culminates in a chapter on the application of SSR to electronic cochlear implants.

Low signal-to-noise ratio systems and sensor networks

Another motivation for this research is the important problem of overcoming the effects of noise in sensors and signal and data processing applications. For example, microelectronics technologies are shrinking, and are beginning to approach the nanoscale level (Martorell *et al.* 2005). At this scale, device behaviour can change and noise levels can approach signal levels. For such small signal-to-noise ratios (SNRs), it may be impossible for traditional noise reduction methods to operate, and it may be that optimal circuit design needs to make use of the effects of SR. Two experimental examples illustrating this possibility include the demonstrations of SR in (i) carbon nanotube transistors for the first time in 2003 (Lee *et al.* 2003, Lee *et al.* 2006), and (ii) a silicon nanoscale mechanical oscillator as reported in 2005 in the journal *Nature* (Badzey and Mohanty 2005, Bulsara 2005).

A second area of much current research is that of distributed sensor networks (Akyildiz *et al.* 2002, Pradhan *et al.* 2002, Chong and Kumar 2003, Iyengar and Brooks 2004, Martinez *et al.* 2004, Xiong *et al.* 2004). Of particular interest to this book is the problem where it is not necessarily a network of complete sensors that is distributed, but it is actually the data acquisition, or compression, that is distributed (Berger *et al.* 1996, Draper and Wornell 2004, Pradhan *et al.* 2002, Xiong *et al.* 2004, Pradhan and Ramchandran 2005). A key aspect of such a scenario is that data are acquired from a number of independently noisy sources that do not cooperate, and are then fused by some central processing unit. In the information theory and signal processing literature, this is referred to as *distributed source coding* or *distributed compression*.

Given that the SSR effect overcomes a serious limitation of all previously studied forms of SR, a complete theoretical investigation of its behaviour may lead to new design approaches to low SNR systems or data acquisition and compression in distributed sensor networks.

1.2 From stochastic resonance to stochastic signal quantization

During the course of writing this book, it became apparent that the SSR effect is equivalent to a noisy, or stochastic, *quantization* of a signal. Consequently, as well as describing its behaviour from the perspective of information transmission, it is equally valid to describe it from the perspective of *information compression* or, more specifically, *lossy compression*. Note that quantization of a signal is a form of lossy compression (Widrow *et al.* 1996, Berger and Gibson 1998, Gray and Neuhoff 1998).

The distinguishing feature of SSR, that sets it apart from standard forms of quantization, is that conventionally the rules that specify a quantizer's operation are considered to be fixed and deterministic. In contrast, when the SSR effect is viewed as quantization, the governing rules lead to a set of parameters that are independent random variables. Hence, we often refer to the SSR model's output as a *stochastic quantization*.

Given this perspective, there are three immediate questions that can be asked:

- Can we describe the SSR effect in terms of conventional quantization and compression theory?
- Given that a central SR question is that of finding the optimal noise conditions, what noise intensity optimizes the performance of the SSR model when it is described as a quantizer?
- How good is the SSR effect at quantization when compared with conventional quantizers?

The underlying theme of this book is to address these three questions.

1.3 Outline of book

This book consists of ten chapters, as follows:

- Chapter 1, the current chapter, provides the background and motivation for the work described in this book.
- Chapter 2 contains an overview of the historical landscape against which this book is set. It defines *stochastic resonance* as it is most widely understood, and gives a broad literature review of SR, with particular emphasis on aspects relevant to quantization. Chapter 2 is deliberately sparse in equations and devoid of quantitative results, but does provide qualitative illustrations of how SR works. It also provides some discussion that is somewhat peripheral to the main scope of this book, but that will prove useful for readers unfamiliar with, or confused about, SR.
- Chapter 3 contains the information-theoretic definitions required for the remainder of this book, and discusses the differences between *dithering* and stochastic resonance.
- Chapters 4 and 5 begin the main focus of the book, by defining the SSR model, giving a detailed literature review of all previous research on SSR, and replicating all the most significant theoretical results to date. These two chapters consider only the original concept of the SSR model as a communications channel. In particular, we examine how the mutual information between the input and output of the SSR model varies with noise intensity. A subset of these results pertains specifically to a large number of individual threshold devices in the SSR model. Chapter 5 is devoted to elaborating on results in this area, as well as developing new results in this area, while Chapter 4 focuses on more general behaviour.
- Chapters 6 and 7 contain work on the description of the SSR model as a quantizer. There are two main aspects to such a description. First, quantizers are specified by two operations: an encoding operation and a decoding operation. The encoding operation assigns ranges of values of the quantizer's input signal to one of a finite number of

output states. The decoding operation approximately reconstructs the original signal by assigning 'reproduction values' to each encoded state. In contrast to conventional quantizers, the SSR model's encoding is stochastic, as the output state for given input signal values is nondeterministic. However, it is possible to decode the SSR output in a similar manner to conventional quantizers, and we examine various ways to achieve this. The second aspect we consider is the performance of the decoded SSR model. Since the decoding is designed to approximate the original signal, performance is measured by the average properties of the error between this original signal and the quantizer's output approximation. Conventionally, mean square error distortion is used to measure this average error, and we examine in detail how this measure varies with noise intensity, the decoding scheme used, and the number of threshold devices in the SSR model. As with Chapters 4 and 5, Chapter 6 focuses on general behaviour, while Chapter 7 is devoted to discussion of the SSR model in the event of a large number of individual threshold devices.

- Chapter 8 expounds work that extends the SSR model beyond its original specification. We relax the constraint that all individual threshold devices must have identical threshold values, and allow each device to have an arbitrary threshold value. We then consider how to optimally choose the set of threshold values as the noise intensity changes. The most important result of this study is the numerical demonstration that the SSR model – where all threshold devices have the same threshold value – is optimal, for sufficiently large noise intensity.

- Motivated by recent neural-coding research, Chapter 9 further extends the SSR model, by including a constraint on the energy available to the system. The performance of a quantizer is characterized by two opposing factors: *rate* and *distortion*. Chapter 9 also explores the SSR model, and its extension to arbitrary thresholds, from the point of view of rate-distortion theory.

- Chapter 10 concludes the technical content of this book, by pointing to a concrete application of SSR theory, in the area of cochlear implants.

- Finally, Chapter 11 contains some speculations on the future of stochastic resonance and suprathreshold stochastic resonance.

2

Stochastic resonance: its definition, history, and debates

Stochastic resonance (SR), being an interdisciplinary and evolving subject, has seen many debates. Indeed, the term SR itself has been difficult to comprehensively define to everyone's satisfaction. In this chapter we look at the problem of defining stochastic resonance, as well as exploring its history. Given that the bulk of this book is focused on suprathreshold stochastic resonance (SSR), we give particular emphasis to forms of stochastic resonance where *thresholding* of random signals occurs. An important example where thresholding occurs is in the generation of action potentials by spiking neurons. In addition, we outline and comment on some of the confusions and controversies surrounding stochastic resonance and what can be achieved by exploiting the effect. This chapter is intentionally qualitative. Illustrative examples of stochastic resonance in threshold systems are given, but fuller mathematical and numerical details are left for subsequent chapters.

2.1 Introducing stochastic resonance

Stochastic resonance, although a term originally used in a very specific context, is now broadly applied to describe any phenomenon where the presence of internal noise or external input noise in a nonlinear system provides a better system response to a certain input signal than in the absence of noise. The key term here is *nonlinear*. Stochastic resonance cannot occur in a linear system – linear in this sense means that the output of the system is a linear transformation of the input of the system. A wide variety of performance measures have been used – we shall discuss some of these later.

The term *stochastic resonance* was first used in the context of noise enhanced signal processing in 1980 by Roberto Benzi, at the NATO International School of Climatology. Since then it has been used – according to the ISI Web of Knowledge database – in around 2000 publications, over a period of a quarter of a century. The frequency of publication, by year, of these papers is shown in Fig. 2.1. This figure

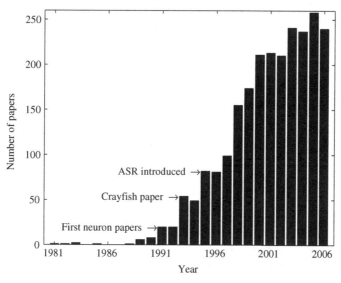

Fig. 2.1. Frequency of stochastic resonance papers by year – between 1980 and 2006 – according to the ISI Web of Knowledge database. There are several epochs in which large increases in the frequency of SR papers occurred. The first of these is between 1989 and 1992, when the most significant events were the first papers examining SR in neural models (Bulsara *et al.* 1991, Bulsara and Moss 1991, Longtin *et al.* 1991), and the description of SR by linear response theory (Dykman *et al.* 1990a). The second epoch is between about 1993 and 1996, when the most significant events were the observation of SR in physiological experiments on neurons (Douglass *et al.* 1993, Levin and Miller 1996), the popularization of array enhanced SR (Lindner *et al.* 1995, Lindner *et al.* 1996), and of aperiodic stochastic resonance (ASR) (Collins *et al.* 1995a). Around 1997, a steady increase in SR papers occurred, as investigations of SR in neurons and ASR became widespread.

illustrates how the use of the term *stochastic resonance* expanded rapidly in the 1990s, and is continuing to expand in the 2000s.

The 'resonance' part of 'stochastic resonance' was originally used because the signature feature of SR is that a plot of output signal-to-noise ratio (SNR) has a single maximum for some nonzero input noise intensity. Such a plot, as shown in Fig. 2.2, has a similar appearance to frequency dependent systems that have a maximum SNR, or output response, for some *resonant frequency*. However, in the case of SR, the resonance is 'noise induced', rather than at a particular frequency – see Section 2.3 for further discussion.

SR has been the subject of many reviews, including full technical journal articles (Jung 1993, Moss *et al.* 1994, Dykman *et al.* 1995, Gammaitoni *et al.* 1998, Wiesenfeld and Jaramillo 1998, Luchinsky *et al.* 1999a, Luchinsky *et al.* 1999b, Anishchenko *et al.* 1999, Hänggi 2002, Harmer *et al.* 2002, Wellens *et al.* 2004,

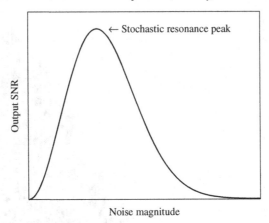

Fig. 2.2. Typical curve of output SNR vs. input noise magnitude, for systems capable of stochastic resonance. For small and large noise values, the output SNR is very small, while some intermediate nonzero noise value provides the maximum output SNR.

Shatokhin *et al.* 2004, Moss *et al.* 2004), editorial works (Bulsara *et al.* 1993, Astumian and Moss 1998, Petracchi *et al.* 2000, Abbott 2001, Gingl 2002), book chapters (Wiesenfeld 1993a), and magazine articles (Wiesenfeld 1993b, Moss and Wiesenfeld 1995, Bulsara and Gammaitoni 1996).

Some of the most influential workshops and conferences on SR include those held in San Diego, USA, in 1992 (published as a special issue of the journal *Il Nuovo Cimento*) (Moss *et al.* 1992) and 1997 (Kadtke and Bulsara 1997), Elba Island, Italy, in 1994 (published as a special issue of *Journal of Statistical Physics*) (Bulsara *et al.* 1994), and Ambleside, UK, in 1999 (Broomhead *et al.* 2000). Various other special issues of journals have been devoted to SR, such as the September 1998 issue of *Chaos*, the September 2000 issue of *Chaos Solitons and Fractals*, and the September 2002 issue of *Fluctuation and Noise Letters*.

There have been articles and letters on SR published in the prestigious journal, *Nature* (Douglass *et al.* 1993, Moss *et al.* 1993, Wiesenfeld and Moss 1995, Moss and Pei 1995, Bezrukov and Voydanoy 1995, Collins *et al.* 1995b, Noest 1995, Collins *et al.* 1995c, Levin and Miller 1996, Bezrukov and Voydanoy 1997b, Astumian *et al.* 1997, Dykman and McClintock 1998, Collins 1999, Russell *et al.* 1999, Moss and Milton 2003, Bulsara 2005, Badzey and Mohanty 2005). A book has been written exclusively on SR (Andò and Graziani 2000), as well as a large section of another book (Anischenko *et al.* 2002), and sections on SR written in popular science books (von Baeyer 2003, Kosko 2006). It has also started to appear in more general textbooks (Gerstner and Kistler 2002).

SR has been widely observed throughout nature – it has been quantified in such diverse systems as climate models (Benzi *et al.* 1982), electronic circuits (Fauve and Heslot 1983, Anishchenko *et al.* 1994), differential equations (Benzi *et al.* 1985, Hu *et al.* 1990), ring lasers (McNamara *et al.* 1988), semiconductor lasers (Iannelli *et al.* 1994), neural models (Bulsara *et al.* 1991, Longtin *et al.* 1991), physiological neural populations (Douglass *et al.* 1993), chemical reactions (Leonard and Reichl 1994), ion channels (Bezrukov and Voydanoy 1995), SQUIDs (Superconducting Quantum Interference Devices) (Hibbs *et al.* 1995), the behaviour of feeding paddlefish (Russell *et al.* 1999, Freund *et al.* 2002), ecological models (Blarer and Doebeli 1999), cell biology (Paulsson and Ehrenberg 2000, Paulsson *et al.* 2000), financial models (Mao *et al.* 2002), psychophysics (Ward *et al.* 2002), carbon-nanotube transistors (Lee *et al.* 2003, Lee *et al.* 2006), nanomechanical oscillators (Badzey and Mohanty 2005, Bulsara 2005), and even social systems (Wallace *et al.* 1997).

The first highly successful application inspired by SR involves the use of electrically generated subthreshold stimuli in biomedical prosthetics to improve human balance control and somatosensation (Priplata *et al.* 2004, Priplata *et al.* 2003, Collins *et al.* 2003, Moss and Milton 2003, Harry *et al.* 2005). This work led to James J. Collins winning a prestigious MacArthur Fellowship in October 2003 (Harry *et al.* 2005).

Prior to this, Collins was also an author on a correspondence to *Nature* (Suki *et al.* 1998) on a noise-enhanced application, described as analogous to SR. The *Nature* paper used a model to verify the experimental results of Lefevre *et al.* (1996). Random noise was introduced into the operation of mechanical life-support ventilators, in order to more closely replicate natural breathing. It was found that added noise enhanced artificial ventilation in several ways. See Brewster *et al.* (2005) for a review.

2.2 Questions concerning stochastic resonance

There are a number of misconceptions and controversies about stochastic resonance. The following list of questions attempts to encapsulate the main points of contention:

(i) What is the definition of stochastic resonance?
(ii) Is stochastic resonance exploited by the nervous system and brain as part of the neural code?
(iii) Does stochastic resonance occur only if a signal's power is weak compared with the power of the noise in a system?
(iv) Can stochastic resonance lead to a signal-to-noise ratio gain?
(v) Was stochastic resonance known prior to the first use in 1981 of the term 'stochastic resonance'?
(vi) How and when is stochastic resonance different from a signal processing technique called *dithering*?

Although question (ii) is quite clearly the interesting scientific question, and seemingly the motivation behind much SR research, it would appear the other questions in the above list have sometimes provided a diversion. The problem is that reaching a consensus on the answers to questions (ii)–(vi) really depends on an agreed answer to question (i).

The broadest possible definition of stochastic resonance is that it occurs when randomness may have some positive role in a signal processing context. Given this definition, *we believe* that the answers to these questions are: (ii) yes, although it is difficult to prove, the brain – or at least, some parts of the nervous system – would almost certainly not function as it does if it were completely deterministic; (iii) no, randomness can have a positive role even if it is only a small amount of randomness; (iv) yes, in the information-theoretic sense, random noise in a system can lead to a less noisy output signal, provided that the system is nonlinear; (v) yes, randomness has been known to have a positive role in many circumstances for decades, if not centuries; and (vi) stochastic resonance occurs when dithering is used – dithering can be described as the exploitation of SR.

On the other hand, if the definition of stochastic resonance is restricted to its original narrow context, then the answers to questions (ii)–(vi) change to: (ii) maybe – this is yet to be conclusively answered, (iii) yes, (iv) no, (v) no, and (vi) dithering is quite different from SR, as is comprehensively explained in Gammaitoni (1995a).

This discussion is intended to illustrate that the debate on the topics listed above can depend crucially on what we mean by *stochastic resonance*. In the remainder of this chapter we discuss some of these issues in more detail in order to bring some clarity to the debate and to illustrate the pitfalls and controversies for readers new to the stochastic resonance field, or who have held only a peripheral interest in the area.

2.3 Defining stochastic resonance

SR is often described as a counter-intuitive phenomenon. This is largely due to its historical background; in the first decade and a half since the coining of the term in 1980, virtually all research into SR considered only systems driven by a combination of a periodic single-frequency input signal and broadband noise. In such systems, a natural measure of system performance is the output signal-to-noise ratio (SNR), or, more precisely, often the ratio of the output power spectral density (PSD) at the input frequency to the output noise floor PSD measured with the signal present. The noise floor is measured with the signal present, rather than absent, as the output noise may change if the signal is not present. This is because the signal and output noise are generally not additive in a nonlinear system, or, in other words, the output noise is signal dependent.

In linear systems driven by periodic input signals – in fact, any linear system – it is well known that the output SNR is maximized in the absence of noise. When systems are analyzed in terms of SNR, it is the norm to assume that noise is a problem, usually with good reason. This means that observations of the presence of noise in a system providing the maximum output SNR are often seen to be highly counter-intuitive – see Kosko (2006, p. 149) for further discussion.

However, although not all noise can be described as a random process – it can, for example, be constant or deterministic (even chaotic) – SR research has tended to focus on the stochastic case. The most common assumption is that the noise is Gaussian distributed, and white – that is, constant in power across all frequencies.

When it is noted that there are many examples of systems or algorithms where randomness is of benefit, SR does not seem quite so counter-intuitive. Such examples include:

- Brownian ratchets (Doering 1995) – mechanical applications of this idea include self-winding (batteryless) wristwatches (Paradiso and Starner 2005);
- dithering in signal processing and analogue-to-digital conversion (Schuchman 1964, Gray and Stockham 1993, Dunay *et al.* 1998, Wannamaker *et al.* 2000b);
- Parrondo's games – the random combination of losing games to produce a winning game (Harmer and Abbott 1999);
- noise-induced linearization (Yu and Lewis 1989, Dykman *et al.* 1994), noise-induced stabilization (Toral *et al.* 1999, Basak 2001), noise-induced synchronization (Neiman 1994), noise-enhanced phase coherence (Neiman *et al.* 1998), and noise-induced order (Matsumoto and Tsuda 1985);
- the use of mixed (probabilistic) optimal strategies in game theory (von Neumann and Morgenstern 1944);
- random switching to control electromagnetic compatibility performance (Allison and Abbott 2000);
- random search optimization techniques, including genetic algorithms (Gershenfeld 1999) and simulated annealing (Kirkpatrick *et al.* 1983);
- random noise radars – that is, radars which transmit random noise waveforms in order to provide immunity from jamming, detection, and interference (Narayanan and Kumru 2005);
- techniques involving the use of Brownian motion for solving nonstochastic partial differential equations, such as the Dirichlet problem (Øksendal 1998);
- stochastic iterative decoding in error control coding applications (Gaudet and Rapley 2003);
- estimation of linear functionals (Plaskota 1996a, 1996b).

Further discussion and other examples appear in Harmer and Abbott (2001), Abbott (2001), Zozor and Amblard (2005), and Kosko (2006).

SR was initially considered to be restricted to the case of periodic input signals. However, now the literature reveals that it is widely used as an all-encompassing term, whether the input signal is a periodic sine-wave, a periodic broadband signal, or aperiodic, and whether stationary, nonstationary (Fakir 1998a), or cyclostationary (Amblard and Zozor 1999). An appropriate measure of output performance depends on the task at hand, and the form of input signal. For example, for periodic signals and broadband noise, SNR is often used, while for random aperiodic signals, SR is usually measured by mutual information or correlation.

Is SR a bona fide resonance?

The definition of SR, and the word 'resonance' itself, have both been objects of debate. In particular, 'resonance' is usually thought of in the sense of a resonant frequency, rather than an optimal noise intensity. For the most conventional kinds of SR, 'resonance' was more or less resolved as being appropriate after a new way – using residence time distributions – of looking at SR found that the effect was a *bona fide* kind of resonance (Gammaitoni *et al.* 1995a), although there was some debate about this too (Choi *et al.* 1998, Giacomelli *et al.* 1999, Marchesoni *et al.* 2000). According to Gammaitoni,

Even if the word resonance has been questioned since the very beginning, it has been recently demonstrated . . . that, for a diffusion process in a double-well system, the meaning of *resonance* as the matching of two characteristic frequencies (or physical time scales) is indeed appropriate for such a phenomenon. (Gammaitoni 1995a)

Stochastic resonance, static systems, and dithering

Although Gammaitoni's work appeared to end the debate for *bistable systems* – see Section 2.4 for discussion about SR in bistable systems – it opened up a new question: Is the term *stochastic resonance* appropriate for systems consisting of simple static threshold[1] nonlinearities? There are at least two reasons why this has been debated.

First, noise-enhanced behaviour in static threshold nonlinearities also occurs in a signal processing technique known as *dithering*. Dithering involves deliberately adding a random – or pseudo-random – signal to another signal, prior to its digitization or quantization (Schuchman 1964). It is most often associated with audio

---

[1] The term 'static threshold' often appears in relation to SR research. It is used to differentiate between *dynamical* systems – such as the bistable potential wells traditionally used in studies of SR – and nondynamical, or *static*, systems (Gingl *et al.* 1995a, Gingl *et al.* 1995b). A system is called static when nonlinear deformation – SR cannot occur in a linear system – of an input signal is not governed by time evolving differential equations, but by simple rules that produce an output signal based on the instantaneous value of the input signal.

or image processing, where the effect of the added noise signal, called the *dither signal*, is to randomize the error signal introduced by quantization. This randomization, although increasing the total power of the noise at the output, reduces undesirable harmonic distortion effects introduced by quantization.

Given that dithering is a way of improving a system using the presence of noise, the question is how to distinguish it from SR? That is, if the system being studied resembles dithering, should noise-enhanced behaviour be called dithering, rather than SR? Since dithering existed for decades prior to the first studies of SR, some SR traditionalists prefer to classify dithering as a completely separate form of noise-enhanced signal processing. However, this requires a quite restrictive definition of SR, where the system must be dynamical, and the presence of noise enables a matching of two time-scales. Such a definition has been superseded in the broader literature. As discussed below, the contemporary definition of SR is such that dithering can be described as a technique that *exploits* SR, and the two terms are not mutually exclusive.

As for determining whether an experiment or application should be labelled as a form of dithering that gives rise to SR, we suggest that the most natural distinction is in the motivation. If we are interested in studying noise-enhanced behaviour in a model or experiment, and how performance varies with noise intensity – for example, examining whether SR occurs – then there is no compulsion to call this dithering. Indeed, in most cases such as this, the signal processing goal is different from that when dither is used in image and audio processing, and therefore different performance measures are appropriate. On the other hand, as implied by Wannamaker *et al.* (2000a), if we are interested in deliberately introducing a random signal prior to digitization, with the sole aim of modifying the effects of quantization noise, then this could naturally be called dithering, while the fact that the presence of noise achieves the aim is the occurrence of SR. Further discussion of this problem is presented in Section 3.4 of Chapter 3.

Secondly, the initial questions about whether noise-enhanced behaviour in static threshold systems should be called *stochastic resonance* relate to whether a *bona fide* resonance occurs, and were concisely expressed by Gammaitoni, who states that:

the use of the term resonance is questionable and the notion of noise induced threshold crossings is more appropriate

and that:

this frequency matching condition, instead, does not apply to the threshold systems we consider here. (Gammaitoni 1995a)

Although these points are certainly fair, given common definitions of the word 'resonance', we take the point of view that such questions of semantic nomenclature are no longer relevant and argue this point in the following.

Stochastic resonance: noise-enhanced signal processing

The term *stochastic resonance* is now used so frequently in the much wider sense of being the occurrence of any kind of noise-enhanced signal processing that this common usage has, by 'weight of numbers', led to a re-definition. Although many authors still define SR only in its original narrow context, where a resonance effect can be considered to be *bona fide*, in line with the evolution of languages, words or phrases often end up with a different meaning from their original roots.

We emphasize here the fact that SR occurs only in the context of *signal* enhancement, as this is the feature that sets it apart from many of the list of randomness-enhanced phenomena above, which could all be described as benefiting in some way from noise, and yet cannot all be defined in terms of an enhanced signal. Furthermore, SR is usually[2] understood to occur in systems where there are both well-defined *input* and *output* signals, and the optimal output signal, according to some measure, occurs for some nonzero level and type of noise.

Indeed, Bart Kosko in his popular science book, *Noise*, succinctly defines SR as meaning 'noise benefit' (Kosko 2006). Implied in this definition – see Kosko (2006, pp. 148–149) – is the caveat that a *signal* should be involved, meaning that SR is 'noise benefit in a signal processing system', or alternatively 'noise-enhanced signal processing'. Put another way, SR occurs when the output signal from a system provides a better representation of the input signal than it would in the complete absence of noise.

With this definition in mind, we now provide a broad history of SR research.

2.4 A brief history of stochastic resonance

The early years: 1980–1992

The term *stochastic resonance* was first used[3] in the open literature – at least, in the context of a noise-optimized system – by Benzi, Sutera, and Vulpiani, in 1981, as a name for the mechanism they suggested is behind the periodic behaviour of the

---

[2] An early paper on SR examined a system in which SR is said to occur that did not have any input signal (Gang *et al.* 1993).

[3] A search in the Inspec database for the phrase 'stochastic resonance' returns a number of published papers prior to 1981, commencing in 1973 (Frisch *et al.* 1973). However, these all use the term 'stochastic resonance' in the context of 'stochastic wave parametric resonance', 'stochastic magnetic resonance', or other stochastic systems, where the term 'resonance' has nothing to do with noise benefits.

Earth's ice ages (Benzi *et al.* 1981, Benzi *et al.* 1982, Benzi *et al.* 1983). The term was apparently also mentioned one year earlier by Benzi (Abbott 2001) in discussions at a workshop on climatic variations and variability (Berger 1980). Climate records show that the period for the Earth's climate switching between ice ages and warmer periods is around 100 000 years. This also happens to be the period of the eccentricity of the Earth's orbit. However, current theories suggest that the eccentricity is not enough to cause such dramatic changes in climate. Benzi *et al.* (1981) – and, independently, Nicolis (1981), Nicolis (1982) – suggested that it is the combination of stochastic perturbations in the Earth's climate, along with the changing eccentricity, which is behind the ice age cycle. Benzi *et al.* (1981) gave this mechanism the name *stochastic resonance*. In its early years, the term was defined only in the very specific context of a bistable system driven by a combination of a periodic force and random noise. Note that Benzi *et al.* (1981) considered the Earth's orbital eccentricity to be a periodic driving signal, and the stochastic perturbations as the random noise. Interestingly, as pointed out in Hohn (2001), this theory for explaining the ice ages is still a subject of debate, even though SR is now well established as a general phenomenon in a huge variety of other systems.

Further mathematical investigations of SR in the following few years included the demonstration of SR in a simple two-state model (Eckmann and Thomas 1982), the quantification of SR in the Landau–Ginzberg equation (Benzi *et al.* 1985) – which is a partial differential equation – and the observation of SR in a system undergoing a Hopf bifurcation (Coullet *et al.* 1985).

Experimental observations of SR in physical systems also came quickly. In 1983, SR was reported in a Schmitt trigger electronic circuit (Fauve and Heslot 1983). Three years later, SR was observed in a bidirectional ring laser, where the deliberate addition of noise was shown to lead to an improved output SNR (McNamara *et al.* 1988, Vemuri and Roy 1989). Both the Schmitt trigger circuit and the bidirectional ring laser are bistable systems; it was originally thought that bistability is a necessary condition for SR (McNamara and Wiesenfeld 1989).

The ring laser paper brought about a large increase in interest in SR – approximately 50 published journal papers from 1989 to 1992 – with a number of theoretical treatments being published in the next few years, for example Gammaitoni *et al.* (1989a), Debnath *et al.* (1989), McNamara and Wiesenfeld (1989), Gammaitoni *et al.* (1989b), and Jung and Hänggi (1991). In particular, Jung and Hänggi (1991) provide a highly cited theoretical study that extends the results of McNamara and Wiesenfeld (1989) to show that SR can lead to peaks in the output PSD at harmonics of the driving frequency. This period also included the important realization that SR, in the limit of relatively small signal and relatively strong noise, can be described using linear response theory (Dykman *et al.* 1990a);

see Dykman *et al.* (1995) for a review. This insight was to lead directly to the realization that neither bistability nor a threshold was necessary for SR to occur, and it
could, for example, occur in monostable systems (Stocks *et al.* 1993).

   Incidentally, it has been suggested that for bistable systems the mechanism of
SR has been known about for over 75 years. According to Dykman *et al.* (1995)
and Luchinsky *et al.* (1998), the work of prolific Nobel Prize-winning chemist,
Peter Debye, on the dielectric properties of polar molecules in a solid (Debye
1929), effectively shows SR behaviour. This fact is not entirely clear from a reading of Debye (1929), as a formula for SNR is not derived, and there is no comment
made that some measure is optimized by a nonzero noise intensity; nevertheless, the relevant page is 105, and further explanation can be found in Dykman *et al.*
(1995, pp. 669–671) and in Luchinsky *et al.* (1998, pp. 930–933). See also
Luchinsky *et al.* (1999a). Dykman *et al.* (1995) also comment that the work of
Snoek in the same field in the 1940s forms part of the 'prehistory of stochastic
resonance'. More recently, Kalmykov *et al.* (2004) pointed out that another work
of Debye can be related to SR, stating that

it is possible to generalize the Debye–Fröhlich model of relaxation over a potential barrier
. . . and so to estimate the effect of anomalous relaxation on the stochastic resonance effect.

However, this does not imply conclusively that Debye knew of SR, only that his
work has been generalized to show SR.

Expansion: 1993–1996

The first important milestone in the period from 1993 to 1996 was the initial investigation into SR in neural and excitable systems. Prior to this time SR was only
observed in bistable systems. The second important milestone was the extension
of SR from periodic to aperiodic driving signals. We shortly discuss both of these
developments in more detail.

   Further extensions that straddle the initial and expansion periods include analysis of different sorts of noise from the standard additive Gaussian white noise,
including multiplicative noise (Dykman *et al.* 1992), coloured noise (Hänggi *et al.*
1993), $1/f$ noise (Kiss *et al.* 1993), and harmonic noise (Neiman and Schimansky-
Geier 1994), as well as the observation that coupling together more than one
SR-capable device can lead to increased output performance (Wiesenfeld 1991,
Jung *et al.* 1992).

Stochastic resonance in excitable systems

An excitable system is one that has only one stable or rest state and a threshold
above which an excited state can occur, but that is not stable. The excited state

eventually decays to the rest state (Gammaitoni *et al.* 1998). Neurons are a significant example of an excitable system. The first papers on the observation of SR in neural models were published by Bulsara *et al.* (1991), Bulsara and Moss (1991), and Longtin *et al.* (1991), although an earlier paper by Yu and Lewis (1989) effectively demonstrates how noise in a neuron model linearizes the system response, a situation later described as SR. However, it was Longtin *et al.* (1991) that brought the initial attention of the broader scientific community to SR, after being featured in a *Nature* 'News and Views' article (Maddox 1991). However, research into SR in neurons and neural models only really took off in 1993, when a heavily cited *Nature* article reported the observation of SR in physiological experiments on crayfish mechanoreceptors[4] (Douglass *et al.* 1993). In the same year a heavily cited paper by Longtin (1993) on SR in neuron models also became widely known and since then many published papers examine stochastic resonance in neurons – whether in mathematical models of neurons, or in biological experiments on the sensory neurons of animals – triggering a large expansion in research into SR. Stochastic resonance in neurons is discussed further in Section 2.9. However, next we discuss the departure of SR research from its original context of periodic input signals – that is, aperiodic stochastic resonance.

Aperiodic stochastic resonance

This important extension of SR was first addressed in Hu *et al.* (1992), who examine a bistable system driven by an aperiodic input signal consisting of a sequence of binary pulses subject to noise. Extending SR to aperiodic signals is significant because, while some important signals in nature and electronic systems are periodic, very many signals are not.

Further discussion of aperiodic SR was not undertaken until 1995, when Collins *et al.* (1995a) popularized the term aperiodic stochastic resonance (ASR) in an investigation of an excitable system – a FitzHugh-Nagumo neuron model – subject to an aperiodic signal. Shortly afterwards, a letter to the journal *Nature* demonstrated the same behaviour in *arrays* of FitzHugh–Nagumo neuron models (Collins *et al.* 1995b). These two studies are amongst the most highly cited of all papers on SR. The *Nature* paper is not without criticism however, and a letter to *Nature* by Noest (1995), followed by a reply from Collins *et al.* (1995c), served only to enhance the exposure of this work to the field.

As an aside, in the abstract of Collins *et al.* (1996a), the term *aperiodic* appears to be equated with the term *broadband*. However, of course, a signal can be periodic

---

[4] Note, however, that it can be convincingly argued that these experiments do not prove that neurons utilize SR in any way. This is because both the signal and the noise were applied *externally* to the mechanoreceptors, and the fact that SR occurs only demonstrates that these cells are nonlinear. It remains an open question as to whether neurons make use of *internally* generated noise and SR effects.

and broadband, for example a periodically repeated radar chirp pulse, or a simple square wave. Equating broadband with aperiodic appears to be due to the early work in SR considering only single frequency periodic signals (Gammaitoni *et al.* 1998), usually a sine wave of the form $x(t) = A \cos(\omega t)$, where $\omega = 2\pi f$ is the frequency of the sine wave in radians per second. In general, an aperiodic signal can be considered to be broadband, but a broadband signal does not need to be aperiodic. The relevance to SR research is that the SNR measure used for single frequency signals is not applicable for either *broadband and periodic* signals or aperiodic signals.

Shortly after these initial two papers by Collins *et al.*, the same authors also demonstrated ASR in three other theoretical models: a bistable-well system, an integrate-and-fire-neuronal model, and the Hodgkin–Huxley neuronal model (Collins *et al.* 1996a); as well as in physiological experiments on rat mechanoreceptors (Collins *et al.* 1996b). In these papers, 'power-norm' measures – both the correlation between the input and output signals, and the normalized power-norm, or correlation coefficient – are used to characterize ASR, instead of SNR.

Almost simultaneously with Collins *et al.*, an alternative approach to measuring SR for aperiodic signals was proposed by Kiss (1996). By contrast, this work employs an SNR-like measure based on cross-spectral densities to show the existence of SR in simple threshold-based systems. The technique is demonstrated on a signal similar to that used by Hu *et al.* (1992), but can theoretically be applied to any broadband input. Although Collins *et al.* thought SNR measures to be inappropriate for aperiodic signals, it was soon demonstrated by Neiman *et al.* (1997) that the correlation measures of Collins *et al.* can be derived from the cross-spectral density, which forms the basis of the SNR measure proposed in Kiss (1996).

Furthermore, the SNR measure used by Kiss can be rewritten in terms of a correlation-like measure. This issue is examined in detail in McDonnell *et al.* (2004a). Kiss considers that one of the advantages of his method is that it is robust to phase shifts between the input and output signal, whereas the correlation-based measures of Collins *et al.* are not (L. B. Kish, personal communication, 2006).[5] However, this only applies to the first papers of Collins *et al.*, since in Collins *et al.* (1996a) – unlike Collins *et al.* (1995a) – the cross-correlation between the input and output signals is calculated as a function of time delay, which is the more conventional way of calculating cross-correlation. Such a measure is indeed robust to

---

[5] Note that Kiss and Kish are the same person. All pre-1999 papers spell his name using 'Kiss,' whereas later papers use the spelling 'Kish.' The pronunciation is the same in both cases – 'Kish.' This book will cite and refer to the spelling used in the corresponding publication.

phase shifts; the maximum cross-correlation simply occurs for a nonzero time lag. In fact, the cross-correlation is an ideal measurement of a time delay, and hence of phase shift.

Of these initial studies of ASR, it is the work by Collins *et al.* that gained the greatest exposure (Collins *et al.* 1995b, Collins *et al.* 1995c), with the result that many expositions on ASR have also used correlation-based measures. However, also of great influence is a paper presented before the work of Collins *et al.* at a 1994 workshop on stochastic resonance (Bulsara *et al.* 1994), which shows that SR can occur for an aperiodic input signal, and the mutual information measure. This paper was subsequently published in a journal (DeWeese and Bialek 1995), and is discussed in detail in Section 2.7

Subsequent to DeWeese and Bialek (1995), the next paper to use mutual information as a measure of ASR is a highly cited paper published in the journal *Nature*, which uses mutual information to experimentally show that SR occurs in the cercal sensory system of a cricket when noise is applied externally (Levin and Miller 1996). In the same year, two articles were published in the same issue of *Physical Review E* – those of Bulsara and Zador (1996) and Heneghan *et al.* (1996) – which theoretically examine the use of mutual information to measure SR for aperiodic signals. These papers paved the way for the use of information theory in SR research. The same year also saw a paper that applies other information-theoretic measures – dynamical entropies and Kullback entropy – to measure SR; however, unlike ASR, the system considered is driven by a periodic signal (Neiman *et al.* 1996).

Further extensions

Some further important extensions of SR in this period included the use of chaotic dynamics to act as the 'noise' source leading to SR (Anishchenko *et al.* 1993), analysis of quantum stochastic resonance (Lofstedt and Coppersmith 1994, Grifoni and Hänggi 1996), the first comprehensive studies of array enhanced stochastic resonance (Lindner *et al.* 1995, Lindner *et al.* 1996) – see Section 2.5 – and the observation of spatio-temporal stochastic resonance (Jung and Mayerkress 1995, Löcher *et al.* 1996, Marchesoni *et al.* 1996, Vilar and Rubi 1997).

Consolidation 1997–2007

Between 1997 and 2000 there was an acceleration in the rate of publication of papers either directly about SR, or listing SR as a keyword, and the number continues to grow. The main development in this period is that a large number of papers examine systems showing ASR for aperiodic input signals. For example some of the papers in 1997–1998 are Chialvo *et al.* (1997), Gailey *et al.* (1997), Eichwald and Walleczek (1997), Neiman *et al.* (1997), Vaudelle *et al.* (1998), Fakir (1998a), Fakir (1998b), and also Godivier and Chapeau-Blondeau (1998). Other developments in more recent years include:

- the first analysis of SR in discrete time rather than continuous time systems (Zozor and Amblard 1999, Zozor and Amblard 2001),
- approaches to controlling stochastic resonance (Mitaim and Kosko 1998, Gammaitoni *et al.* 1999, Löcher *et al.* 2000, Kosko and Mitaim 2001, Mitaim and Kosko 2004),
- comprehensive theoretical studies into the limits of when SR can occur, leading to the 'forbidden interval theorem', which states that SR will occur unless a simple condition relating the threshold value to the mean of the noise is violated (Kosko and Mitaim 2003, Mitaim and Kosko 2004, Patel and Kosko 2005, Lee *et al.* 2006),
- the observation of SR in carbon nanotube transistors (Lee *et al.* 2003, Lee *et al.* 2006),
- the observation of an SR-like effect called 'diversity-induced resonance' (Tessone *et al.* 2006) and
- a comprehensive signal-processing-based approach to understanding the detection capabilities of SR from a binary hypothesis testing perspective (Chen *et al.* 2007).

As discussed already, the popularization of SR as a phenomenon that is not restricted to periodic signals has seen its original definition expanded to encompass almost any system in which input and output signals can be defined, and in which noise can have some sort of beneficial role. This period of growth however appears to have slowed a little between 2001 and 2007. The general consensus appears to be that the most recent highly significant result in SR research is its expansion to aperiodic input signals. It could be said however, due to the number of papers investigating SR in neurons, that the most significant discovery on SR may be yet to come – see Section 2.9.

2.5 Paradigms of stochastic resonance

Stochastic resonance in bistable systems

As mentioned above, in the early years of the SR community, it was thought that SR effects were restricted to bistable systems (Fauve and Heslot 1983, Fox 1989, Dykman *et al.* 1990a, Gammaitoni *et al.* 1990, Dykman *et al.* 1990b, Zhou and Moss 1990, Jung and Hänggi 1991, Gammaitoni *et al.* 1991, Dykman *et al.* 1992). Here, following Harmer *et al.* (2002), a simple bistable system consisting of a periodically driven double-well potential is described.

A classic one-dimensional nonlinear system that exhibits stochastic resonance is the damped harmonic oscillator. Following Lanzara *et al.* (1997), this can be modelled with the Langevin equation of motion in the form

$$m\frac{d^2x(t)}{dt^2} + \gamma\frac{dx(t)}{dt} = -\frac{dU(x)}{dx} + \sqrt{D}\xi(t). \tag{2.1}$$

This equation describes the motion of a particle of mass m moving in the presence of friction, γ. The restoring force is expressed as the gradient of some bistable or multistable potential function $U(x)$. In addition, there is an additive stochastic force $\xi(t)$ with intensity D. Typically, this is white Gaussian noise with mean and autocorrelation given respectively by

$$\langle\xi(t)\rangle = 0 \quad \text{and} \quad \langle\xi(t)\xi(t')\rangle = \delta(t - t'). \tag{2.2}$$

This implies that $\xi(t)$ and $\xi(t')$ are statistically independent for $t \neq t'$. The angled brackets $\langle\cdot\rangle$ denote an ensemble average.

When $U(x)$ is a bistable potential, Eq. (2.1) may model several physical processes, ranging from the dynamics of a nonlinear elastic mechanical oscillator to the transient dynamics of a laser. A simple symmetric bistable potential has the form of a standard quartic

$$U(x) = -a\frac{x^2}{2} + b\frac{x^4}{4}. \tag{2.3}$$

If the system is heavily damped, the inertial $m\frac{d^2x(t)}{dt^2}$ term can be neglected. Rescaling the system in Eq. (2.1) with the damping term γ gives the stochastic overdamped Duffing equation

$$\frac{dx(t)}{dt} = -\frac{dU(x)}{dx} + \sqrt{D}\xi(t), \tag{2.4}$$

which is frequently used to model nonequilibrium critical phenomena. For $a > 0$ the potential is bistable as shown in Fig. 2.3. From simple algebra, there is an unstable state at $x = 0$ and two stable states at $x_s^{\pm} = \pm\sqrt{a/b}$, separated by a barrier of height $\Delta U = a^2/4b$ when the noise, $\xi(t)$, is zero. The position of the particle $x(t)$ is considered to be the output of the system and has a power spectral density (PSD) $S(\omega)$.

Two examples (Bulsara and Gammaitoni 1996) of nonlinear dynamic systems are (i) the analogue Hopfield neuron, for which the potential is

$$U(x(t)) = \alpha x(t)^2 - \beta \ln(\cosh x(t)), \tag{2.5}$$

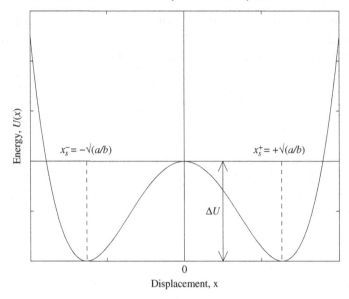

Fig. 2.3. The quartic bistable potential of Eq. (2.3), without any noise, with minima at $x_s^{\pm}$ and barrier height ΔU. Reprinted figure with permission after Harmer *et al.* (2002), © 2002 IEEE.

where the state point $x(t)$ denotes a cell membrane voltage; and (ii) the SQUID loop, for which the potential is

$$U(x(t)) = \alpha x(t)^2 - \beta \cos(2\pi x(t)), \tag{2.6}$$

where $x(t)$ denotes the magnetic field flux in the loop.

 By itself, the bistable system is stationary as described by the motion of the particle. That is, if the particle is in one of the two wells, it will stay there indefinitely. By adding a periodic input signal, $A \sin(\omega_s t)$, to the bistable system, the dynamics are governed by the following equation

$$\frac{dx}{dt} = \left(-\frac{dU(x)}{dx} + A \sin(\omega_s t) \right) + \sqrt{D} \xi(t). \tag{2.7}$$

The bistable potential, which is now time dependent, becomes

$$U(x, t) = U(x) - Ax \sin(\omega_s t)$$

$$= -a\frac{x^2}{2} + b\frac{x^4}{4} - Ax \sin(\omega_s t), \tag{2.8}$$

where A and ω_s are the amplitude and the frequency of the periodic signal respectively.

 It is assumed that the signal amplitude is small enough that, in the absence of any noise, it is insufficient to force a particle to move from one well to another.

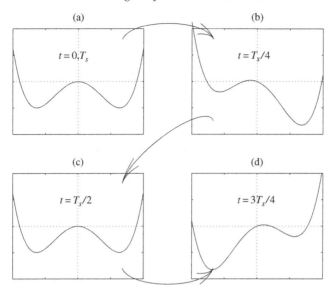

Fig. 2.4. Quartic bistable potential: fixed points. The periodic driving signal causes the double well potential to be tilted back and forth, antisymmetrically raising and lowering the left and right potential wells. Reprinted figure with permission after Harmer *et al.* (2002), © 2002 IEEE.

It is also assumed that the signal period is longer than some characteristic intrawell relaxation time for the system.

Due to the presence of the modulating signal, the double well potential $U(x, t)$ is periodically tilted back and forth with the same frequency, $\omega_s = 2\pi/T_s$. Relating this to the potentials in Fig. 2.4 the effect is to weakly tilt the potential barriers of the right and left sides respectively. In one period of the signal the potential is cycled through Fig. 2.4 (a)–(d). The maxima and minima of the signal correspond to when the potential barrier is at its lowest, which is shown in Fig. 2.4 (b) and (d).

While this tilting on its own is not enough to allow a particle to move from one well to another, the presence of the white noise process in Eq. (2.7) allows noise-induced transitions of the particle. For a certain range of values of the noise intensity, D, transitions will occur in synchronization with the periodic signal. One way of measuring how well the position of the particle represents the frequency of the input is to measure the PSD of the position, and determine the signal-to-noise ratio at ω_s. This will have a peak at a nonzero value of D, and hence SR occurs. The optimal value of D is the one that provides the best time-scale matching between ω_s and the residence times of the particle in each well (Gammaitoni *et al.* 1998, Wellens *et al.* 2004).

A natural simplification of the double-well problem is the discrete two-state system, in which the dynamical variable can take on only two discrete values in these systems. The Schmitt trigger is an example of such a discrete system. Another two-state system in common use is the simple threshold-based system, as we now consider.

Stochastic resonance in threshold systems

The prime objective of this book is to examine *stochastic quantization* of a signal.[6] We define stochastic quantization to mean:

The quantization of a signal by threshold values that are independent and identically distributed random variables.

The starting point of this work is a particular form of SR known as suprathreshold stochastic resonance (SSR). To set the context for SSR, it is necessary to briefly review SR in threshold-based systems – a story that began in 1994. In a decade, the phenomenon is now known to be so widespread that a recent paper has made the point that almost all threshold systems – and noise distributions – exhibit stochastic resonance (Kosko and Mitaim 2003, Kosko and Mitaim 2004).

A single threshold

The first paper to consider SR in a system consisting of a simple threshold, which when crossed by an input signal gives an output pulse, was published in April 1994 under the curious title of *Stochastic Resonance on a Circle* (Wiesenfeld *et al.* 1994). In this paper, the authors note that all previous treatments of SR are based on the classical motion of a particle confined in a monostable or multistable potential. This appears to not be strictly true, as in 1991 and 1993 Bulsara *et al.* (1991), Bulsara and Moss (1991), Longtin *et al.* (1991), Longtin (1993), Douglass *et al.* (1993), and Moss *et al.* (1993) all analyzed SR in neurons, which as excitable systems are not bistable systems in the classical sense. Additionally, although the work on SR in Schmitt trigger circuits (Fauve and Heslot 1983, Melnikov 1993) has been put into the bistable system category, a Schmitt trigger is closely related to the simple threshold crossing system – see the discussion on page 1219 of Luchinsky *et al.* (1999b).

Nevertheless, the paper by Wiesenfeld *et al.* (1994) does appear to be the first paper to show the presence of SR in a system consisting of a simple threshold and

---

[6] Note that this is a different usage of the term 'stochastic quantization' from that prevalent in areas of quantum physics, stochastic differential equations, and path-integrals such as the Parisi–Wu stochastic quantization method (Damgaard and Huffel 1988).

the sum of a subthreshold signal and noise. Unlike previous work on SR, this paper explores SR with:

a different class of dynamical systems based not on bistability but rather an excitable dynamics. (Wiesenfeld *et al.* 1994)

The system considered consists of a weak subthreshold periodic signal, a potential barrier, and zero-mean Gaussian white noise. The output is a 'spike', which is a short duration pulse, with a large amplitude and deterministic refractory time. Note that a short refractory time for the pulse is important; the refractory time must be much less than the correlation time of the noise, or threshold crossings will occur that cannot give an output pulse. This system is described as:

a simple dynamical process based on a single potential well, for which SR can be observed. (Wiesenfeld *et al.* 1994)

It is also observed to represent the process of action potential events in sensory neurons.

Figure 2.5 illustrates qualitatively why SR occurs in such a system. In each subfigure, the lower plot indicates the input signal's amplitude against increasing time, with the straight dotted line being the threshold value. The upper trace is the output signal plotted against the same time-scale as the input signal. This output signal is simply a narrow pulse, or 'spike', every time the input signal crosses the threshold with positive slope.

Figure 2.5(a) shows a noiseless subthreshold periodic input signal. Such a signal never induces a threshold crossing, and the output remains constant and spike-less. Clearly nothing at all can be said about the input signal by examining the output signal, except that the input is entirely subthreshold. Figure 2.5(b) indicates how in the absence of a signal, threshold crossings will occur randomly. The input in this plot is only bandlimited white Gaussian noise – that is, noise with a frequency spectrum that is flat within its bandwidth. The statistics of this case are required for obtaining output SNRs when a signal is present, although such formulas will only hold when the input signal is small compared to the noise. In Fig. 2.5(c), the lower trace shows the sum of a periodic input signal, and bandlimited Gaussian noise. Unlike Fig. 2.5(a), for this signal some threshold crossings do occur. The probability of an output pulse is dependent on the amplitude of the noiseless input signal. This is clear visually and intuitively; a threshold crossing is more likely to occur when the input signal is close to the threshold than when it is not. Compared with the absence of noise, some information about the original noiseless input signal is now available at the output, and removing the noise is counterproductive in this system. Note that the output pulse 'train' does not 'look' anything like the input signal, but in fact encodes the input by a form of stochastic frequency modulation.

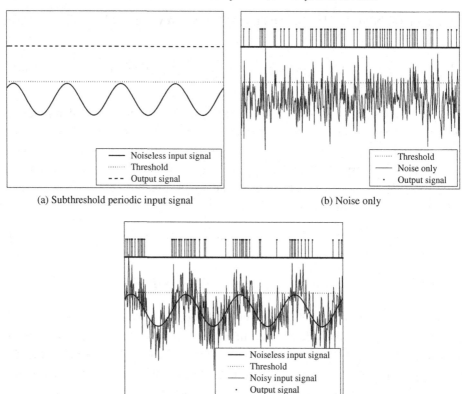

(c) Subthreshold but noisy periodic input signal

Fig. 2.5. Qualitative illustration of SR occurring in a simple threshold-based system. In each subfigure, the lower trace indicates the input signal's amplitude against increasing time, with the dotted line being the threshold value. The upper trace is the output signal plotted against the same time-scale as the input signal. The output signal is a 'spike' – that is, a short duration pulse – every time the input crosses the threshold with positive slope, that is from subthreshold to suprathreshold but not *vice versa*. Figure 2.5(a) shows an entirely subthreshold periodic input signal. Such a signal will never cause an output spike. Figure 2.5(b) shows the case of threshold crossings occurring randomly due to bandlimited Gaussian noise. Figure 2.5(c) illustrates how the presence of noise enables some information about a subthreshold input signal to be present at the output. In the absence of noise, the output will always remain constant. When noise is added to the input, threshold crossings can occur, with a probability related to the amplitude of the input signal.

If the input signal is to be recovered, we must obtain an ensemble average to make sense of this, as is demonstrated shortly, and described in Sulcs *et al.* (2000).

Another way of understanding this effect is to realize that adding noise to a subthreshold periodic signal, and then thresholding the result, is equivalent to thresholding the noiseless signal with a time-varying stochastic threshold value. This is equivalent to having a signal that is thresholded at a random amplitude, where sometimes the random amplitude is greater than the signal's maximum amplitude. This means the periodic signal is thresholded at a random phase in its period, and sometimes not at all.

Returning to Wiesenfeld's work, his paper notes that such a threshold-based system can be realized by a Josephson junction biased in its zero voltage state. He then provides a general theoretical approach for what he calls 'generic threshold-plus-reinjection dynamics' and shows how it can lead to SR. The main theoretical result given is a derivation of the SNR at the output of the generic system, using the autocorrelation of the output. This formula for SNR is shown to have a maximum for a nonzero value of noise strength, which shows that SR exists in such a system.

Shortly after this initial paper, several other authors also analyzed SR in threshold systems subject to periodic input signals and additive noise. The first of these is an excellent paper by Jung (1994), which extends the theory of Wiesenfeld *et al.* (1994) by removing an assumption of Poisson statistics, and using the decades-old work of S. O. Rice, who analyzed the expected number of threshold crossings by either broadband noise alone, or the sum of a sinusoid and broadband noise (Rice 1944, 1945, 1948). Jung (1994) shows mathematically that the correlation coefficient between a subthreshold periodic input signal and its corresponding random output pulse train possesses a noise-induced maximum.

Jung (1994) also gives a convincing argument, which has been much used in SR research, for why such simple threshold crossing systems can be used to provide simple models of neurons, particularly in studies of neural networks, with large numbers of neurons. As noted by Jung (1994), such a simple two-state model of a neuron was first proposed over 50 years ago by McCulloch and Pitts (1943).

Another observation made is that SR occurs for subthreshold signals since, in the absence of noise, no threshold crossings can occur, whereas

in the presence of noise, there will be noise-induced threshold crossing(s), but at preferred instants of time, that is, when the signal is larger. (Jung 1994)

In addition, the conclusion is made that suprathreshold signals will never show stochastic resonant behaviour. We shall see later in this chapter that this is not true, once some of the assumptions used by Jung are discarded. Jung (1994) was followed up shortly afterwards in two more papers on the same topic (Jung and Mayer-Kress 1995, Jung 1995).

Simultaneously with Jung, an alternative approach was taken by Gingl *et al.* (1995b) – aspects of which also appeared in Kiss (1996) and Gingl *et al.* (1995a) – who consider a threshold-based system to be a 'level crossing detector' (LCD), that is a system that detects whether or not an input signal has crossed a certain voltage level. Such a system is described as 'non-dynamical', to differentiate it from the bistable systems used in 'classical' – that is, pre 1994 – SR studies.

Like Jung (1994), this work also uses the established work of Rice. Formulas first given in Rice (1944) are applied to derive a formula for the SNR in the linear response limit at the output of an LCD, a formula that is quite similar to that of Wiesenfeld *et al.* (1994). The equation obtained is verified by simulation in Gingl *et al.* (1995b).

Other early papers on SR in threshold-based systems appeared in 1996 from Bulsara and Zador (1996) and Bulsara and Gammaitoni (1996), who use mutual information to measure SR for a subthreshold aperiodic input signal. Shortly after this, Chapeau-Blondeau and co-authors published several papers examining threshold-based SR from a wide variety of new angles (Chapeau-Blondeau 1996, Chapeau-Blondeau and Godivier 1996, Chapeau-Blondeau and Godivier 1997, Chapeau-Blondeau 1997b, Godivier and Chapeau-Blondeau 1997, Chapeau-Blondeau 1997a, Godivier *et al.* 1997).

Array enhanced stochastic resonance

The bulk of this book concerns a form of SR that occurs in arrays of parallel threshold devices. We briefly mention some of the history of what is some-times known as array enhanced stochastic resonance (AESR). There have been many observations that coupling together more than one SR-capable device can lead to increased output performance. Perhaps the first to demonstrate this were Wiesenfeld (1991), Jung *et al.* (1992), and Bulsara and Schmera (1993). However, Lindner *et al.* (1995) were the first to use the term AESR, and show that a chain of coupled nonlinear oscillators can provide an enhanced SR effect when compared with a single oscillator. A very similar effect is discussed in Collins *et al.* (1995b), where ASR is studied in an array of FitzHugh–Nagumo neuron models. Each neuron is considered to receive the same subthreshold aperiodic input sig-nal, but independent noise, and the overall output is the summed response from all neurons. This result is also discussed in Moss and Pei (1995).

Other early papers showing the effect of AESR include Bezrukov and Voydanoy (1995), Lindner *et al.* (1996), Pei *et al.* (1996), Gailey *et al.* (1997), Neiman *et al.* (1997), and Chialvo *et al.* (1997). An unpublished preprint gives a more detailed history (P. F. Góra, arXiv:cond-mat/0308620). The main point in these works is that the magnitude of SR effects can be enhanced by combining the outputs of more than one single SR-capable component.

Multiple thresholds and soft thresholds

The first paper to consider systems consisting of more than one static threshold is Gammaitoni (1995b), which considers SR in threshold-based systems to be equivalent to dithering. A comparison between dithering and stochastic resonance is given in Section 3.4 of Chapter 3. The second is Gailey *et al.* (1997), which analyzes an ensemble of N threshold elements, using the classical theory of nonlinear transformations of a Gaussian process to obtain cross correlations.

Another extension was the realization of the existence of a:

class of non-dynamical and threshold-free systems that also exhibit stochastic resonance. (Bezrukov and Voydanoy 1997b, Bezrukov and Voydanoy 1997a)

This work demonstrates that a 'hard' threshold – that is, a threshold that divides its inputs into exactly two states, rather than a continuum of states – is not a necessary condition for SR to occur in nondynamical systems. Further development of this approach can be found in Bezrukov (1998).

Forbidden interval theorems

Although demonstrations of SR in threshold systems since the initial 1994 paper are many and varied, detailed theoretical proofs of the fact that SR effects should be expected in nearly all threshold systems were not published until much later (Kosko and Mitaim 2003). A number of theorems, collectively known as the *forbidden interval theorems*, give simple conditions predicting when SR will occur for random binary (Bernoulli distributed) signals, and static threshold systems (Kosko and Mitaim 2003, Kosko and Mitaim 2004). Specifically, the theorems prove that SR will occur (i) for all finite variance noise distributions, if and only if the mean of the noise is outside a 'forbidden interval', and (ii) all infinite variance *stable* noise distributions, if and only if the location parameter of the noise is outside the forbidden interval. The forbidden interval is the region $[\theta - A, \theta + A]$, where θ is the threshold value, and the binary signal takes values from $\{-A, A\}$.

More recently, similar forbidden interval theorems were published that partially extend their validity from the basic static threshold to spiking neuron models (Patel and Kosko 2005).

When the consequences of the forbidden interval theorem are combined with the observation that SR occurs for impulsive (infinite variance) noise, as pointed out by Kosko and Mitaim (2001), it is clear that SR is actually very robust. This fact highlights that, although published SR research usually includes an assumption of finite variance noise – usually Gaussian – this is not a necessary condition for SR to occur.

The validity of the forbidden interval theorem has been tested experimentally using carbon nanotube transistors (Lee *et al.* 2003, Lee *et al.* 2006), both with finite variance and with impulsive noise. This demonstration that SR effects are extremely robust is potentially important for future applications making use of novel transistor technologies, as it means that SR could be utilized even with little control over the distribution of SR-inducing noise signals.

To date, the forbidden interval theorem has not been extended beyond binary input signals. It is an open question whether conditions exist stating whether SR can or cannot occur when the input signal is a continuously valued random variable. This latter situation is the one studied in the remaining chapters of this book.

2.6 How should I measure thee? Let me count the ways . . .

The reason for the name of this section[7] is to draw attention to the fact that SR has been measured in many different ways. Examples include SNR (Benzi *et al.* 1981), spectral power amplification (Jung and Hänggi 1991, Rozenfeld and Schimansky-Geier 2000, Imkeller and Pavlyukevich 2001, Drozhdin 2001), correlation coefficient (Collins *et al.* 1995a), mutual information (Levin and Miller 1996), Kullback entropy (Neiman *et al.* 1996), channel capacity (Chapeau-Blondeau 1997b), Fisher information (Greenwood *et al.* 1999), ϕ–divergences (Inchiosa *et al.* 2000, Robinson *et al.* 2001), and mean square distortion (McDonnell *et al.* 2002a). Stochastic resonance has been analyzed in terms of residence time distributions – see Gammaitoni *et al.* (1998) for a review – as well as Receiver Operating Characteristic (ROC) curves (Robinson *et al.* 1998, Galdi *et al.* 1998, Zozor and Amblard 2002), which are based on probabilities of detecting a signal to be present, or falsely detecting a nonexisting signal (Urick 1967).

The key point is that the measure appropriate to a given task should be used. Unfortunately, due to historical reasons, some authors tend to employ the original measure used, SNR, in contexts where it is effectively meaningless. This section gives a brief history of the use of SNR in SR research, and a discussion of some of the criticisms of its use.

It was first thought that SR occurs only in bistable dynamical systems, generally driven by a periodic input signal, $A \sin(\omega_0 t + \phi)$, and broadband noise. Since the input to such systems is a simple sinusoid, the SNR at the output is a natural measure to use to determine how well the output signal can reflect the input periodicity, with the following definition most common

$$\text{SNR} = \frac{P(\omega_0)}{S_N}. \tag{2.9}$$

In Eq. (2.9), $P(\omega_0)$ is the output power spectral density (PSD) at the frequency of the input signal, ω_0, and S_N is the PSD of the output background noise, as measured with a signal present. This definition assumes that the overall output PSD is the superposition of a constant noise background, corresponding to white noise, and a

---

[7] Apologies to Elizabeth Barrett Browning (1850).

delta-function spike at the input frequency. Stochastic resonance occurs when the SNR is maximized by a nonzero value of input noise intensity.

It is well known in electronic engineering that nonlinear devices cause output frequency distortion – that is, for a single frequency input, the output will consist of various harmonics of the input (Cogdell 1996). This means that basic circuit design requires the use of filters that remove unwanted output frequencies. For example, this *harmonic distortion* in audio amplifiers is very undesirable. On the other hand, high frequency oscillators make use of this effect by starting with a very stable low frequency oscillator and sending the generated signal through a chain of frequency multipliers. The final frequency is harmonically related to the low frequency source.

For more than one input frequency, the output of the nonlinear device will contain the input frequencies, as well as integer multiples of the sum and difference between all frequencies (Cogdell 1996). This effect of creating new frequencies is known as *intermodulation distortion*. In the field of optics this phenomenon can be used to generate lower frequency signals – for example, T-rays (that is, terahertz radiation) – from different optical frequencies by a method known as *optical rectification* (Mickan *et al.* 2000, Mickan and Zhang 2003).

A study of such higher harmonics generated by a nonlinear system exhibiting SR has been published (Bartussek *et al.* 1994), and the phenomenon is also discussed in subsequent works (Bulsara and Inchiosa 1996, Inchiosa and Bulsara 1998). However, much research into SR has only been interested in the output frequency component that corresponds to the fundamental frequency of the periodic input signal, in which case the output SNR is given by Eq. (2.9) and ignores all other output harmonics.

More recently, attempts have been made to overcome this, by defining the output SNR as a function of all frequencies present at the input, even if not in a narrow band around the fundamental (Kiss 1996, Gingl *et al.* 2001, Mingesz *et al.* 2005). However, while such formulations may have some uses (McDonnell *et al.* 2004a), there has been much discussion regarding the inadequacies of SNR as an appropriate measure for many signal processing tasks (DeWeese and Bialek 1995, Galdi *et al.* 1998).

One of the most important objections can be illustrated as follows. Consider a periodic, but broadband input signal, such as a regularly repeated radar chirp signal. The use of SNR as the ratio of the output power of the fundamental frequency to the background noise PSD is meaningless for signal recovery here, unless only the fundamental period is of interest. This output SNR measure only provides information about the period of the signal – the output SNR at that frequency – and nothing about the shape of the chirp in the time domain.

This inadequacy was recognized when researchers first turned their attention to ASR, which ushered in the widespread use of cross-correlation and information-theoretic measures, which can, in some sense, describe how well the shape of the output signal is related to the input signal. An excellent description of the issue is that:

a nonlinear signal processor may output a signal that has infinite SNR but is useless because it has no correlation with the input signal. Such a system would be one which simply generates a sine wave at the signal frequency, totally ignoring its input. (Inchiosa and Bulsara 1995)

While many SR researchers realized that studying aperiodic input signals substantially increases the relevance of SR to applications such as studies of neural coding, some did not get past the need to move on to measures other than SNR in such circumstances. This has led to a somewhat strange debate about whether or not 'SNR gains' can be made to happen in a nonlinear system by the addition of noise. Next we discuss the main questions on this issue.

The SNR gain debate

In the last decade, a number of researchers have reported results claiming that it is possible to obtain an SNR gain in some nonlinear systems by the addition of noise (Kiss 1996, Loerincz *et al.* 1996, Vilar and Rubí 1996, Chapeau-Blondeau 1997a, Chapeau-Blondeau and Godivier 1997, Chapeau-Blondeau 1999, Gingl *et al.* 2000, Liu *et al.* 2001, Gingl *et al.* 2001, Makra *et al.* 2002, Casado-Pascual *et al.* 2003, Duan *et al.* 2006). There has been some criticism of these works, for example the comment on Liu *et al.* (2001) given in Khovanov and McClintock (2003), and in the SR community, such results have been seen as fairly controversial, for two reasons, as discussed in the remainder of this section.

Can SNR gains occur at all?

Initially it seemed that SNR gains contradicted proofs that SNR gains cannot occur. For example DeWeese and Bialek (1995) – see also Dykman *et al.* (1995) – show for stationary Gaussian noise and a signal that is small compared to the noise, that for nonlinear systems the gain, $G = \text{SNR}_{\text{out}}/\text{SNR}_{\text{in}}$, must be less than or equal to unity, and that no SNR gain can be induced by utilizing SR. This proof is based on the use of linear response theory, where, since the signal is small compared to the noise, both the signal and noise are transferred linearly to the output, and, as in a linear system, no SNR gain is possible. Much attention has been given to this fact, since most of the earlier studies on SR were kept to cases where the linear response limit applies, to ensure that the output is not subject to the above-mentioned harmonic distortion (Gingl *et al.* 2000).

Once this fact was established, researchers still hoping to be able to find systems in which SNR gains due to noise could occur turned their attention to situations not covered by the proof – that is, the case of a signal that is not small compared to the noise, or broadband signals or non-Gaussian noise.

For example, Kiss (1996) considers a broadband input signal, and, being broadband, the conventional SNR definition cannot be used. Instead, a new frequency dependent SNR measure is derived, a measure with which an SNR gain is shown to occur. Further examples are Chapeau-Blondeau and Godivier (1997) and Chapeau-Blondeau (1997a), which use the conventional SNR definition, but the large signal regime to show the existence of SNR gains. Furthermore, Chapeau-Blondeau (1999) also considers the case of non-Gaussian noise.

However, the interpretation in some of the papers on this topic can be a little fuzzy. For example, it is sometimes implied that an SNR gain is *due* to the addition of more noise to an already noisy signal.

Instead, the SNR gains reported are caused by an increase in input noise in order to find the optimal point on the system's SR curve. The side effect of this is a decrease in input SNR. This leads to an increase in SNR gain, simply due to the same mechanism that causes SR itself. This means that the gain is due to the characteristics of the system itself rather than the addition of noise.

Nevertheless, the main point emphasized is that the SNR gain can be greater than one, which does not occur for the linear response regime, and is a valid point. Our conclusion is that the answer to 'can SNR gains occur at all?' is 'yes, SNR gains can occur'. The more important question is whether such gains are meaningful.

Are SNR gains meaningful?

By looking outside the conditions of the proof that SNR gains cannot occur in the linear response limit, SNR gains can be found. However, the second reason that an emphasis on SNR gains due to SR are seen to be controversial is that the definitions of SNR used in cases where SNR gains occur are not always particularly meaningful. Taking the approach of looking outside the parameters of the proof assumes that SNR is still a useful measure outside these parameters. A strong argument against this assumption, and for the use of information theory, rather than SNRs, is given in DeWeese and Bialek (1995), as discussed in Section 2.7.

Useful discussions of this point, and discussions of signal detection theory in the context of SR are given in Inchiosa and Bulsara (1995), Galdi *et al.* (1998), Robinson *et al.* (1998), Petracchi (2000), Hänggi *et al.* (2000), Robinson *et al.* (2001) and Chen *et al.* (2007). For example, Hänggi *et al.* (2000) give a general investigation of SNR gains due to noise, are highly critical of the use of SNR in such systems, and indicate more appropriate measures to use, at least for signal detection or estimation problems. Our conclusion is that the answer to the title of this subsection is 'probably not, for most tasks'.

Recall the quote above: Inchiosa and Bulsara (1995) recognize what is well-known to electronic engineers – that an SNR gain is not in itself a remarkable thing, and that SNR gains are routinely obtained by filtering – for example, the bandpass filter. The reason that more is made of such phenomena in the SR literature is that the reported SNR gains are said to be due to the *addition of noise* to an already noisy signal, rather than a deliberately designed filter. Another paper by the same authors also discusses this topic (Inchiosa and Bulsara 1996). As discussed above, the view that SNR gains occur *due to* SR can be misleading.

An associated problem is that of relating SNRs to information theory. For example, it has sometimes been stated that an SNR gain in a periodic system is analogous to an increase in information. We now investigate such a claim.

2.7 Stochastic resonance and information theory

The previous section highlighted that problems can occur if SNR measures are used in situations where they are not appropriate. Here, we indicate why this is the case. The crux of the matter is that SNR measures are, for example, appropriate when the goal is to decide if a signal is present or not – that is, the problem of *signal detection*. In the original context of SR, the detection of a weak periodic signal was certainly the goal, and a small SNR can make it very difficult to detect the signal. This is quite a different matter from other signal processing problems, such as estimation, compression, error-free transmission, and classification.

Rather than attempting to observe SR effects in a particular measure, it makes more sense to first define the signal processing objective of a system. This leads to an appropriate measure of performance quality, which can then be analyzed to see whether SR can occur.

One influential paper on SR that takes this approach is now discussed.

Signal detection vs. information transmission

The first paper to discuss stochastic resonance in the context of information theory was DeWeese and Bialek (1995). In this paper, it is considered that the signal processing objective of a neuron is to transmit as much information about its input as possible. The measure used is mutual information (Cover and Thomas 1991).

There are several reasons why this paper has been significant for SR researchers:

• The point is made that one potentially universal characteristic of neural coding is that the:

SNR (is) of order unity over a broad bandwidth. (DeWeese and Bialek 1995)

Since this means that the environment in which sensory neural coding takes place appears to be very noisy, it is highly plausible that neural coding makes use of SR. This point is also made in Bialek *et al.* (1993) and DeWeese (1996).

- It is pointed out that measuring information transfer for single frequency sine waves by SNR is only really applicable in linear systems. Since SR cannot occur in linear systems, the use of SNR only really applies to the case of small input signals so that the output exhibits a linear response.

- A proof is given of the fact that, for small signals in Gaussian noise, it is impossible for the output SNR to be greater than the input SNR. This fact has led some researchers to search for – and find (see Section 2.6) – circumstances in which the proof does not apply and that SNR gains can occur. It could be argued that any such work that does not have a detection goal does not pay attention to the previous point above, and is possibly not really proving much.

- It is pointed out that sine waves do not carry information that increases with the time of observation:

No information can be carried by the signal unless its entropy is an extensive quantity. In other words, if we choose to study a signal composed of a sine wave, the information carried by the signal will not grow linearly with the length of time we observe it, whether or not the noise is present. In addition to this, we would like to compare our results to the performance of real neurons in as natural conditions as possible, so we should use ensembles of broadband signals, not sine waves. (DeWeese and Bialek 1995)

- It is demonstrated for the case of a subthreshold signal in a single threshold system that the information transferred through such a system can be optimized by modifying the threshold setting. With the optimal value for the threshold, the mutual information is strictly decreasing for increasing noise, and SR does not occur.

So it seems that if you adapt your coding strategy, you discover that *stochastic resonance* effects disappear ... More generally, we can view the addition of noise to improve information transmission as a strategy for overcoming the *incorrect* setting of the threshold. (DeWeese and Bialek 1995)

The conclusions drawn are that single-frequency periodic signals are not relevant for information transfer, and particularly not relevant for neurons. Indeed, it has been pointed out that rather than using SNR:

It is the total information encoded about a signal that is the biologically relevant quantity to consider. (Levin and Miller 1996)

At the time, with a growing interest into SR in neurons, such a realization led researchers away from studying conventional single frequency SR in neurons, to more realistic broadband and aperiodic signals. For most authors, this naturally led to using measures other than SNR.

One exception to the above reasoning is that there are some neurons that act as binary switches; that is, they are essentially simple signal detectors asking the question: 'is there a signal present at my input or not?'

Perhaps another important exception to these ideas is in the encoding of sound by the cochlear nerve. The mechanism by which audio signals are encoded has been likened to a biological Fourier transform; different spatial regions in the *organ of Corti* – that is, the part of the inner ear that contains sensory neurons – are sensitive to different frequencies of sound waves (Kandel *et al.* 1991).

The conclusion that SR in threshold systems is simply a way to overcome the incorrect threshold setting seems to have led many to think that making use of noise is a suboptimal means of designing a system. The contrasting viewpoint is that noise is ubiquitous; since it is virtually impossible to remove all noise completely from systems, design methods should consider the effects of SR, and that various design parameters, such as a threshold value, may in some circumstances need to be set in ways that make use of the inherent noise to obtain an optimal response. We discuss exactly this situation in Chapter 8.

Furthermore, the analysis in DeWeese and Bialek (1995) assumes that the input signal to a neuron is random with an ideal white spectrum. In other words, there is no time correlation in the signal. As is argued by illustration in Section 2.8, time correlation can lead to a noise-enhanced benefit in a single threshold system, even if the threshold is optimally set.

Information theory and SNR gains

The search for SNR gains due to SR in the case of periodic input signals naturally led some authors to look for an analogy to compare input performance to output performance for aperiodic stochastic resonance (ASR). As mentioned, Kiss (1996) defines a frequency dependent SNR measure based on cross-spectral densities, which he considers to be valid for such aperiodic input signals. This method is discussed further in McDonnell *et al.* (2004a).

An alternative approach for measuring ASR is mutual information. A special case of mutual information is known as *channel capacity* (Cover and Thomas 1991). Channel capacity is simply defined as the maximum possible mutual information through a 'channel' or system. It is usually defined in terms of the input probability distribution that provides the maximum mutual information, subject to certain constraints. For example, the input signal may be restricted to two states – that is, a binary signal – or to be a continuously valued random variable, but with a specified power.

The most widely known formula describing channel capacity is the Shannon–Hartley formula, which gives the channel capacity for the transmission of a power-limited and band-limited signal through an additive, signal-independent, Gaussian white noise channel. As mentioned in Berger and Gibson (1998), this formula is often misused in situations where it does not apply, including, one could argue, in the SR literature.

Channel capacity as a measure of SR is discussed in Chapeau-Blondeau (1997b), Godivier and Chapeau-Blondeau (1998), Goychuk and Hänggi (1999), Kish *et al.* (2001), Goychuk (2001), and Bowen and Mancini (2004), all of which show that the right level of noise can provide the maximum channel capacity. However, in general, this means only that the right level of input noise optimizes the channel; that is, SR occurs.

Of more interest to us here is a way of comparing the input signal to the output signal in a way analogous to SNR gains for periodic signals. For example, in Chapeau-Blondeau (1999) it is considered that comparing the channel capacity at the input and the output of a system for an aperiodic input signal is analogous to a comparison of the input and output SNRs for periodic input signals. Here we investigate the use of channel capacity in simple threshold-based systems where SR can occur, and show by use of a well-known theorem of information theory that such an analogy is a false one.

The data processing inequality

The data processing inequality (DPI) of information theory asserts that no more information can be obtained out of a set of data than is there to begin with. It states that given random variables X, Y, and Z that form a Markov chain in the order $X \rightarrow Y \rightarrow Z$, then the mutual information between X and Y is greater than or equal to the mutual information between X and Z (Cover and Thomas 1991). That is

$$I(X, Y) \geq I(X, Z). \tag{2.10}$$

In practice, this means that no signal processing on Y can increase the information that Y contains about X.

It should be noted that the terminology *Markov chain* used in connection with the DPI is somewhat more inclusive than that prevalent in applied probability. DPI usage requires only the basic Markov property that Z and X are conditionally independent given Y. By contrast the usage in applied probability requires also that X, Y, and Z range over the same set of values and that the distribution of Z given Y be the same as that of Y given X.

Generic nonlinear noisy system

To illustrate the arguments we now present, consider a generic system where a signal, $s(t)$, is subject to independent additive random noise, $n(t)$, to form another random signal, $x(t) = s(t) + n(t)$. The signal $x(t)$ is then subjected to a nonlinear transformation, $T[\cdot]$, to give a final random signal, $y(t) = T[x(t)]$. A block diagram of such a system is shown in Fig. 2.6.

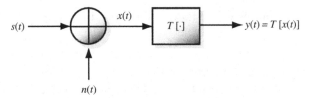

Fig. 2.6. Schematic diagram of a generic noisy nonlinear system. The input signal, $s(t)$, is subject to additive noise, $n(t)$. The sum of the signal and noise, $x(t)$, is subjected to the nonlinear transfer function, $T[\cdot]$, to give $y(t) = T[x(t)]$.

As noted above, many papers have demonstrated that SNR gains can occur due to SR. Such a system as that of Fig. 2.6 describes many of those reported to show SNR gains for both periodic or aperiodic input signals.

Channel capacity

Ignoring for now the question over whether SNR measures have much relevance in such cases, the observation of SNR gains can appear on the surface to contradict the DPI. The reason for this is that one could be led to believe that information can always be related to SNR by the Shannon–Hartley channel capacity formula[8]

$$C = 0.5 \log_2 (1 + \text{SNR}) \quad \text{bits per sample.} \tag{2.11}$$

Clearly, when this formula applies, an increasing SNR leads to an increase in the maximum possible mutual information through a channel. Suppose that Eq. (2.11) does apply in Fig. 2.6 and that the SNR of $s(t)$ in $x(t)$ is SNR_1. This means that the maximum mutual information between $s(t)$ and $x(t)$ is $I(s, x) = 0.5 \log_2 (1 + \text{SNR}_1)$ bits per sample.

Suppose also that the operation $T[\cdot]$ filters $x(t)$ to obtain $y(t)$, such that the filtering provides an output SNR for $s(t)$ in $y(t)$ of SNR_2. If the filtering provides an SNR gain, then $\text{SNR}_2 > \text{SNR}_1$. Consider the overall system that has input, $s(t)$, and output, $y(t)$. If Eq. (2.11) applies for this whole system, the mutual information between $s(t)$ and $y(t)$ is $I(s, y) = 0.5 \log_2 (1 + \text{SNR}_2) > I(s, x)$. This is clearly a violation of the DPI, and an SNR gain either cannot occur in a system in which Eq. (2.11) applies, or Eq. (2.11) does not apply. If we believe that Eq. (2.11) always applies, then scepticism about the occurrence of SNR gains can be forgiven.

However, it is instead the validity of Eq. (2.11) that needs consideration. As mentioned, this formula is often widely misused (Berger and Gibson 1998), as it applies only for additive Gaussian white noise channels, where the signal is independent of the noise. Of particular relevance here is the *additive noise* part. No SNR gain such

---

[8] Some references instead refer to this formula as the 'Hartley–Shannon formula', the 'Shannon–Hartley–Tuller law', or simply as 'Shannon's channel capacity formula'. It is also variously known as a 'theory', 'law', 'equation', 'limit', or 'formula'.

as that from SNR_1 to SNR_2 can be achieved in such an additive noise channel. This means that Eq. (2.11) can never apply to the situation mentioned above between signals $s(t)$ and $y(t)$, since even if it applies between $s(t)$ and $x(t)$, the SNR gain required in the filtering operation rules it invalid. The conclusion of this reasoning is that there is no reason to be sceptical about SNR gains, except for cases where the Shannon–Hartley channel capacity formula is actually valid.

However, such a discussion does indicate that any analogy between SNR gains and mutual information is fraught with danger. For example, it is shown using simple examples in McDonnell *et al.* (2003b) and McDonnell *et al.* (2003c) that such an analogy is generally false. From these investigations, it is clear that although SNR gains may exist due to SR for periodic input signals, no information-theoretical analogy exists for random noisy aperiodic signals. The simplest illustration of this is to threshold a noisy binary pulse train at its mean. For uniform noise with a maximum value less than half the pulse amplitude, the mutual information between input and output remains constant, regardless of the input SNR.

Furthermore, since the DPI holds, the addition of more noise to a noisy signal cannot be of benefit as far as obtaining an input–output mutual information gain is concerned. Such a result does not rule out the fact that the addition of noise at the input to a channel can maximize the mutual information at the output; in other words, the effect of SR for aperiodic signals is perfectly valid. When this occurs, an optimal value of input noise means minimizing the information lost in the channel.

2.8 Is stochastic resonance restricted to subthreshold signals?

The common aspect of the cited works in the section on SR in threshold systems is that, for a single threshold, SR is shown to occur only for subthreshold input signals. It has been discussed several times that SR cannot occur in a threshold-based system for an optimally placed threshold. However, this is true only for certain situations. One such situation is where only the output SNR at the frequency of a periodic signal is measured. For example, consider the case of a threshold system where the output is a pulse whenever the input signal crosses the threshold with positive slope. In the absence of noise, placing the threshold at the mean of the input signal will cause output pulses to occur once per input period. In the presence of noise, the output signal will be noisy, since spurious pulses, or jitter in the timing of the desired pulse, will occur, and the absence of noise is desirable in this case. This situation is illustrated in Fig. 2.7.

Another such situation, as already discussed, is given in DeWeese and Bialek (1995). When the input signal is aperiodic, with no time correlation, and the measure used is mutual information, an optimally placed threshold also precludes SR from occurring.

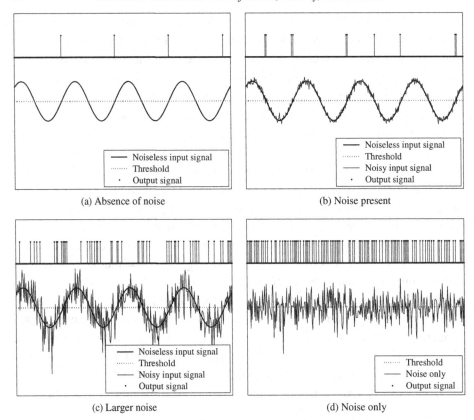

Fig. 2.7. Thresholding a periodic signal at its mean. This plot illustrates that for an optimally placed threshold, at least as far as determining the frequency of a sine wave, the optimal output occurs in the absence of noise. In each subfigure, the lower trace indicates the input signal's amplitude against increasing time, with the straight line being the threshold value. The upper plot is the output signal plotted against the same time-scale as the input signal. The output signal is a 'spike' – that is, a short duration pulse – every time the input crosses the threshold with positive slope – that is, from subthreshold to suprathreshold but not *vice versa*. Figure 2.7(a) shows that in the absence of noise, there is exactly one threshold crossing per period. Figure 2.7(b) shows that as soon as some small amount of additive noise is present, extra threshold crossings occur. This results in some noise being present in the output, as there is no longer exactly one output pulse per period. Figure 2.7(c) shows that as the noise becomes larger, many more output pulses occur, although with a higher frequency when the input signal is close to the threshold than when it is near its maximum or minimum. This indicates that averaging should recover the signal period. Figure 2.7(d) shows that, in the absence of a signal, the occurrence of output pulses is completely random.

However, one example that shows that SR can indeed occur for a signal that is not entirely subthreshold[9] is when averaging is allowed for a periodic – but not

---

[9] In this text, the term 'suprathreshold' is not intended to mean that a signal is always entirely greater than a certain threshold. Instead, it refers to a signal that is allowed to have values that are both above and below a threshold. This includes the special case of entirely suprathreshold signals.

single frequency – signal. Figs 2.8 and 2.9 illustrate that the ensemble average of a noisy thresholded periodic signal – with an optimal threshold – can be better in a certain sense in the presence of noise than without it.

Figures 2.8(b) and 2.8(c) show a periodic but *not single frequency* signal being thresholded at its mean. Figure 2.8(a), which shows the same signal completely subthresholded, is provided for comparison. In Fig. 2.8(b) where noise is absent, the output is identical to that of the single frequency sine wave of Fig. 2.7(a), and no amount of averaging can differentiate between the two output signals. However, in Fig. 2.8(c), where noise is present, the shape of the periodic signal can be recovered upon averaging. This is illustrated in Figs 2.9(a) and 2.9(b), which show that the shape of both subthreshold and suprathreshold input signals can be recovered in the presence of noise by ensemble averaging the output, albeit with a certain amount of distortion.

The use of ensemble averaging to increase the SNR of a noisy signal is a well-known technique in areas such as sonar signal processing – see also Section 6.2 in Chapter 6. When N independently noisy realizations of the same signal are averaged, the resulting SNR is known to be increased by a factor of N. This is illustrated in Figs 2.9(c) and 2.9(d), where the input signals shown in Figs 2.5 and 2.8 respectively have been ensemble averaged for 1000 different noise realizations. The difference between this technique and the situation described in Figs 2.9(a) and 2.9(b) is that in one case the signal being averaged is continuously valued, whereas in the thresholded signal case, the signal being averaged is binary. This distinction is crucial. Different techniques must be used to properly analyze each case. Furthermore, in practice, ensemble averaging a continuously valued signal to increase its SNR is simply not practical, due to the difficulty of precisely storing an analogue quantity. Instead, such a signal is converted to a digital signal by quantization, prior to averaging, giving rise to a situation highly analogous to the scenario in Figs 2.9(a) and 2.9(b). The difference is that quantization is performed by a quantizer with a number of different threshold values, on each signal realization. In the case of Figs 2.9(a) and 2.9(b), a one-bit quantization is performed. The presence of noise allows ensemble averaging to improve the resulting output SNR.

It seems that, at least in terms of efficiency, if the input SNR is very large, it might be worthwhile to perform the one-bit quantization N times, provided the noise is independent in each realization. This is in contrast with trying to obtain a high precision quantization of a very noisy signal – which is stored in a multi-bit binary number – and then averaging the result N times. Such a technique has indeed been performed in sonar signal processing, in a method known as DIgital MUltibeam Steering (DIMUS), employed in submarine sonar arrays (Rudnick 1960) – see Section 4.2 of Chapter 4.

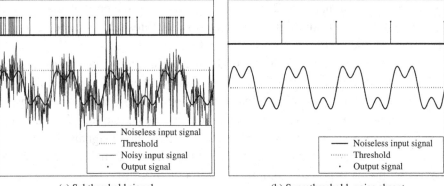

(a) Subthreshold signal   (b) Suprathreshold, noise absent

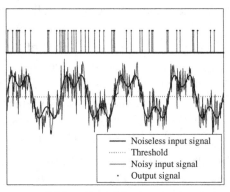

(c) Suprathreshold, noise present

Fig. 2.8. Threshold SR for a periodic but not sinusoidal signal. This figure
shows an input signal that, while periodic, is not a single frequency sine wave.
Instead, the input is the sum of two sine waves of different frequencies. In each
subfigure, the lower trace indicates the input signal's amplitude against increasing
time, with the straight line being the threshold value. The upper plot is the output
signal plotted against the same time-scale as the input signal. The output signal
is a 'spike' – that is, a short duration pulse – every time the input crosses the
threshold with positive slope – that is, from subthreshold to suprathreshold but
not *vice versa*. Figure 2.8(a) shows the output signal for the case where the input
signal is subthreshold, but additive noise causes threshold crossings. As with the
single frequency input signal of Fig. 2.5, the noise causes output pulses to occur.
The probability of a pulse is higher when the signal is closer to the threshold.
Figure 2.8(b) illustrates how, in the absence of noise, thresholding this signal at
its mean will provide exactly the same output as a single frequency sine wave.
Figure 2.8(c) shows the output signal when the input is thresholded at its mean
and additive noise is present. Unlike in the absence of noise, more than one pulse
per period can occur. This means output noise is created, if it is only the input
signal's period that is to be recovered. On the other hand, the presence of noise
allows the shape of the input signal to be recovered by ensemble averaging – see
Fig. 2.9.

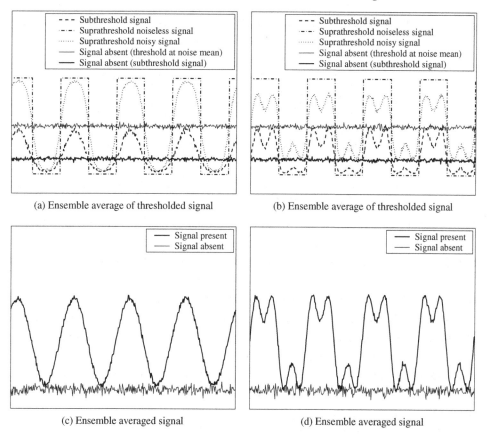

(a) Ensemble average of thresholded signal

(b) Ensemble average of thresholded signal

(c) Ensemble averaged signal

(d) Ensemble averaged signal

Fig. 2.9. Ensemble averages of thresholded and unthresholded signals. This figure shows the result of ensemble averaging 1000 realizations of the outputs shown in Figs 2.5, 2.7, and 2.8, as well as ensemble averages of the input signals. Figure 2.9(a) shows ensemble averages when the input is the single frequency sine wave of Figs 2.5 and 2.7. In the absence of noise, the ensemble average is a square wave. However, in the presence of noise, the ensemble average is clearly closer in shape to the original input signal, although also somewhat distorted. This is true for both subthreshold and suprathreshold input signals. The average of the noise is also shown for comparison. Figure 2.9(b) shows the ensemble averages for the periodic but not single frequency input signal of Fig. 2.8. The absence of noise gives the same signal as for the single frequency input, whereas again the presence of noise provides a signal that has a shape close to that of the input signal. It is well known that averaging a noisy signal N times reduces the SNR by a factor of N. This is illustrated for the unthresholded input signals in Figs 2.9(c) and 2.9(d). Clearly the ensemble averaging of the thresholded signals in the presence of noise provides a similar effect to this, although since distortion is introduced due to the discrete nature of the output, the output SNR is greater.

In light of the above discussion, it can be said that SR can occur for non-subthreshold signals; we simply need to clarify what it is that is being measured! This fact was perhaps overlooked due to an ingrained emphasis on measuring SR by the output SNR at the fundamental frequency of the input periodic signal. When this was starting to be questioned, the emphasis switched to aperiodic input signals. As discussed below, the ensemble averaging performed here cannot be carried out for aperiodic signals in the same way.

Suprathreshold stochastic resonance

The fact that ensemble averaging a thresholded periodic signal can provide a better response in the presence of noise rather than its absence leads, almost, but not quite, to the concept of suprathreshold stochastic resonance (SSR) (Stocks 2000c).

In the situations illustrated above, it is the periodicity of the input – regardless of the shape – that allows ensemble averages, taken over a period of time, to increase the output SNR. Since we know that the signal is periodic, provided that separate ensembles of the input are all mutually in phase, each signal segment that is averaged can be collected at any point in time. This allows independent amplitudes of noise to be added to each amplitude of the signal, since each amplitude of signal is periodically repeated, whereas the noise is not. This situation is described in Gammaitoni *et al.* (1998). In such a situation, if the input signal is aperiodic, ensemble averaging in this fashion would not work.

By contrast, we shall see in Chapter 4 that SSR refers to the *instantaneous* averaging of the outputs from an *array* of independently noisy threshold devices that all receive the same signal.

We note briefly that although it is normally considered that SR does not occur in a single device for suprathreshold signals, it has in fact been reported that an SR effect, known as *residual* stochastic resonance, can occur in a bistable system for a weakly suprathreshold signal (Gammaitoni *et al.* 1995b, Apostolico *et al.* 1997, Duan *et al.* 2004).

SSR has also received some criticism because of its apparent similarity to a situation where independently noisy versions of the same signal are each passed through linear amplifiers, and then averaged. However, this view misses a crucial point. The result of this procedure would remain an analogue signal. In the SSR model discussed in this book, the output signal is a quantized – that is, digital – signal. This fact should be considered to be equally important to the fact that the output signal is less noisy than the input, as digitizing a signal – that is, *compressing it* – provides the possibility of immunity from the effects of more noise in subsequent propagation and computation. This is the case whether we are considering artificial digital technology – where nearly all modern systems are digital – or biology, where the senses communicate analogue stimuli that they observe to the brain predominantly using 'all-or-nothing' electrical action potentials.

Further discussion of SSR is deferred until Chapters 4–9. We demonstrate in these chapters that SSR can be described as a form of nondeterministic quantization. As an aid to those unfamiliar with quantization in the signal processing context, the next chapter discusses quantization theory. Before then, however, we give a brief pointer to some of the most important results on SR in neurons and neural systems.

2.9 Does stochastic resonance occur *in vivo* in neural systems?

According to the ISI Web of Knowledge database, at the time of writing, about 20% of SR papers also contain a reference in the title, abstract, or keywords to the words *neuron* or *neural*.

As mentioned previously, the first papers investigating SR in neuron models appeared in 1991 (Bulsara *et al.* 1991, Bulsara and Moss 1991, Longtin *et al.* 1991) with such research accelerating – for example Longtin (1993), Chialvo and Apkarian (1993) and Longtin *et al.* (1994) – after the 1993 observation of SR in physiological experiments where external signal and noise were applied to crayfish mechanoreceptors (Douglass *et al.* 1993). A good history of this early work on SR in neurons is given in Hohn (2001), and a recent summary of progress in the field was published in Moss *et al.* (2004).

However, there are some other published works that indicate that the positive role of noise in neurons was noticed prior to 1991.

For example, in 1971 the first comprehensive analytical studies of the effects of noise on neuron firing demonstrated that noise 'smoothes' the firing response of neurons (Lecar and Nossal 1971a, Lecar and Nossal 1971b). Later, Horsthemke and Lefever (1980) discuss noise-induced transitions in neural models, and in particular Yu and Lewis (1989) advocate noise as being an important element in signal modulation by neurons.

Crucially, none of the above cited papers has been able to prove that neurons use SR in a natural setting – the evidence for neurons 'using SR' is only indirect. What has been observed is that neurons are nonlinear dynamical systems for which SR effects occur when signal and noise are both added externally. A direct observation of SR would require an application of an external signal, and measurements of *internal* neural noise.

If it can be established that SR plays an important role in the encoding and processing of information in the brain, and that it somehow provides part of the brain's superior performance to computers and artificial intelligence in some areas, then using this knowledge in engineering systems may revolutionize the way we design computers, sensors, and communications systems.

One of the most intriguing proposed applications inspired by SR, as first suggested by Morse and Evans (1996), is that of enhanced cochlear implant signal encoding. Various authors have since advocated the exploitation of SR in this area (Morse and Evans 1996, Morse and Roper 2000, Morse and Meyer 2000, Hohn and Burkitt 2001, Stocks *et al.* 2002, Chatterjee and Robert 2001, Rubinstein and Hong 2003, Behnam and Zeng 2003, Moore 2003, Chatterjee and Oba 2005, Morse *et al.* 2007). The basic idea is that several sources of substantial randomness are known to exist in healthy functioning inner ears (Hudspeth 1989, Lewis *et al.* 2000, Henry and Lewis 2001, Robles and Roggero 2001). Well-controlled noise in the output of cochlear implant electrical signals would therefore stimulate nerve fibres in a more natural way, and hopefully improve hearing in deaf cochlear-implant patients. The same principle of making the output of biomedical prosthetics more like biology – including any random aspects – has previously been applied in mechanical ventilators (Suki *et al.* 1998). The subject of cochlear implants is discussed further in Chapter 10.

2.10 Chapter summary

In this chapter a historical perspective is used to review SR, and the main sub-areas of SR important to this book. In particular, we define SR, as it is most widely understood, and discuss its occurrence in simple threshold-based systems.

Chapter 2 in a nutshell

This chapter includes the following highlights:

- A discussion of the evolution of the term 'stochastic resonance'.
- A historical review and elucidation of the major epochs in the history of SR research.
- A qualitative demonstration that SR can occur in a single threshold device, where the threshold is set to the signal mean. SR will not occur in the conventional SNR measure in this situation, but only in a measure of distortion, after ensemble averaging.
- A discussion of some of the confusing and controversial aspects of SR research, and a critique of the application of SNR and information-theoretic measures.

This concludes Chapter 2, which sets the historical context for this book. The next chapter presents some information-theoretic definitions required in the remainder of this book, and overviews signal quantization theory.

3

Stochastic quantization

By definition, signal or data quantization schemes are noisy in that some information about a measurement or variable is lost in the process of quantization. Other
systems are subject to stochastic forms of noise that interfere with the accurate
recovery of a signal, or cause inaccuracies in measurements. However stochastic noise and quantization can both be incredibly useful in natural processes or
engineered systems. As we saw in Chapter 2, one way in which noisy behaviour
can be useful is through a phenomenon known as stochastic resonance (SR). In
order to relate SR and signal quantization, this chapter provides a brief history of
standard quantization theory. Such results and research have come mainly from
the electronic engineering community, where quantization needs to be understood
for the very important process of analogue-to-digital conversion – a fundamental
requirement for the plethora of digital systems in the modern world.

3.1 Information and quantization theory

Analogue-to-digital conversion (ADC) is a fundamental stage in the electronic stor-
age and transmission of information. This process involves obtaining samples of a
signal, and their quantization to one of a finite number of levels.

According to the *Australian Macquarie Dictionary*, the definition of the word
'quantize' is

1. *Physics*: **a.** to restrict (a variable) to a discrete value rather than a set of continuous
values. **b.** to assign (a discrete value), as a quantum, to the energy content or level of a
system. **2.** *Electronics*: to convert a continuous signal waveform into a waveform which
can have only a finite number (usually two) of values. (Delbridge *et al.* 1997)

One of the aims of this book is to consider theoretical measures of the performance of a *stochastic* quantization method – that is, a method that assigns discrete
values in a nondeterministic fashion – and compare its performance with some of
the conventional quantization schemes that are often used in ADCs.

This chapter describes the basic ideas of quantization and then lists some important results in quantization theory. However, first we touch briefly on information theory, and the concepts of *entropy* and *mutual information*. These ideas are required for most of the other chapters in this book.

3.2 Entropy, relative entropy, and mutual information

Entropy

Consider a continuous and stationary random variable, X, with probability density function (PDF) $f_X(x)$, and support S. The *entropy* of X is defined (Cover and Thomas 1991) as

$$H(X) = - \int_S f_X(x) \log_2 (f_X(x)) dx. \tag{3.1}$$

An analogous expression to this holds for a discrete random variable, Z, with probability mass function, $P_Z(i)$, $i = 1, \ldots, n$

$$H(Z) = - \sum_{i=1}^{n} P_Z(i) \log_2 (P_Z(i)). \tag{3.2}$$

Note that the entropy of a continuous random variable is subtly different from the discrete case, and is more properly known as *differential entropy* (Cover and Thomas 1991). For example, discrete entropy is upper bounded by $\log_2 (n)$, and is always nonnegative. In contrast, differential entropy can be positive, zero, or negative. The reasons for this are discussed in Cover and Thomas (1991).

Relative entropy

Consider a second continuous random variable, Y, with PDF $f_Y(y)$, and the same support, S. The *relative entropy* – or Kullback–Liebler divergence – between the distributions of the two random variables is defined as (Cover and Thomas 1991)

$$D(f_X \| f_Y) = \int_{\eta \in S} f_X(\eta) \log_2 \left(\frac{f_X(\eta)}{f_Y(\eta)} \right) d\eta. \tag{3.3}$$

Mutual information

Suppose X and Y may be correlated, and have joint PDF $f_{XY}(x, y)$. Shannon's mutual information between the random variables, X and Y, is defined as the relative entropy between the joint PDF and the product of the marginal PDFs (Cover and Thomas 1991)

$$I(X, Y) = \int_x \int_y f_{XY}(x, y) \log_2 \left(\frac{f_{XY}(x, y)}{f_X(x) f_Y(y)} \right) dx\, dy \quad \text{bits per sample.} \quad (3.4)$$

It can be shown (Cover and Thomas 1991) that mutual information can be expressed as the difference between the entropy of X, $H(X)$, and the average conditional entropy, or equivocation, $H(Y|X)$, as

$$I(X, Y) = H(Y) - H(Y|X). \quad (3.5)$$

Unlike entropy, mutual information is nonnegative for both continuous and discrete random variables.

Conditional entropy

Later, we shall consider the conditional entropy of a discrete random variable, Z, given a continuous random variable, X. This is given by

$$H(Z|X) = \sum_{i=1}^n \int_x P_{X,Z}(x, i) \log_2 (P_{Z|X}(i|x)) dx, \quad (3.6)$$

where $P_{X,Z}(x, i)$ is the joint probability density of X and Z and $P_{Z|X}(i|x)$ is the conditional probability mass function of Z given X.

3.3 The basics of lossy source coding and quantization theory

Comprehensive reviews of the history of quantization theory, and the closely related topic of lossy source coding, can be found in Gray and Neuhoff (1998) and Berger and Gibson (1998). These papers provide a detailed history of early practically motivated quantization work, such as pulse code modulation in the 1950s, as well as early theoretical work on lossy source coding, including Shannon's initial formulation of the problem of minimizing the rate required to achieve a given distortion. As well as setting the historical context for these fields, Gray and Neuhoff (1998) and Berger and Gibson (1998) also work through the state of the art, and future directions for research. Another excellent reference is the textbook Gersho and Gray (1992).

We shall now briefly discuss some of the aspects of quantization theory most pertinent to the topic of this book.

Quantization of a signal or 'source' consists of the partitioning of every sample of the signal into a discrete number of intervals, or cells. Certain rules specify which range of values of the signal gets assigned to each cell. This process is known as *encoding*. If an estimate of the original signal is required to be made from this encoding, then each cell must also be assigned a reproduction value. This process is known as *decoding*.

Encoding

Although quantization is an integral part of an ADC, it is of course not restricted to such a narrow scope; the input does not need to be an electronic signal or a continuously valued variable. A basic example of quantization is in the formation of a histogram for some real valued data set. For example, consider a study that measures the heights of 1000 people. The researcher may decide to divide her measurements up into ten bins, of which eight are equally spaced with length 5 cm starting from 160 cm. The other two bins – overflow bins – are for measured heights of less than 160 cm, and more than 200 cm. To obtain a histogram, the measurements that fall into each bin are then counted and plotted against the index of the bin.

What information can be gleaned from the histogram about the heights of the people in the study? The researcher will look at statistical measures such as percentage frequency of each bin, mean, mode, median, and variance. If all the measurements fall in just one or two bins, the researcher will realize that the bin spacing, and the difference between the maximum and minimum bins is too wide to obtain any detail about the distribution of heights, or that more bins are required. Another similar problem will occur if most of the measured heights are in one or both of the overflow bins.

Decoding

Such questions of bin number, size, and placement are precisely those faced by the designer of a quantizer. The difficulty of finding a good design may also be compounded by the fact that the signal being quantized may not be as stationary as the heights of people. Furthermore, not only does the quantization of a signal usually require the *encoding* into bins, it also requires the *decoding* operation to be specified. In the histogram binning analogy, the need for decoding is probably not all that interesting to the anatomy researcher measuring people's heights, but could be understood as follows. Select randomly one of the 1000 people whose heights were measured and ask that person to specify which bin his height falls in, without specifying his exact height. What can the researcher say about the height? If the bin is the one from 180 to 185 cm, only that the person is no shorter than 180 cm and no taller than 185 cm. If asked to guess the height, the researcher would probably guess 182.5 cm, knowing that the maximum error in the guess would be 2.5 cm.

It is this question of assigning an estimated height to a bin that is exactly the problem of decoding in quantization. In the case of a signal being quantized, the value assigned as the decoding for each quantization bin is sometimes known as the *reproduction point*.

Another example of quantization is the representation of real numbers in a computer's architecture. Examples of such quantization schemes include the IEEE floating-point and fixed-point standards (Widrow *et al.* 1996).

Measures of a quantizer's performance

One important measure of a quantizer is its *rate*. In this context, 'rate' does not necessarily refer to a quantity that is defined in terms of 'per unit time'. For example, mutual information, entropy, or the number of output bits have all been used as the definition of 'rate'. The idea is that rate provides a measure of how many bits – that is, how many binary symbols – are required to represent information. This means that the rate of a quantizer is usually meant as the (average) number of bits per sample that the quantizer output consists of, or contains about the input. If mutual information is used as the definition of rate, then, in general, the rate will depend on the statistics of the input signal as well as the encoding process. However, for deterministic encoding, the rate is simply the average entropy of the output encoding, which will be the same as the mutual information, and is given by

$$I(X, Y) = H(Y) = -\sum_{i=0}^{N} P_Y(i) \log_2 P_Y(i), \tag{3.7}$$

where $P_Y(i)$ is the probability of output state i occurring. The maximum rate occurs when all output states are equally likely and is given by $I(X, Y)_{\max} = \log_2(N + 1)$, that is the number of bits at the output of the quantizer.

Information theory also tells us that the quantization of a signal will always cause some error in a reproduction of the original signal. This error is known as the distortion, and is most commonly measured by the mean square error between the original signal and the reproduced signal (Gray and Neuhoff 1998). If the encoding is decoded to a signal, z, the error is given by

$$\epsilon = x - z, \tag{3.8}$$

and the mean square distortion is

$$D_{\mathrm{ms}} = \mathrm{E}[(x - z)^2]. \tag{3.9}$$

A commonly used measure of a quantizer's performance is its signal-to-quantization-noise ratio (SQNR), which is the ratio of the input signal's power to the mean square distortion power. If the input signal has power σ_x^2, this can be expressed as

$$\text{SQNR} = 10 \log_{10} \left(\frac{\sigma_x^2}{D_{\text{ms}}} \right).\tag{3.10}$$

Optimal quantization

We have seen that the design of a quantizer reduces to choosing how to partition an input into bins, and the selection of reproduction points for each bin. Selecting the bins requires choosing the number of bins and the values of the thresholds that define which input values go in to each bin. Reproduction points should be selected that provide a good estimate of all the values represented by the corresponding bin.

So how should a quantizer be optimally designed? Suppose that quantization is to result in $\log_2 (N)$ bits. Then $N - 1$ threshold values and N reproduction points are required.

Given a measure of distortion, the reproduction points can be chosen as the values that, on average, minimize that distortion measure, for a given input signal. For continuously valued random source distributions – other than the uniform distribution – analytic expressions for the optimal thresholds are rare, at least for the mean square error distortion. However, standard numerical algorithms such as the Lloyd–Max method (Max 1960, Lloyd 1982, Gersho and Gray 1992) can be applied to find the optimal partition and reproduction points for a given specified source distribution, or a set of training data.

However, these algorithms appear never to have been extended to consider the situation where the threshold values that form the partitions are independently noisy or random variables. Such a situation is discussed in this book, and we call it *stochastic quantization*. As the final part of this chapter, we outline why stochastic quantization should be considered as a different phenomenon from dithering. This requires discussion of the relationship between dithering and SR.

3.4 Differences between stochastic quantization and dithering

We discussed in Section 2.3 how there has been some debate over whether SR in static threshold systems should be called dithering. Dithering is seen by some to be equivalent to SR in threshold systems, since it is a technique where a random signal is deliberately added to another signal to achieve a desired effect. Dithering has a specific goal: it is used in the process of digitization in order to reduce undesirable distortion resulting from quantization. We concluded in Chapter 2 that the definition of SR has evolved sufficiently from its initial use such that SR can be said to occur in the process of dithering, and that the two terms are not mutually exclusive.

However, we want to be very clear that when we discuss stochastic quantization in this book, although it bears some similarities to dithering, it is fundamentally different from any previous work on dithering that we are aware of. Before illustrating this further, it is worth considering some of the prior work that explicitly discusses the similarities and differences between SR and dithering (Gammaitoni 1995a, Gammaitoni 1995b, Wannamaker *et al.* 2000a, Andò and Graziani 2000, Lim and Saloma 2001, Andò 2002, Zozor and Amblard 2005), as well as summarizing dithering as it is understood by engineers.

Dithering in the context of stochastic resonance

The initial studies on SR in simple static threshold systems (Wiesenfeld *et al.* 1994, Jung 1994, Jung and Mayer-Kress 1995, Jung 1995, Gingl *et al.* 1995a, Gingl *et al.* 1995b) led Gammaitoni to publish two separate papers illustrating his view that SR in threshold systems is equivalent to the effect of *dithering* (Gammaitoni 1995a, Gammaitoni 1995b).

Both papers consider a periodic subthreshold input signal subject to additive white noise. In the initial paper, only a single threshold is considered at first. In subsequent sections and in the follow-up paper, a system in which more than one threshold is used to quantize a subthreshold signal is examined. However in both cases, the main point made is that SR in such systems can be considered as a special case of dithering.

Also relevant to later chapters of this book is a formula for measuring the dithering effect proposed in Gammaitoni (1995a). This can be written as

$$D = \sqrt{\int_x (\mathrm{E}[y|x] - x)^2 \, dx},   (3.11)$$

where x is the input signal to a threshold system, y is the output signal, and $\mathrm{E}[\cdot]$ indicates the expected value. This means that D is the root mean square (rms) error between the input signal and the average output signal. An undiscussed assumption built into this equation is that a uniform weighting is given to each possible input signal value. In general, if the input signal is taken from a random distribution with PDF $f_x(x)$, Eq. (3.11) can be rewritten to take into account the varying probabilities of each value of x occurring as

$$D = \sqrt{\int_x (\mathrm{E}[y|x] - x)^2 \, f_x(x) dx},   (3.12)$$

where the integration is over the support of $f_x(x)$. This formula now gives the *root mean square (rms) bias* of the system, a term that will be discussed in Chapter 6, where we shall see that a performance measure of a quantizer should, in contrast to Gammaitoni (1995b) and Gammaitoni (1995a), also take into account the average conditional error variance as well as the bias.

Some years after this, Andò and Graziani tackled the question head-on, suggesting that SR and dithering be differentiated as follows:

Stochastic resonance is a phenomenon that is observed in the natural world, whereas dithering is an artificial technique. (Andò and Graziani 2001)

Although *prima facie* this is a suitable way of distinguishing SR from dithering, such a mutually exclusive distinction does not leave room for the possibility for unintentional noise enhancement that acts identically to dither – there is plenty of anecdotal evidence from engineers that this can occur. Moreover, Andò and Graziani's statement also ignores that SR has been observed in many kinds of artificial systems.

Furthermore, Andò and Graziani (2001) refers to SR as a noise-added 'technique' – see also Andò (2002). Calling SR a technique is somewhat misleading, as this implies that any manifestation of SR requires some deliberate artificial procedure to be carried out. Instead, SR is a phenomenon that is observed, when an experiment, measurement, or simulation is undertaken. We do not say 'stochastic resonance is carried out', or the like; when measurements indicate that a nonzero level of noise optimizes performance, then SR is said to *occur*. In contrast, dithering is usually understood to be a technique where noise is deliberately introduced to a system, with a specific aim in mind, and SR can be said to occur if the dither signal provides the desired enhancement. Indeed, this viewpoint was put by Wannamaker *et al.*, who compared SR in threshold systems with nonsubtractive dithering, arguing that

many static systems displaying stochastic resonance are forms of dithered quantizers, and that the existence or absence of stochastic resonance in such systems can be predicted from the effects of 'dither averaging' upon their transfer characteristics. (Wannamaker *et al.* 2000a)

This view that SR occurs in a dithered system contradicts that of Andò and Graziani (2001). Further illustration of the occurrence of 'noise-enhanced processing effects' in dithered sigma-delta ADCs appears in Zozor and Amblard (2005).

Rather than attempt to classify noise-enhanced behaviour in threshold systems as either dithering or SR, the question that might be asked instead is whether dithering has been carried out in order for SR to occur. To answer this requires discussion of what dithering actually means to those who use it in engineering.

What does dithering mean to an engineer?

It is difficult to find a comprehensive definition of *dither* and what it achieves. It is treated slightly differently in different contexts, for example digital audio (Pohlmann 2005), image processing (Roberts 1962), analogue-to-digital converter circuits (Kikkert 1995), instrumentation and measurement (Dunay *et al.* 1998, Carbone and Petri 2000), control theory (Zames and Shneydor 1976), and information theory (Schuchman 1964, Gray and Stockham 1993). There is no text or reference book solely on dithering, and perhaps the most comprehensive and readable introductory references, which also contain a concise history of the topic, are Wannamaker (1997) and Wannamaker *et al.* (2000b). A textbook with a large section on dithering is Jayant and Noll (1984).

The aspects of dithering that are agreed upon are:

- a *dither signal* is a random, or pseudo-random, signal that is added to a signal prior to quantization;
- the purpose of adding a dither signal is to modify the statistical properties of the quantization error signal.

The most common kind of dithering is subtractive dithering, in which a copy of the dither signal is subtracted after quantization. This subtraction immediately makes dithering different from SR in threshold systems where subtraction is not carried out (Wannamaker *et al.* 2000a). However a less common kind of dithering called nonsubtractive dithering exists, and, as pointed out in Wannamaker *et al.* (2000b), there has been some confusion in the literature about the different properties of subtractive and nonsubtractive dithering, for example Jayant and Noll (1984, p. 170).

In either case, the addition of a dither signal prior to quantization is recognized as a tradeoff – a small amount of additional noise power at the output of the quantizer is a small price to pay for a large decrease in undesirable 'distortion' effects resulting from the quantization error. What are these effects, and why are they undesirable? Given dithering's close association with audio and image processing, the 'undesirable distortion' is usually described in terms of how it is perceived by our ears and eyes. This subjective description can be quantified in terms of how the statistics of the quantization error after dithering can be made independent of the input signal (Jayant and Noll 1984, Pohlmann 2005, Wannamaker *et al.* 2000b).

However, not only does dithering reduce undesirable distortion effects for suprathreshold signals, it also allows the possibility of encoding low-level signals, that is parts of a signal where the amplitude is always less than the least threshold (Pohlmann 2005). In this event, the absence of a dither signal would mean the input is always coded as being completely absent. When dither is present, the least threshold value will sometimes be crossed, with a probability related to how close the input signal is to that threshold. This is identical to the situation described in Section 2.5 of Chapter 2, and is the aspect of dithering most like studies of SR in threshold-based systems.

Recall that a subthreshold signal subject to a thresholding operation is nondetectable at the output. Adding noise to the input signal to allow threshold crossings is effectively the same as a dither signal when the signal amplitude is smaller than the quantizer's bin size. See, for example, Wannamaker *et al.* (2000a) and Lim and Saloma (2001) for discussions on this.

Why dithering and SR are not mutually exclusive

We now attempt to summarize and rebut several possible ways to differentiate between dithering and SR, in light of the above.

SR occurs in dynamical systems, dither occurs in static threshold systems

This is the view expressed by Gammaitoni (1995a), and follows from an interpretation that SR needs to somehow result in a *bona fide* resonance, that is the matching of two characteristic time-scales. As discussed in Section 2.3, contemporary widespread usage of the term *stochastic resonance* has seen its definition evolve beyond this, and others agree that:

this approach seems to be unsuitable for a clear classification. (Andò 2002)

SR occurs in the natural world, while dither is artificial

While it is true that dither signals are usually artificial, this view, expressed by Andò and Graziani (2001), ignores the fact that SR, as it is usually understood, is an *observed* effect that is well known to occur in either natural or artificial systems. In other words, SR *occurs* in dithered systems (Wannamaker *et al.* 2000a).

SR enhances subthreshold signals, dither affects suprathreshold signals

Given the historical emphasis in the SR literature for analyzing weak subthreshold signals, it is no surprise that the papers discussing the differences between SR and dithering have all restricted their attention to small periodic input signals, whose amplitudes are not sufficient to allow threshold crossings to occur without additive noise (Gammaitoni 1995a, Gammaitoni 1995b, Wannamaker *et al.* 2000a). In general, dithering is applied to a signal that is then quantized by multiple thresholds, rather than a single threshold, and which has a dynamic range that does not require noise to induce threshold crossings, that is the signal is suprathreshold. Given this difference, perhaps it could be said that dithering occurs for suprathreshold signals, while SR occurs for subthreshold signals? This is more or less the suggestion made by Andò (2002) – that is, that SR is equivalent to threshold reduction via noise, while dithering linearizes a system's characteristics.

However, this ignores the fact that dithering is widely recognized as providing two desirable effects. In addition to making the error signal less correlated with the input signal, it also allows coding of amplitudes smaller than the smallest threshold value. Hence, to say that dither only affects suprathreshold signals would not be correct.

Dithering is a technique that provides conditions in which SR can occur

In line with the definition of SR, *noise-enhanced signal processing*, which we believe reflects its widespread contemporary usage, rather than classifying some studies as dithering and others as SR, it is logical to suggest that dithering and SR are not mutually exclusive, but that SR occurs when dithering is used. If a noise signal such as a dither signal is introduced to exploit SR, with an aim that is different from that of dithering as it is commonly understood, then the term dithering is best avoided.

Stochastic quantization

We have defined 'stochastic quantization' to mean quantization by thresholds that are independent random variables. This makes it clear that stochastic quantization is not the same as dithering, since dither signals are added to the signal to be quantized prior to quantization. Although such a dither signal could be viewed as causing stochastic quantization, it is different in that all threshold values are made to vary by the same amount; for a given source sample, the sample of the dither signal will modify all threshold values equally. This is very different from the case where N independent noise sources act to independently randomize each threshold value.

There are also several other crucial differences between dithering as it is usually understood and the concept we have called 'stochastic quantization'. These are that dither signals:

- are usually considered to have a dynamic range smaller than the width of a quantizer's bin size, and therefore are small compared to the signal's dynamic range;
- are usually applied to quantizers with widely spaced thresholds;
- and are usually distributed on a finite range, such as the uniform distribution, rather than having PDFs with infinite tails, such as the Gaussian distribution.

By contrast, we shall consider stochastic quantization in the following scenarios:

- large dither – or noise – amplitudes, compared to the signal's amplitude;
- the case where all thresholds in a quantizer have identical values, but become independent random variables due to the addition of noise;
- and noise signals with PDFs with infinite tails.

3.5 Estimation theory

We shall also in this book touch on areas of *point estimation theory*. An excellent technical reference encompassing this field is Lehmann and Casella (1998). The main estimation topic we shall look at is that of minimizing mean square error distortion between the input and output signals of a nonlinear system. Such a goal also appears in quantization theory; however, we shall find it useful to use ideas from estimation theory, such as Fisher information and the Cramer–Rao bound, which are not generally used in conventional quantization theory.

There are also a number of papers in the SR literature that tackle SR from this point of view (Greenwood *et al.* 1999, Greenwood *et al.* 2000, Chapeau-Blondeau and Rojas-Varela 2001, Greenwood *et al.* 2003, Chapeau-Blondeau and Rousseau 2004).

3.6 Chapter summary

This chapter introduces the concept of signal quantization, briefly indicates some of the most commonly used measures of a quantizer's performance, and points out references in the literature that discuss quantization in full technical detail.

Chapter 3 in a nutshell

This chapter includes the following highlights:

- Definitions of information-theoretic quantities.
- The definition of signal quantization.
- A discussion on the differences between SR and dithering.

This concludes Chapter 3. The next chapter begins the main topic of this book – that of suprathreshold stochastic resonance (SSR).

4

Suprathreshold stochastic resonance: encoding

In many of the systems and models in which stochastic resonance has been observed, the essential nonlinearity is effectively a single threshold. Usually SR occurs when an entirely subthreshold signal is subjected to additive noise, which allows threshold crossings to occur that otherwise would not have. In such systems, it is generally thought that when the input signal is suprathreshold, then the addition of noise will not have any beneficial effect on the system output.

However, the 1999 discovery of a novel form of SR in simple threshold-based systems showed that this is not the case. This phenomenon is known as suprathreshold stochastic resonance, and occurs in arrays of identical threshold devices subject to independent additive noise. In such arrays, SR can occur regardless of whether the signal is entirely subthreshold or not, hence the name *suprathreshold* SR. The SSR effect is quite general, and is not restricted to any particular type of signal or noise distribution.

This chapter reviews the early theoretical work on SSR. Recent theoretical extensions are also presented, as well as numerical analysis of previously unstudied input and noise signals, a new technique for calculating the mutual information by integration, and an investigation of a number of channel capacity questions for SSR. Finally, this chapter shows how SSR can be interpreted as a *stochastic quantization* scheme.

4.1 Introduction

Suprathreshold stochastic resonance (SSR) is a form of stochastic resonance (SR) that occurs in arrays of identical threshold devices. A schematic model of the system is shown in Fig. 4.1, and is described in detail in Section 4.3. The discovery of SR in such a system was made by Stocks in 1999 (Luchinsky *et al.* 1999b, Stocks

59

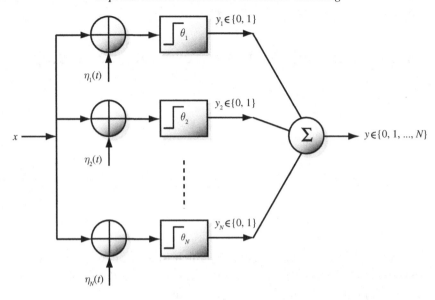

Fig. 4.1. Array of N noisy threshold devices. This schematic diagram shows the model in which SSR occurs, sometimes known as a *stochastic pooling network*. There are N identical threshold devices. The input signal is taken to be a random signal, consisting of a sequence of discrete time uncorrelated samples drawn from a continuously valued probability distribution. Each device receives the same input signal sample, x, and is subject to independent additive noise, η_i. The noise is also a random signal, and is independent from the input signal, x. The output from the nth device, y_i, is unity if the sum of the signal and noise at its input is greater than the corresponding threshold, θ_i, and zero otherwise. The overall output, y, is the sum of the individual outputs, y_i.

2000c).[1] Stocks showed – using an aperiodic input signal, meaning that SSR is a form of aperiodic stochastic resonance (ASR) – that SR can occur in Shannon's average mutual information measure between the input and output of the array, under the constraint that all thresholds are set to the same value. Most importantly, SR occurs regardless of whether the input signal is entirely subthreshold or not, which is the first known occurrence of such behaviour in threshold devices. Stocks named this effect *suprathreshold stochastic resonance*, to distinguish it from the occurrence of SR in previous studies of single-threshold systems, and subsequently showed that the effect is maximized when all threshold values are set to the signal mean (Stocks 2001a). Figure 4.1 also serves to describe the more general case when all thresholds are not identical, a case which is examined in detail in Chapter 8. In this chapter we shall only consider the case of all thresholds set to the same value.

---

[1] Note that the first description of SSR was given in a 1999 paper reviewing nonconventional forms of SR (Luchinsky *et al.* 1999b). However, although this paper was published first, it references a paper submitted to *Physical Review Letters*, which was subsequently published in 2000 (Stocks 2000c).

The mathematical treatment in this chapter follows along the lines of Stocks (2000c), Stocks (2001c), Stocks (2001a), McDonnell *et al.* (2001), McDonnell *et al.* (2002a), McDonnell and Abbott (2004b), and McDonnell *et al.* (2007).

Chapter structure

This chapter provides, in Section 4.2, a brief literature review that outlines the main results of all previous studies of SSR. Then, in Section 4.3, we reproduce the most important results from this previous work, provide some recent extensions to theory, and numerically examine a number of hitherto unconsidered signal and noise distributions. In contrast with the comment in Kosko (2006, p. 222), it is shown that the occurrence of SSR is not dependent on a particular signal or noise distribution. This analysis is followed in Section 4.4 by an investigation into a number of channel capacity questions for the SSR model. Finally, an interpretation of SSR as *stochastic quantization* is introduced in Section 4.5. A number of mathematical results relevant to this chapter are derived in more detail in Appendix 1.

4.2 Literature review

Information-theoretic studies of stochastic resonance in threshold systems

As discussed in Chapter 2, although the term *aperiodic stochastic resonance* was popularized by work using correlation-based measures (Collins *et al.* 1995a), the first papers to study ASR using mutual information (Levin and Miller 1996, Bulsara and Zador 1996, Heneghan *et al.* 1996) are of more relevance to SSR. These papers, along with DeWeese and Bialek (1995), paved the way for the use of information theory in SR research.

Of particular importance to SSR was the first study using mutual information to analyze a single threshold-device system (Bulsara and Zador 1996). This system is very similar to that discussed in Section 2.7 of Chapter 2, where the input signal to the system is a random binary signal, subject to continuously valued noise. The initial section of Bulsara and Zador (1996) calculates the input-output mutual information for this system for various threshold values, and shows that SR occurs for entirely subthreshold signals. Subsequently, Bulsara and Zador (1996) also consider mutual information in a leaky integrate-and-fire neuron model.

Also of relevance to SSR are the final paragraphs of Bulsara and Zador (1996), which discuss how future work might extend the system they considered. Specifically, Bulsara and Zador (1996) propose allowing continuously valued input signals, and an output signal which is the sum of the outputs of N binary threshold devices. It is stated that the mutual information in such a model would be of the order of $0.5 \log_2 (N)$. These extensions are exactly what is carried out in the first work on SSR.

Original work on suprathreshold stochastic resonance

The earliest work published in the open literature in which the phenomenon of SSR is presented is in the second part of a two-part review of SR in electrical circuits (Luchinsky *et al.* 1999b). Although this work was published first, it references a paper submitted to *Physical Review Letters*, which was subsequently published in 2000 (Stocks 2000c). This latter paper contains the same results as Luchinsky *et al.* (1999b), as well as some further analysis. In this original work of Stocks, the input signal, x, is taken to be a Gaussian random variable, and the noise on each threshold is likewise a Gaussian random variable. No reduced analytical expressions are obtained for the mutual information, which is instead calculated by numerical integration and validated by digital simulation.

A later paper (Stocks 2001c) presents an analytical expression for the mutual information in the specific case of a uniformly distributed signal, and uniformly distributed threshold noise. Another longer paper (Stocks 2001a) examines the behaviour of SSR in the event of large N, as well as providing an exact expression for the mutual information in the event that both the signal and noise distributions are identical, that is they have the same distribution and moments. Furthermore, Stocks (2001a) finds that the SSR effect is optimized in the case of all thresholds being set equal to the signal mean, as well as presenting an analysis of the output signal-to-noise ratio (SNR). Several conference and book chapters published a combination of the material presented in Stocks (2000c), Stocks (2001c), and Stocks (2001a). These include Stocks (2000b), Stocks (2000a), and Stocks (2001b). In Stocks (2001b), an interesting new measure of the performance of SSR is briefly described, that of a coding efficiency. This measure is briefly explored in Section 4.4. Furthermore, in both Stocks (2000a) and Stocks (2001b), an analysis of the optimal threshold configuration for the system in Fig. 4.1 is given. Such calculations are the subject of Chapter 8, and will not be considered here.

SSR in neurons and cochlear implant encoding

The original work on SSR contained brief comments – see, in particular, Stocks (2000a) – that the model shown in Fig. 4.1 has similarities to ensembles of sensory neurons. This, combined with the fact that SR is well known to occur in neurons (Levin and Miller 1996) – although whether *in vivo* is an open question – is one of the motivations for studying such a system. In order to further investigate the possibility that SSR can occur in real neurons, Stocks and Mannella (2001) – see also the book chapter, Stocks and Mannella (2000) – replaces the simple threshold device building block that makes up the system shown in Fig. 4.1, with the FitzHugh–Nagumo neuron model. It is shown that SSR can still occur despite this change in the model. Calculations of the mutual information in such a case are more complicated than the simple model analyzed in depth in this chapter, as the output is no longer discretely valued. Further discussion of this topic is left for Chapter 10.

Further justification of the SSR model's relevance to neural coding is discussed in Hoch *et al.* (2003b), Morse *et al.* (2007), McDonnell *et al.* (2007), and Nikitin *et al.* (2007), and by other extensions of the model to include more biologically realistic neural features. For example, the parallel array has been modified to consist of parallel leaky integrate-and-fire neuron models (Hoch *et al.* 2003b)[2] and Hodgkin–Huxley models (Hoch *et al.* 2003b), and for the case of signal-dependent (multiplicative) noise (Nikitin *et al.* 2007). In all cases the same qualitative results as for the simple threshold model were obtained.

Cochlear implants[3] are prosthetic devices that enable profoundly deaf people to hear (Dorman and Wilson 2004, Clark 2000). The operation of cochlear implants requires direct electrical stimulation of the cochlear nerve. Such stimulation requires sophisticated methods of signal encoding, and, although hearing can be restored, patients still have difficulty perceiving speech in a noisy room, or music. An interesting and potentially very important proposed application of both SR in general, and the SSR effect, is to incorporate its effects in the design of cochlear implant encoding. The idea is based on the fact that people requiring cochlear implants are missing the natural sensory hair cells that a functioning inner ear uses to encode sound in the cochlear nerve. It is known that these hair cells undergo significant Brownian motion – that is, randomness (Jaramillo and Wiesenfeld 1998, Bennett *et al.* 2004, Lindner *et al.* 2005). The hypothesis is that employing the principles of SSR to re-introduce this natural randomness to the encoding of sound could improve speech comprehension in patients fitted with cochlear implants (Morse *et al.* 2002, Stocks *et al.* 2002, Allingham *et al.* 2003, Allingham *et al.* 2004, Morse and Stocks 2005, Morse *et al.* 2007). Theoretical studies related to this have included analysis of SSR-like effects for multiplicative – that is, signal-dependent – noise, rather than additive input noise (Nikitin *et al.* 2005, Morse and Stocks 2005, Nikitin *et al.* 2007). Investigation and further discussion of this work is presented in Chapter 10.

---

[2] As yet unpublished work by S. J. Durrant and J. Feng has also discussed this.

[3] Historical note: the world's first *multi-channel* cochlear implant was invented in Australia by Graeme Clark – the first patient to receive the implant was in Melbourne, Australia, 1978 (Clark 1986, Clark 2000).

Other work on SSR

An increasing number of authors have now published results on SSR. On the theoretical side, Hoch *et al.* (2003b) – identical work is given in Hoch *et al.* (2003b) and Wenning (2004) – derives an approximation for the mutual information through the system of Fig. 4.1 that applies in the case of large N. Using this result, it is shown that for the case of a Gaussian signal and independent Gaussian noise – as studied in Stocks (2000c) and Stocks (2001a) – the value of noise intensity at which the mutual information reaches its maximum converges to a fixed value as N increases. More recently, Amblard *et al.* (2006) and Zozor *et al.* (2007) have also considered the large N situation for a generalization of the SSR model that they describe as *stochastic pooling networks*. This framework has been used to extend analysis of SSR to detection scenarios, and is currently an ongoing area of research.

As mentioned, Hoch *et al.* (2003b) also examine SSR in a neural coding context, including an investigation into the effects of SSR under energy efficiency coding constraints. We consider such a question from a different perspective in Chapter 9.

An analysis of the SSR model in terms of Fisher information is given in Rousseau *et al.* (2003). This paper also discusses for the first time the case of deterministic time-varying input signals, rather than random input signals. Such an input signal allows analysis of the performance of the SSR model in terms of SNR in the conventional SR sense. A detailed investigation of exactly this topic is reported in Rousseau and Chapeau-Blondeau (2004). Subsequently, Wang and Wu (2005) also use Fisher information to measure the performance of the SSR model, by examining how the 'thickness' of the 'tail' of the noise distribution affects the output response.

The SSR model shown in Fig. 4.1 is mentioned in Sato *et al.* (2004); however, this work considers the input signal to always be entirely sub-threshold. In such a case, SR can still occur; however, in this chapter, unlike Sato *et al.* (2004), we shall not restrict attention to entirely subthreshold input signals.

More practically oriented work includes an investigation of the potential use of SSR in motion detection systems (Harmer 2001, Harmer and Abbott 2001, McDonnell and Abbott 2002). There has also been a proposal to use the entire SSR model as the comparator component of a sigma-delta modulator (Oliaei 2003), which is a specific type of analogue-to-digital converter that makes use of feedback. Very recently, a heuristic method for analogue-to-digital conversion based on the SSR model has been proposed (Nguyen 2007). In this method, the input signal is transformed so that it has a large variance and is bimodal.

Similarity to SSR in other work

As already mentioned, the model shown in Fig. 4.1 can also serve as a model for a population of neurons. Furthermore, we shall see in Chapters 6–8 that it also serves as a model for quantization and analogue-to-digital conversion (ADC). However, in quantizers and ADCs it is certainly not conventional to assume that all thresholds have the same value, as is the case in this chapter. There are, however, some other systems in which all thresholds are identical, and so the SSR model shares similarities with work previously considered in the literature. The remainder of this subsection briefly points out references to such work. Finally, it discusses how the results on SSR first given in Stocks (2000c) differ from these works.

DIMUS sonar arrays

DIMUS sonar arrays (Rudnick 1960) are arrays of spatially distributed hydrophones that each 'clip' their inputs to provide a digital binary signal. If there are N hydrophones, then the overall output is the sum of the N individual binary signals. Such a method for acquiring acoustic sonar signals was first used prior to the 1960s, and provided the introduction of digital signal processing to the field. Due to the spatially distributed nature of the hydrophones, it is usually assumed that noise at the input to each hydrophone is mutually independent of every other hydrophone, while the signal at each remains identical. As noted in Stocks (2001a), such a situation is virtually identical to the SSR model. However, unlike the SSR model, DIMUS has never been analyzed from an information-theoretic perspective. The reason for this is that the signal processing goal of a DIMUS array is to detect whether or not a signal is present in background noise. Analysis of detection or false-alarm probabilities does not require information theory. Since we use information theory, we implicitly assume that a signal is present in the background noise. Rather than detection, we are interested in how much information about the signal can be obtained as the noise and N vary. Furthermore, no results have been published indicating the performance of the DIMUS sonar array as the noise intensity increases from zero to input SNRs less than zero decibels (dB), as we discuss here. The interested reader is referred to the following references (Anderson 1960, Rudnick 1960, Wolff *et al.* 1962, Remley 1966, Kanefsky 1966, Berndt 1968, Fitelson 1970, Wang 1972, Bershad and Feintuch 1974, Wang 1976, Anderson 1980, Tuteur and Presley 1981).

Semi-continuous channels

Although it is not immediately apparent on a first reading of a paper by Chang and Davisson (1988), which considers algorithms for calculating channel capacity for 'an infinite input and finite output channel', it actually is highly relevant to the SSR scenario. Both the channel considered by Chang and Davisson (1988) – see also Davisson and Leon-Garcia (1980) – and the SSR model of Fig. 4.1 have an input signal consisting of a continuously valued random signal, and an output signal that consists of a finite number of states. As will be discussed in Section 4.4, calculations of channel capacity values for various N given in Chang and Davisson (1988) are in fact equivalent to channel capacity values for SSR, under certain assumptions on the noise distribution.

Detection scenarios

A summation of independent binary values is also used in Kay (2000) in a detection scenario. This work describes the occurrence of SR in a suboptimal detector. However, unlike SSR, the input signal is only ever a single value, and is not considered to be a random variable.

Discussion on the use of information theory

All of the situations above – apart from Chang and Davisson (1988) – are considered from the viewpoint of signal detection, and the SNR of the output signal. As in Chang and Davisson (1988), if we assume that a signal is present in the background noise, all these systems can be analyzed in terms of information theory. In each case, the system in which a signal is propagated can be described as a *channel*. An important branch of information theory is concerned with the transmission of information through channels. Often, it is assumed that the signal is discrete time in nature. The basis for this assumption is that the input signal can be described as a random signal, and has a certain finite bandwidth. Under such an assumption, the sampling theorem guarantees that a continuous time signal can be sampled at the Nyquist rate. Subsequently, the original continuous time signal can be perfectly reconstructed from those samples. In such a situation, or indeed any case where a signal is composed of *iid* – independent and identically distributed – samples from some probability distribution, Shannon's mutual information measure provides a measure of the *average* number of bits per sample that can be transmitted by a single sample in some channel. When combined with a sample rate, measured in samples per second – often called bandwidth – an *information rate* through a channel can be stated, by multiplying the sample rate by the mutual information, to obtain the number of bits per second that are transmitted through the channel.

The remainder of this chapter studies the system in Fig. 4.1 from this information-theoretic viewpoint. All input signals are assumed to be discrete-time sequences of samples drawn from some stationary probability distribution. The result is that this work differs further from the detection scenario of Kay (2000), which considers a constant signal. Such a signal does not convey new information with an increasing number of samples, and cannot be considered from an information-theoretic viewpoint.

Finally, note that another branch of information theory is concerned with the problem of *lossy source coding*. This includes studies of quantization theory, where a continuously valued signal is compressed to a discretely valued encoding. Although combined source–channel coding research is increasing, information theory traditionally separates source coding and channel coding into two separate independent components of a communications system. An interesting feature of the SSR model is that it can be interpreted both as a channel model and as a source coding model. This is because the channel is semi-continuous; it has a continuously valued input signal, but a discretely valued output signal. This viewpoint leads to a natural interpretation of the SSR model as a quantization scheme. This subject is further studied in Chapters 6 and 7. Alternatively, since there is channel noise in the system, and the output is a noisy version of the input signal, the system can also naturally be seen as a channel. The remainder of this chapter looks at this topic, and analyzes SSR from a channel coding viewpoint.

4.3 Suprathreshold stochastic resonance

Notation

Throughout, we denote the probability mass function (PMF) of a discrete random variable, α, as $P_\alpha(\cdot)$, the probability density function (PDF) of a continuous random variable, β, as $f_\beta(\cdot)$, and the cumulative distribution function (CDF) of β as $F_\beta(\cdot)$.

The SSR model

Figure 4.1 shows a schematic diagram of the model system in which SSR occurs. This system consists of N threshold devices – we shall also refer to these as comparators – which all receive the same sample of a random input signal, x. This random signal is assumed to consist of a sequence of independent samples drawn from a distribution with PDF, $f_x(x)$. For such a situation to apply in a real system, the independence of each sample usually requires that a random continuous time signal is bandlimited and sampled at the Nyquist rate (Proakis and Salehi 1994), prior to input into this system as a sequence of discrete-time samples.

The ith device in the model is subject to continuously valued *iid* additive noise, $(\eta_i, \ i = 1, \ \ldots, N)$, drawn from a probability distribution with PDF $f_\eta(\eta)$. Each noise signal is required to also be independent of the signal, x. For each individual comparator, the output signal, y_i, is unity if the input signal, x, plus the noise on that comparator's thresholds, η_i, is greater than the threshold value, θ_i. The output signal is zero otherwise. The output, y_i from comparator i is summed to give the overall system output signal, y. Hence, y is a discrete signal, which can take integer values between zero and N.

The output of device i is then given by

$$y_i = \begin{cases} 1 & \text{if} \quad x + \eta_i \geq \theta_i, \\ 0 & \text{otherwise.} \end{cases} \tag{4.1}$$

The overall output of the array of comparators is $y = \sum_{i=1}^{N} y_i$. This can be expressed as a function of x in terms of the signum (sign) function as

$$y(x) = \frac{1}{2} \sum_{i=1}^{N} \text{sign}[x + \eta_i - \theta_i] + \frac{N}{2}. \tag{4.2}$$

As mentioned in Section 4.2, Kay (2000) has previously published work describing the occurrence of SR in a suboptimal detector. This work uses an expression similar to Eq. (4.2) in terms of the signum function, indicating some similarities between SSR and detection problems. However, the problem of deciding whether a constant signal has been detected, as carried out in Kay (2000), is beyond the scope of this text; the main task considered here is that of signal transmission and quantization, rather than detection.

Note that Eq. (4.2) completely describes the *transfer function* for the model system of Fig. 4.1. However, the output of the array of comparators is nondeterministic – except in the complete absence of noise – and represents a lossy encoding of the input signal. This means that the transfer function is not deterministic, since for a given input value, x, the output, $y(x)$, depends on the set of random variables, $\{\eta_i\}$, $i = 1, \ldots, N$.

This chapter considers only the case of all threshold values being identical. We let $\theta_i = \theta$ for all i and, without loss of generality, in this chapter the subscript, i, is dropped from all references to threshold values. It is also dropped for references to noise signals, η, but this does not imply that the noise on each threshold is no longer independent.

Since the encoding given by the transfer function of Eq. (4.2) is not deterministic, probabilistic measures are required for mathematical analysis of the system. The key function required is therefore the joint PDF between the input and output signals, $f_{xy}(x, y)$. Denoting the probability mass function of the output signal as $P_y(n)$ and making use of Bayes' theorem (Shiryaev 1996, Yates and Goodman 2005) gives the joint PDF as

$$f_{xy}(x, y) = P_{y|x}(y = n|x) f_x(x) \tag{4.3}$$
$$= f_{x|y}(x|y = n) P_y(n). \tag{4.4}$$

We shall describe the conditional distribution of the output given the input – denoted by $P_{y|x}(y = n|x)$ – as the *transition probabilities*, since $P_{y|x}(y = n|x)$ gives the probability that the encoding for a given input value, x, is encoded by output state n. From here on we shall abbreviate the notation $P_{y|x}(y = n|x)$ to $P_{y|x}(n|x)$. The transition probabilities can be used to obtain $P_y(n)$ as

$$P_y(n) = \int_{-\infty}^{\infty} P_{y|x}(n|x) f_x(x) dx \quad n = 0, \ldots, N. \tag{4.5}$$

In this chapter we shall always assume that the PDF, $f_x(x)$, is known, except briefly in Section 4.4, where channel capacity is considered. To progress further requires a method for calculating the transition probabilities. Using the notation of Stocks (2000c), let $P_{1|x}$ be the probability of any comparator being 'on' – that is, the sum of the signal, x, and noise, η, exceeds the threshold value, θ, for a single comparator – *given* that the input signal value, x, is known, that is

$$P_{1|x} = P(x + \eta > \theta|x). \tag{4.6}$$

This probability depends on the noise PDF, $f_\eta(\eta)$, as

$$P_{1|x} = \int_{\theta-x}^{\infty} f_\eta(\eta) d\eta = 1 - F_\eta(\theta - x), \tag{4.7}$$

where $F_\eta(\cdot)$ is the cumulative distribution function (CDF) of the noise. If $f_\eta(\eta)$ is an even function of η, then

$$P_{1|x} = F_\eta(x - \theta). \tag{4.8}$$

Since all thresholds have the same value, then – as noted in Stocks (2000c) – the transition probabilities are given by the binomial distribution (Shiryaev 1996) as

$$P_{y|x}(n|x) = \binom{N}{n} P_{1|x}^n (1 - P_{1|x})^{N-n} \quad n = 0, \ldots, N. \tag{4.9}$$

In many applications of the binomial distribution, the probability of a single event is a constant. Instead, here $P_{1|x}$ is a function of x and the transition probabilities are binomially distributed for a given value of x.

However, for a given value of n, the nature of the transition probabilities for a given n as a function of x is not as easily characterized. We can, however, easily find the value of x at which the maximum value, or peaks, of $P_{y|x}(n|x)$ for a given n occurs, by differentiating $P_{y|x}(n|x)$ with respect to x, setting to zero and solving for x. The details of this procedure are given in Section A1.1 of Appendix 1, which shows that the peak occurs when

$$P_{1|x} = \begin{cases} 1 & n = 0, \\ \frac{n}{N} & n = 1, \ldots, N - 1, \\ 0 & n = N. \end{cases} \tag{4.10}$$

In all cases we examine, the CDF will be sigmoidal rather than peaked for $n = 0$ and $n = N$, which means that the peaks occur at $\pm\infty$. For the context of determining the peaks, we shall ignore these values of n in this discussion. Substituting for Eq. (4.7) in Eq. (4.10) and solving for x gives

$$x = \theta - F_\eta^{-1}\left(1 - \frac{n}{N}\right) \quad n = 1, \ldots, N-1, \tag{4.11}$$

where $F_\eta^{-1}(\cdot)$ is the inverse cumulative distribution function (ICDF) of the noise distribution. For even noise PDFs

$$x = \theta + F_\eta^{-1}\left(\frac{n}{N}\right) \quad n = 1, \ldots, N-1. \tag{4.12}$$

These values of x at which the maximum value of $P_{y|x}(n|x)$ occurs are known as the *mode* of the conditioned random variable, y, given x (Yates and Goodman 2005).

Measuring SSR with mutual information

We shall refer to the model described above as the *SSR model*.

The mutual information – see Section 3.2 in Chapter 2 for a definition – between the input and output signals of the SSR model can be expressed as

$$I(x, y) = H(y) - H(y|x)$$

$$= -\sum_{n=0}^{N} P_y(n) \log_2 P_y(n) -$$

$$- \int_{-\infty}^{\infty} f_x(x) \sum_{n=0}^{N} P_{y|x}(n|x) \log_2 P_{y|x}(n|x)dx. \tag{4.13}$$

As noted in Stocks (2001a), since the input signal is continuously valued and the output is discretely valued, the mutual information is that of a semi-continuous channel. Since $P_y(n)$ is a function of $f_x(x)$ and $P_{y|x}(n|x)$ – see Eq. (4.5) – the mutual information can be expressed in terms of only $f_x(x)$ and $P_{y|x}(n|x)$. The transition probabilities, $P_{y|x}(n|x)$, are given by Eq. (4.9), which for a given N depends only on $P_{1|x}$ and therefore only on the noise PDF, $f_\eta(\eta)$, and the threshold values, θ.

Stocks (2000c) simplifies this expression for the mutual information by use of Eq. (4.9). In particular, consider the entropy of the output for a given value of the input

$$\hat{H}(y|x) = -\sum_{n=0}^{N} P_{y|x}(n|x) \log_2 P_{y|x}(n|x). \tag{4.14}$$

Note that the quantity, $\hat{H}$, is labelled with a caret to make explicit that this formula is not the *average* conditional entropy, $H(y|x)$ – see Eq. (4.16). Substituting from Eq. (4.9) into Eq. (4.14) and simplifying gives

$$\hat{H}(y|x) = - N \left(P_{1|x} \log_2 P_{1|x} + (1 - P_{1|x}) \log_2 (1 - P_{1|x}) \right)$$

$$- \sum_{n=0}^{N} P_{y|x}(n|x) \log_2 \binom{N}{n}. \tag{4.15}$$

This simplification uses the facts that $\sum_{n=0}^{N} P_{y|x}(n|x) = 1$ and that the expected value of the binomial distribution is $\sum_{n=0}^{N} n P_{y|x}(n|x) = N P_{1|x}$. Equation (4.15) is identical to one expressed by Davisson and Leon-Garcia (1980) and Chang and Davisson (1990), although their use of this equation is under far more specific conditions.

The average conditional entropy, or equivocation, is

$$H(y|x) = \int_x f_x(x) \hat{H}(y|x) dx$$

$$= - \sum_{n=0}^{N} P_y(n) \log_2 \binom{N}{n}$$

$$- N \int_x f_x(x) \left(P_{1|x} \log_2 P_{1|x} + (1 - P_{1|x}) \log_2 (1 - P_{1|x}) \right) dx, \tag{4.16}$$

where the simplification comes from use of Eq. (4.5). Notice that the integral on the right-hand side (rhs) is N times the conditional entropy of the output of a single comparator. Since $\sum_{n=0}^{N} P_y(n) \log_2 \binom{N}{n}$ is always less than zero, the overall average conditional entropy is less than the sum of the average conditional entropy of its parts, by this factor. This means that the average uncertainty about the input signal, x, is reduced by observing the added outputs of more than one single-threshold device.

Note that in the absence of noise, the system is deterministic, and $P_{1|x}$ can be only zero or unity. This means that the only possible overall output states are zero or N, and the conditional output entropy is always zero, since there is never any uncertainty about the output given the input. The result is that the mutual information is always exactly one bit per sample, provided the threshold is set to the signal mean, so that $P_y(0) = P_y(N) = 0.5$.

Substituting from Eq. (4.16) into Eq. (4.13) gives the mutual information as

$$I(x, y) = - \sum_{n=0}^{N} P_y(n) \log_2 B_y(n)$$

$$+ N \int_x f_x(x) \left(P_{1|x} \log_2 P_{1|x} + (1 - P_{1|x}) \log_2 (1 - P_{1|x}) \right) dx, \tag{4.17}$$

where $B_y(n) = P_y(n)/\binom{N}{n}$, so that $B_y(n) = \int_x f_x(x) P_{1|x}^n (1 - P_{1|x})^{N-n} dx$. The above derivation is given in Stocks (2001a).

Change of variable

We now introduce a mathematically convenient change of variable. The function $P_{1|x}$ is a function of x. The range of x is the range of allowable values of the input, and for infinite tail input PDFs, $x \in [-\infty, \infty]$ – an example of this is the Gaussian distribution. We shall see that certain equations can be simplified if integrations are performed over the interval $[0, 1]$ rather than $(-\infty, \infty)$. This can be achieved by a change of variable as follows.

Let $\tau = P_{1|x}$ so that from Eq. (4.7), $\tau = 1 - F_\eta(\theta - x)$. Then, differentiating τ with respect to x – assuming τ is differentiable for all x – gives

$$\frac{d\tau}{dx} = f_\eta(\theta - x). \tag{4.18}$$

Multiplying both sides by $f_x(x)$ and rearranging leaves

$$f_x(x)dx = \frac{f_x(x)}{f_\eta(\theta - x)} d\tau. \tag{4.19}$$

Substituting for $x = \theta - F_\eta^{-1}(1 - \tau)$ gives

$$f_x(x)dx = \frac{f_x(\theta - F_\eta^{-1}(1 - \tau))}{f_\eta(F_\eta^{-1}(1 - \tau))} d\tau. \tag{4.20}$$

The left-hand side (lhs) of the above expression is the probability that the input lies between x and $x + dx$. The rhs is $d\tau$ times the ratio of the signal PDF to the noise PDF, calculated at the values x and $(\theta - x)$ respectively. Let this rhs expression be $f_\varrho(\tau)d\tau$ so that

$$f_\varrho(\tau) = \frac{f_x(x)}{f_\eta(\theta - x)}\bigg|_{x=\theta - F_\eta^{-1}(1-\tau)}. \tag{4.21}$$

Provided the support of $f_x(x)$ is contained in the support of $f_\eta(\theta - x)$, since otherwise division by zero occurs, then $f_\varrho(\tau)$ is a PDF defined on the interval $[0, 1]$. We shall see later in this chapter that the case of uniform signal and uniform noise, with the signal variance larger than the noise variance, is an example where $f_\varrho(\tau)$ is not a PDF.

Making a change of variable in Eq. (4.16) gives

$$H(y|x) = -\sum_{n=0}^{N} P_y(n) \log_2 \binom{N}{n}$$

$$- N \int_{\tau=0}^{\tau=1} f_\varrho(\tau) \left(\tau \log_2 \tau d\tau + (1 - \tau) \log_2 (1 - \tau) \right) d\tau, \tag{4.22}$$

and in Eq. (4.17) gives

$$I(x, y) = -\sum_{n=0}^{N} P_y(n) \log_2 B_y(n)$$

$$+ N \int_{\tau=0}^{\tau=1} f_Q(\tau) \left(\tau \log_2 \tau d\tau + (1 - \tau) \log_2 (1 - \tau)\right) d\tau, \quad (4.23)$$

where

$$B_y(n) = \int_{\tau=0}^{\tau=1} f_Q(\tau)\tau^n(1 - \tau)^{N-n} d\tau. \quad (4.24)$$

Input and noise PDFs that are even functions about identical means

Further simplification can be made in the case where both the signal and noise PDFs are even functions, with identical means, and all thresholds are set equal to that mean. Without loss of generality, assume the mean is zero. If $f_\eta(\eta)$ is even and $\theta = 0$, from Eq. (4.8), $P_{1|x} = F_\eta(x) = 1 - P_{1|-x}$. Considering the integral

$$A = \int_{x=-\infty}^{x=\infty} f_x(x)(1 - P_{1|x}) \log_2 (1 - P_{1|x}) dx, \quad (4.25)$$

and putting $s = -x$, then making a change of variable and letting $f_x(x)$ be even so that $f_x(x) = f_x(-x)$, gives

$$A = -\int_{s=\infty}^{s=-\infty} f_x(-s)(1 - P_{1|-s}) \log_2 (1 - P_{1|-s}) ds$$

$$= \int_{s=-\infty}^{s=\infty} f_x(s) P_{1|s} \log_2 P_{1|s} ds. \quad (4.26)$$

Equation (4.16) reduces to

$$H(y|x) = -\sum_{n=0}^{N} P_y(n) \log_2 \binom{N}{n} - 2N \int_x f_x(x) P_{1|x} \log_2 P_{1|x} dx \quad (4.27)$$

and Eq. (4.17) reduces to

$$I(x, y) = -\sum_{n=0}^{N} P_y(n) \log_2 B_y(n) + 2N \int_x f_x(x) P_{1|x} \log_2 P_{1|x} dx. \quad (4.28)$$

Also, $f_Q(\tau)$ simplifies when $\theta = 0$ and $f_x(x)$ and $f_\eta(\eta)$ are even to yield

$$f_Q(\tau) = \frac{f_x(F_\eta^{-1}(1 - \tau))}{f_\eta(F_\eta^{-1}(1 - \tau))} = \frac{f_x(F_\eta^{-1}(\tau))}{f_\eta(F_\eta^{-1}(\tau))}. \quad (4.29)$$

It turns out that besides being mathematically convenient, the function $f_Q(\tau)$ has further significance, and assists greatly in intuitively understanding the behaviour of SSR. We shall return to this point in Section 4.3 and Chapter 5.

The average conditional entropy can now be written in terms of the function $f_Q(\tau)$ as

$$H(y|x) = -\sum_{n=0}^{N} P_y(n) \log_2 \binom{N}{n} - 2N \int_{\tau=0}^{\tau=1} f_Q(\tau)\tau \log_2 \tau d\tau, \qquad (4.30)$$

and the mutual information as

$$I(x, y) = -\sum_{n=0}^{N} P_y(n) \log_2 B_y(n) + 2N \int_{\tau=0}^{\tau=1} f_Q(\tau)\tau \log_2 \tau d\tau. \qquad (4.31)$$

Identical signal and noise distributions

Further analytical simplification of Eq. (4.13) is possible in the case where the signal's PDF is 'matched' with the noise PDF, so that $f_x(x) = f_\eta(\theta - x) \; \forall \; x$. This means that the signal and noise must have identical even moments.

Note that for some signal and noise distribution pairs, it is not possible for $f_x(x) = f_\eta(\theta - x) \; \forall \; x$. Examples include (i) if $f_x(x)$ and $f_\eta(\eta)$ are uniform, and $E[x] \neq \theta - E[\eta]$ and (ii) if both $f_x(x)$ and $f_\eta(\eta)$ have only positive support, with infinite tails, that is $f_x(x) > 0 \; \forall \; x \in [0, \infty]$.

We begin with a result due to Stocks (2001a) showing that all output states are equiprobable in this situation, that is

$$P_y(n) = \frac{1}{N+1} \quad n = 0, \ldots, N. \qquad (4.32)$$

This result can be proved as follows. Consider $f_Q(\tau)$ as given by Eq. (4.21). If $f_x(x) = f_\eta(\theta - x) \; \forall \; x$, then it is clear that $f_Q(\tau) = 1 \; \forall \; \tau$, and from Eq. (4.24)

$$B_y(n) = \int_{\tau=0}^{\tau=1} \tau^n (1 - \tau)^{N-n} d\tau. \qquad (4.33)$$

This is simply a beta function (Spiegel and Liu 1999), which evaluates as

$$B_y(n) = \frac{n!(N-n)!}{(N+1)!}, \qquad (4.34)$$

so that

$$P_y(n) = \binom{N}{n} B_y(n) = \frac{1}{N+1}. \qquad (4.35)$$

Interestingly, this means that the output entropy is maximized, since maximum entropy occurs when all states are equally likely (Cover and Thomas 1991). The output entropy is

$$H(y) = -\sum_{n=0}^{N} \frac{1}{N+1} \log_2 \left(\frac{1}{N+1} \right) = \log_2 (N+1). \tag{4.36}$$

This does not, however, mean that mutual information is maximized, since it depends also on the average conditional entropy, $H(y|x)$. It is possible for a discrete signal such as y to have a uniform distribution, yet still have a very large average uncertainty about y given x, whereas a smaller output entropy might provide a far smaller $H(y|x)$.

Stocks (2001a) calculates the integral on the rhs of Eq. (4.16), to obtain an exact expression for the average conditional entropy. This integral can be solved equivalently from Eq. (4.22), by using the fact that $f_Q(\tau) = 1$ to get

$$\int_{\tau=0}^{\tau=1} f_Q(\tau) \left(\tau \log_2 \tau d\tau + (1-\tau) \log_2 (1-\tau) \right) d\tau = 2 \int_{\tau=0}^{\tau=1} \tau \log_2 \tau d\tau$$

$$= -\frac{1}{2 \ln 2}. \tag{4.37}$$

In order to further simplify the average conditional entropy, Stocks (2001a) uses the identity (see Section A1.2 in Appendix 1 for a proof)

$$-\sum_{n=0}^{N} \log_2 \binom{N}{n} = \sum_{n=1}^{N} (N+1-2n) \log_2 n. \tag{4.38}$$

Substituting from Eqs (4.37) and (4.38) into Eq. (4.22) gives

$$H(y|x) = \frac{N}{2 \ln 2} + \frac{1}{N+1} \sum_{n=1}^{N} (N+1-2n) \log_2 n \tag{4.39}$$

and the mutual information is

$$I(x, y) = \log_2 (N+1) - \frac{N}{2 \ln 2} - \frac{1}{N+1} \sum_{n=2}^{N} (N+1-2n) \log_2 n. \tag{4.40}$$

This is plotted against increasing N in Fig. 4.2. What is quite remarkable about this result is that it is independent of the nature of the PDFs of the signal and noise, other than that $f_x(x) = f_\eta(\theta - x)$ for all valid values of x, so that both PDFs have the same shape, but may possibly have different means, and be mutually reversed along the x-axis about their means.

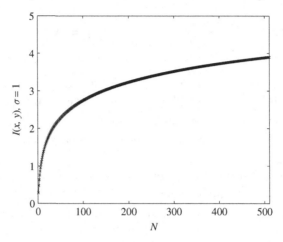

Fig. 4.2. Mutual information against an increasing number of array elements, $N = 0, \ldots, 511$, for the case of $f_x(x) = f_\eta(\theta - x)$, calculated from Eq. (4.40). Notice that the mutual information increases with increasing N, although the rate of increase decreases with increasing N.

Further simplification of Eq. (4.40) is insightful. The second term of Eq. (4.39) can be expressed as

$$\frac{1}{N+1} \sum_{n=1}^{N} (N + 1 - 2n) \log_2 n = \log_2 (N!) - \frac{2}{N+1} \sum_{n=1}^{N} n \log_2 n$$

$$= \log_2 (N!) - N \log_2 (N + 1)$$

$$- 2 \sum_{n=1}^{N} \frac{n}{N+1} \log_2 \left(\frac{n}{N+1} \right). \qquad (4.41)$$

The average conditional entropy can be written as

$$H(y|x) = \log_2 (N!) - N \log_2 (N + 1) + \frac{N}{2 \ln 2} - 2 \sum_{n=1}^{N} \frac{n}{N+1} \log_2 \left(\frac{n}{N+1} \right). \qquad (4.42)$$

Note that the first two terms scale with $N \log N$, since $\log_2 (N!)$ approaches $N \log N$ for very large N. Such large N behaviour is discussed in Chapter 5. The term inside the summation in the final term is always between 0 and 0.5, so the summation is always less than N. Hence the average conditional entropy consists of two terms of the order of $N \log N$ and two terms of the order of N.

Stocks (2001a) shows also that as N becomes large, Eq. (4.42) reduces to $I(x, y) \simeq 0.5 \log_2 (N + 1)$, and that this implies that a maximum must occur in the mutual information for a nonzero noise intensity for any $N > 1$. The large N behaviour of SSR is further examined in Chapter 5. Additional analytical progress on calculating the mutual information for small N can be made only if the signal and/or noise distributions are specified. This is the focus of the following paragraphs.

SSR for specific signal and noise distributions

We now examine whether SSR can occur for a range of different signal and noise distributions. Before doing this, we first describe seven different probability distributions, and also make some comments on the function, $f_Q(\tau)$.

Seven different probability distributions

Table 4.1 gives expressions for the PDF, CDF, and ICDF of a number of different probability distributions. Section A1.3 in Appendix 1 gives specific details of employing each of these distributions as the signal or noise distribution. The PDF, CDF, and ICDF of each distribution are plotted in Fig. 4.3, where the variance is $\sigma = 1$ for all distributions except Cauchy, and the *dispersion* parameter $\lambda = 1$ for the Cauchy distribution. For the CDF, z is limited to the values of support of the corresponding PDF, and for the ICDF $w \in [0, 1]$. Note that $P_{1|x}$ can be obtained from the CDF by use of Eq. (4.7) and the mode of $P_{y|x}(n|x)$ for each n by use of Eq. (4.11).

The first five PDFs can be seen to be even functions about their means. The Gaussian and logistic PDFs are almost identical in shape, and have infinite tails, whereas the uniform PDF is nonzero on a finite range, and is flat. The Laplacian PDF has infinite tails, but is more sharply peaked than either the Gaussian or logistic distributions. The Cauchy distribution – also known as the Lorentzian distribution – differs from the other distributions considered, in that it has an infinite variance. We discuss the Cauchy distribution in this chapter as a way of illustrating that SSR is not restricted to finite variance signal or noise distributions, just as SR is not (Kosko and Mitaim 2001). Instead of characterizing the Cauchy distribution's width by its standard deviation, a parameter known as its dispersion, or full width at half maximum (FWHM), is usually used, as explained in Section A1.3 of Appendix 1. The FWHM is the distance between points on the curve at which the function reaches half its maximum value. This is given for the Cauchy distribution by the parameter, λ_x. The Rayleigh and exponential distributions exist only for positive support. This fact ensures that no PDF, $f_Q(\tau)$, can be found for these cases for all τ, since there will always exist points of support for which $f_x(x)$ is nonzero, yet $f_\eta(\theta - x) = 0$.

For the CDFs, in each case there is a substantial range of z for which the CDF is approximately linearly increasing. The support of the CDF is the same as the

Table 4.1. *List of probability distributions and corresponding PDFs, CDFs, and ICDFs*

Distribution	support	mean	variance	PDF, $f_x(x)$	CDF, $F_\eta(z)$	ICDF, $F_\eta^{-1}(w)$
Gaussian	$(-\infty, \infty)$	0	σ_x^2	$\dfrac{1}{\sqrt{2\pi\sigma_x^2}}\exp\left(-\dfrac{x^2}{2\sigma_x^2}\right)$	$0.5 + 0.5\,\mathrm{erf}\left(\dfrac{z}{\sqrt{2}\sigma_\eta}\right)$	$\sqrt{2}\sigma_\eta\,\mathrm{erf}^{-1}(2w-1)$
Uniform	$\left[-\dfrac{\sigma_x}{2}, \dfrac{\sigma_x}{2}\right]$	0	$\dfrac{\sigma_x^2}{12}$	$\dfrac{1}{\sigma_x},\quad x\in\left[-\dfrac{\sigma_x}{2},\dfrac{\sigma_x}{2}\right],$ $0,\quad$ otherwise	$0,\quad z < -\dfrac{\sigma_\eta}{2},$ $\dfrac{z}{\sigma_\eta}+\dfrac{1}{2},\quad z\in\left[-\dfrac{\sigma_\eta}{2},\dfrac{\sigma_\eta}{2}\right],$ $1,\quad z > \dfrac{\sigma_\eta}{2}.$	$\sigma_\eta\left(w-\dfrac{1}{2}\right)$
Laplacian	$(-\infty, \infty)$	0	σ_x^2	$\dfrac{1}{\sqrt{2}\sigma_x}\exp\left(\dfrac{-\sqrt{2}|x|}{\sigma_x}\right)$	$\dfrac{1}{2}\exp\left(\dfrac{\sqrt{2}z}{\sigma_\eta}\right),\quad z\le 0,$ $1-\dfrac{1}{2}\exp\left(\dfrac{-\sqrt{2}z}{\sigma_\eta}\right),\quad z\ge 0.$	$\dfrac{\sigma_\eta}{\sqrt{2}}\ln(2w),\quad w\in\left[0,\dfrac{1}{2}\right],$ $\dfrac{-\sigma_\eta}{\sqrt{2}}\ln(2(1-w)),\quad w\in\left[\dfrac{1}{2},1\right].$
Logistic	$(-\infty, \infty)$	0	$\sigma_x^2 = \dfrac{\pi^2 b_x^2}{3}$	$\dfrac{\exp\left(-\dfrac{x}{b_x}\right)}{b_x\left(1+\exp\left(-\dfrac{x}{b_x}\right)\right)^2}$	$\dfrac{1}{1+\exp\left(-\dfrac{z}{b_\eta}\right)}$	$b_\eta\ln\left(\dfrac{w}{1-w}\right)$
Cauchy	$(-\infty, \infty)$	$-$	∞	$\dfrac{\lambda_x}{\pi}\dfrac{1}{\lambda_x^2+x^2}$	$\dfrac{1}{2}+\dfrac{1}{\pi}\arctan\left(\dfrac{z}{\lambda_\eta}\right)$	$\lambda_\eta\tan\left(\pi\left(w-\dfrac{1}{2}\right)\right)$
Exponential	$[0, \infty)$	σ_x	σ_x^2	$\dfrac{1}{\sigma_x}\exp\left(-\dfrac{x}{\sigma_x}\right)$	$1-\exp\left(-z/\sigma_\eta\right)$	$\sigma_\eta\ln\left(\dfrac{1}{1-w}\right)$
Rayleigh	$[0, \infty)$	$\sigma_x\sqrt{\dfrac{\pi}{2}}$	$\left(2-\dfrac{\pi}{2}\right)\sigma_x^2$	$\dfrac{x}{\sigma_x^2}\exp\left(-\dfrac{x^2}{2\sigma_x^2}\right)$	$1-\exp\left(-\dfrac{z^2}{2\sigma_\eta^2}\right)$	$\sigma_\eta\sqrt{2\ln\left(\dfrac{1}{1-w}\right)}$

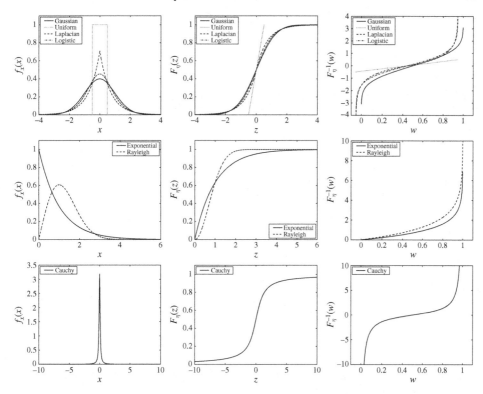

Fig. 4.3. The PDF, $f_x(x)$, the CDF, $F_\eta(z)$, and the ICDF, $F_\eta^{-1}(w)$, for seven different distributions. The left-hand panels show the PDFs, the middle panels the CDFs, and the right-hand panels the ICDFs. The first row shows the four double-sided, zero-mean, finite variance distributions, the second row shows the two one-sided finite variance distributions, and the third row shows the infinite variance Cauchy distribution. All distributions have been plotted with a variance of $\sigma = 1$, except for the Cauchy distribution, which has been plotted with a FWHM of $\lambda = 1$.

support for the corresponding PDF. The ICDF can be seen to have support between zero and unity. This reflects the fact that probabilities must be between zero and unity.

Observations about $f_Q(\tau)$

It will prove to be convenient to parameterize the forthcoming results in terms of the ratio of noise standard deviation to signal standard deviation, $\sigma = \sigma_\eta/\sigma_x$. Likewise, for the infinite variance Cauchy distribution, let $\sigma_\lambda = \lambda_\eta/\lambda_x$. Note from Appendix A1.3 that – apart from the Cauchy case – the variance of the noise is

Table 4.2. *List of ratios of signal PDF to noise PDF, $f_Q(\tau)$, for $\theta = 0$. Also shown is $H(Q)$ – that is, the entropy of the random variable, Q, with PDF $f_Q(\tau)$. The label 'NAS' indicates that there is no analytical solution for the entropy. Table adapted with permission from McDonnell et al. (2007). Copyright 2007 by the American Physical Society.*

Distribution	$f_Q(\tau)$	$H(Q)$
Gaussian	$\sigma \exp\left((1-\sigma^2)\left(\mathrm{erf}^{-1}(2\tau-1)\right)^2\right)$	$-\log_2(\sigma) - \frac{1}{2\ln 2}\left(\frac{1}{\sigma^2} - 1\right)$
Uniform, $\sigma \geq 1$	$\begin{cases} \sigma, & -\frac{1}{2\sigma}+0.5 \leq \tau \leq \frac{1}{2\sigma}+0.5, \\ 0, & \text{otherwise.} \end{cases}$	$\log_2 \sigma$
Laplacian	$\begin{cases} \sigma(2\tau)^{(\sigma-1)} & \text{for } 0 \leq \tau \leq 0.5, \\ \sigma(2(1-\tau))^{(\sigma-1)} & \text{for } 0.5 \leq \tau \leq 1. \end{cases}$	$-\log_2(\sigma) - \frac{1}{2\ln 2}\left(\frac{1}{\sigma} - 1\right)$
Logistic	$\sigma\frac{(\tau(1-\tau))^{(\sigma-1)}}{(\tau^\sigma+(1-\tau)^\sigma)^2}$	NAS
Cauchy	$\sigma_\lambda \frac{1+\tan^2(\pi(\tau-0.5))}{(1+\sigma_\lambda^2\tan^2(\pi(\tau-0.5)))}$	NAS

always a function of σ_η, and the variance of the signal is a function of σ_x. The reason that this parameterization is convenient is that the mutual information turns out – in the cases examined here, where θ is equal to the signal mean – to be a function of the ratio, σ, so that it is invariant to a change in σ_x provided σ_η changes by the same proportion.

This result can be seen by deriving the function $f_Q(\tau)$ for specific signal and noise distributions. When the signal and noise share the same distribution – but with not necessarily equal variances – the function $f_Q(\tau)$ is listed in Table 4.2 for $\theta = 0$. In all cases, $f_Q(\tau)$ is a function of the ratio σ. Hence, by inspection of Eqs (4.24), (4.22), and (4.23) it is clear that $P_y(n)$, $H(y)$, $H(y|x)$, and $I(x, y)$ will also be functions of σ, and not σ_η or σ_x in isolation. Note that if $\theta \neq 0$, then $f_Q(\tau)$ will depend on the ratio $\frac{\theta}{\sigma_x}$, as well as θ and the ratio, σ, and therefore so will the mutual information.

The PDFs, $f_Q(\tau)$, are also plotted for various values of σ in Fig. 4.4 for $\theta = 0$. Table 4.2 also lists the entropy, $H(Q)$ – which is equivalent to the negative of the relative entropy between $f_x(x)$ and $f_\eta(x)$ – for two cases. In this table, NAS indicates that no analytical solution can be found for the entropy.

We are now in a position to comment further on interpreting the significance of the function $f_Q(\tau)$. Recall that $f_x(x)dx = f_Q(\tau)d\tau$. Assume that $f_Q(\tau)$ is a PDF – which will be the case if the support of $f_x(x)$ is contained in the support of

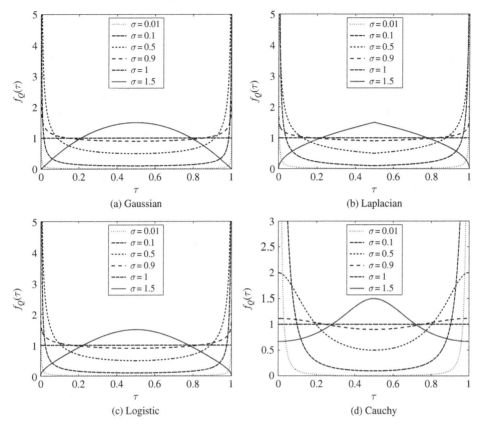

Fig. 4.4. The PDF, $f_Q(\tau)$, for various values of σ for four even, infinite tail PDFs, where both $f_\eta(\cdot)$ and $f_x(\cdot)$ have identical PDFs and $\theta = 0$. For $\sigma < 1$, $f_Q(\tau)$ is convex, for $\sigma = 1$, $f_Q(\tau)$ is uniform, and for $\sigma > 1$, $f_Q(\tau)$ is concave. For the Gaussian, Laplacian, and logistic cases, $f_Q(\tau)$ is infinite at $\tau = 0$ and $\tau = 1$ for $\sigma < 1$, and is equal to zero at $\tau = 0$ and $\tau = 1$ for $\sigma > 1$. For the Cauchy case, $f_Q(0) = f_Q(1) = \frac{1}{\sigma_\lambda}$ $\forall \sigma$.

$f_\eta(\theta - x)$, and is given by Eq. (4.21). The entropy of the corresponding random variable Q is

$$H(Q) = -\int_{\tau=0}^{\tau=1} f_Q(\tau) \log_2(f_Q(\tau)) d\tau$$

$$= -\int_{x=-\infty}^{x=\infty} f_x(x) \log_2\left(\frac{f_x(x)}{f_\eta(\theta - x)}\right) dx, \qquad (4.43)$$

which is the negative of the *relative entropy* between $f_x(x)$ and $f_\eta(x - \theta)$, that is $H(Q) = -D(f_x(x)||f_\eta(\theta - x))$ – see Section 3.2 in Chapter 2 for the definition of relative entropy. If $f_\eta(\eta)$ is even about a mean of zero, and θ is equal to the signal mean of zero, then the entropy, $H(Q)$, is the negative of the relative entropy, $D(f_x||f_\eta)$. Since relative entropy is always positive, this means that the entropy corresponding to f_Q is always negative or zero. Since this is the entropy of a continuously valued random variable, $H(Q)$ is differential entropy, and negative entropy is allowable.

The relative entropy for the four cases of matched Gaussian, logistic, Laplacian and Cauchy distributions is shown for $\theta = 0$ in Fig. 4.5, against increasing σ. The Gaussian case was calculated from the exact formula given in Table 4.2. The other cases were found numerically. Clearly, the relative entropy is zero at $\sigma = 1$, which is as expected, since at $\sigma = 1$, $f_x(x) = f_\eta(x) \forall x$. For $\sigma < 1$, the relative entropy becomes very large as σ gets smaller, and for $\sigma > 1$ gets larger, but more slowly with σ. The Gaussian case always gives the largest relative entropy for the same value of σ.

Also, the first two moments of the random variable, τ, are

$$E[\tau] = \int_{\tau=0}^{\tau=1} \tau f_Q(\tau) d\tau, \tag{4.44}$$

$$E[\tau^2] = \int_{\tau=0}^{\tau=1} \tau^2 f_Q(\tau) d\tau. \tag{4.45}$$

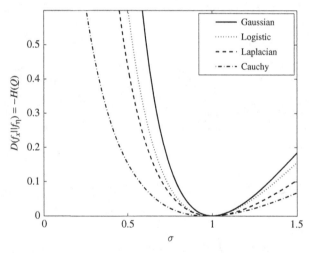

Fig. 4.5. Relative entropy between $f_x(x)$ and $f_\eta(x)$. This figure shows the relative entropy – which is equivalent to the negative of the entropy, $H(Q)$ – plotted against σ for four cases of matched signal and noise distributions, for $\theta = 0$. The Gaussian case was calculated from the exact formula given in Table 4.2. The other cases were found numerically. Clearly, the relative entropy is zero at $\sigma = 1$, which is as expected, since at $\sigma = 1$, $f_x(x) = f_\eta(x) \forall x$. For $\sigma < 1$, the relative entropy becomes large as σ gets smaller, and for $\sigma > 1$ it becomes larger, but more slowly with σ. The Gaussian case always gives the largest relative entropy for the same value of σ.

Notice also that the term that appears on the rhs in Eq. (4.30) and Eq. (4.31) is $E[\tau \log_2 (\tau)]$.

For the specific distributions listed in this section that have infinite tails, $f_Q(\tau)$ has support $\tau \in [0, 1]$. For the uniform case, when $\sigma > 1$, τ has support $[-\frac{1}{2\sigma} + 0.5, \frac{1}{2\sigma} + 0.5]$. Note that for uniform signal and noise, $f_Q(\tau)$ is valid only for $\sigma \geq 1$, since for $\sigma \leq 1$, $\frac{d\tau}{dx} = 0$ for $x < -\frac{\sigma_\eta}{2}$ and $x > \frac{\sigma_\eta}{2}$. In all the cases in Table 4.2, $f_Q(\tau)$ is a PDF, as one would expect from Eq. (4.20). This is proven in Section A1.4 in Appendix 1 for the cases of Gaussian and Laplacian signal and noise.

Note also from Table 4.2 and Fig. 4.4 that, apart from the Cauchy case, when $\sigma = 1$, $f_Q(\tau) = 1 \,\forall\, \tau$, that is $f_Q(\tau)$ is uniform. Furthermore, for $\sigma > 1$, $f_Q(0) = f_Q(1) = 0$, so that $f_Q(x)$ is concave, and for $\sigma < 1$, $f_Q(x)$ is convex and $f_Q(0) = f_Q(1) = \infty$. For the Cauchy case, a finite limit exists at $\tau = 0$ and $\tau = 1$, such that $f_Q(0) = f_Q(1) = \frac{1}{\sigma_\lambda} \,\forall\, \sigma$.

That a PDF may have values that approach infinity may at first seem surprising. However, to be a PDF, a function only needs to be nonnegative for its entire support, and have a total area of unity when integrated with respect to its support. Infinite values do not preclude this, no more than the fact that a Gaussian PDF is never equal to zero, even for infinite values of its support. An example of a relatively well-known PDF with infinite values is the PDF of a sine wave with a random phase and amplitude A (Damper 1995, Wannamaker *et al.* 2000a)

$$f_x(x) = \begin{cases} \frac{1}{\pi\sqrt{A^2-x^2}}, & x \in [-A, A], \\ 0 & \text{otherwise.} \end{cases} \tag{4.46}$$

This PDF is plotted in Fig. 4.6 for $A = 0.5$.

Interpreting $f_Q(\tau)$

For uniform signal and noise with $\sigma < 1$, define $g'_Q(\tau) = f_Q(\tau)$ to be valid for $x \in [-\frac{\sigma_\eta}{2}, \frac{\sigma_\eta}{2}]$. Hence, $g'_Q(\tau)$ is not a PDF for $\sigma < 1$, since $\int_{\tau=0}^{\tau=1} \sigma d\tau = \sigma$. Therefore any integrations over τ in this case require extra consideration of $x < -\frac{\sigma_\eta}{2}$ and $x > \frac{\sigma_\eta}{2}$. This will not be the case for the integral on the rhs of Eq. (4.28) since in this range of x, $P_{1|x}$ is either zero or unity and therefore $P_{1|x} \log_2 P_{1|x} = 0$. Likewise, it will not be the case in calculating $P_y(n)$, since $P_{1|x}(1 - P_{1|x}) = 0$, except when $n = 0$ or $N = 1$. A simple integration in these cases finds that

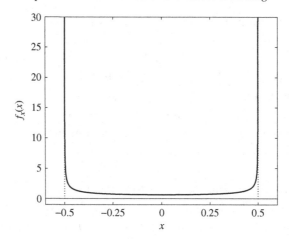

Fig. 4.6. The PDF of a random phase sine wave that has an amplitude of $A = 0.5$. Notice that the PDF approaches infinity at $x = \pm A$, just as the PDF $f_Q(\tau)$ does for small values of σ.

$$P_y(0) = P_y(N) = \int_{\tau=0}^{\tau=1} g'_Q(\tau)\tau^N d\tau + \int_{x=\sigma_\eta/2}^{x=\sigma_x/2} \frac{1}{\sigma_x} 1^N dx$$

$$= \frac{\sigma}{N+1} + \frac{1}{2} - \frac{\sigma}{2}. \tag{4.47}$$

Note that the integral in the region $x \in [\sigma_\eta/2, \sigma_x/2]$ is the region of the support of $f_x(x)$ for which all thresholds are always 'on'. This occurs since for these values of x, the noise is never negative enough to cause any threshold devices to be 'off', that is we have $x + \eta > \theta \; \forall \, x \in [\sigma_\eta/2, \sigma_x/2]$.

A possible interpretation of the significance of $f_Q(\tau)$ is that it is related to the output distribution, $P_y(n)$. As previously noted, for $\sigma < 1$, $f_Q(\tau)$ is concave, and for $\sigma > 1$, $f_Q(\tau)$ is convex. For $\sigma = 1$, $f_Q(\tau)$ is uniform. Recall also that $\sigma = 1$ corresponds to the case of optimal output entropy, so that $P_y(n) = \frac{1}{N+1}$, that is the output probability mass function is uniform. Furthermore, for $\sigma = 0$, only output states zero and N are available, each with probability one half. Also, for uniform signal and noise and $\sigma < 1$, where $g'_Q(\tau)$ is not a PDF, there is a contribution to $P_y(0)$ and $P_y(N)$ that cannot be explained by this interpretation of $g'_Q(\tau)$.

The obvious possibility is to suggest that

$$P_y(n) = \int_{\tau=\frac{n}{N+1}}^{\tau=\frac{n+1}{N+1}} f_Q(\tau)d\tau. \tag{4.48}$$

However, this does not agree with Eq. (4.24). In the case of large N however, it will be seen in Chapter 5 that the output distribution does indeed approach $f_Q(\tau)$.

Hence, given that $f_Q(\tau)$ is a PDF, an intuitive explanation of its significance is that it is the PDF of the average fraction of thresholds that are 'on', given x, that is $f_Q(\tau)d\tau$ is the probability that 100τ is the average percentage of times that $x + \eta > \theta$. Hence, since there are N thresholds, the average number of thresholds crossed is $N\tau$, and $f_Q(\tau)$ is the probability that $N\tau$ thresholds are crossed. We discuss this in more detail in Section 5.4.

By contrast with the probability mass function, $P_y(n)$, of the output states of the discrete random variable, y, the random variable of the average number of thresholds crossed is a continuous random variable. The reason why the average value of y is a continuous random variable is that it is being conditioned on x, which is also a continuous random variable. Hence, the average value of y given x – that is, $E[y|x]$ – is described by a PDF, which we are interpreting as $f_Q(\tau)$. The CDF of $f_Q(\tau)$ evaluated at x is the probability that $E[y|x]/N$ is less than x.

Furthermore, the expected value of τ, as given by Eq. (4.44), is then the actual expected value of the output y (not conditioned on x) and the expected value of τ^2, as given by Eq. (4.45), is the mean square value of the output (not conditioned on x).

We have presented the distributions we shall study for the signal and/or noise, as well as commenting on the function $f_Q(\tau)$, and giving an interpretation for its significance in the SSR model. Next we present an exact result for the mutual information for the specific case of signal and noise both being uniformly distributed, and $\sigma \leq 1$.

Exact result for uniform signal and noise, and $\sigma \leq 1$

Exact analytical results for the SSR model are rare, and in most cases quantities need to be found numerically. However, Stocks (2001c) gives an exact expression for the mutual information in the case of a uniform signal and uniform noise when $\sigma \leq 1$ and $\theta = 0$.

In this case, $P_y(0)$ and $P_y(N)$ have already been derived, as given by Eq. (4.47). For other n, from Eq. (4.24) for $x \in [-\frac{\sigma_\eta}{2}, \frac{\sigma_\eta}{2}]$

$$B_y(n) = \sigma \int_0^1 \tau^n (1 - \tau)^{N-n} d\tau. \tag{4.49}$$

For x outside this range, there is no additional contribution to $B_y(n)$, as already discussed. Equation (4.49) is the same expression as Eq. (6) in Stocks (2001c). As noted in Stocks (2001a), this integral is simply a beta function. The solution to such a beta function (Spiegel and Liu 1999) is

$$B_y(n) = \sigma \frac{\Gamma(n+1)\Gamma(N-n+1)}{\Gamma(N+2)}, \tag{4.50}$$

where $\Gamma(\cdot)$ is the Gamma function (Spiegel and Liu 1999). For integer k, $\Gamma(k+1) = k!$ and hence, since n and N are integers

$$B_y(n) = \sigma \frac{n!(N-n)!}{(N+1)!} = \frac{\sigma}{(N+1)\binom{N}{n}}. \tag{4.51}$$

Therefore

$$P_y(n) = \binom{N}{n} B_y(n) = \frac{\sigma}{N+1}. \tag{4.52}$$

The complete output probability mass function for uniform signal and noise and $\sigma \leq 1$ is

$$P_y(n) = \begin{cases} \frac{\sigma}{N+1} + \frac{1}{2} - \frac{\sigma}{2} & n = 0, N \\ \frac{\sigma}{N+1} & n = 1, \ldots, N-1. \end{cases} \tag{4.53}$$

It is clear that when $\sigma = 0$, $P_y(0) = P_y(N) = 0.5$, which is expected, as in the absence of noise either all comparators are switched on, or all are off. It is also clear for $\sigma = 1$ that $P_y(n) = \frac{1}{N+1} \ \forall \ n$, which is verified by Eq. (4.32).

The output entropy is

$$H(y) = -2P_y(0) \log_2 P_y(0) - (N-1)P_y(1) \log_2 P_y(1)$$
$$= -\left(\frac{2\sigma}{N+1} + 1 - \sigma\right) \log_2 \left(\frac{\sigma}{N+1} + \frac{1}{2} - \frac{\sigma}{2}\right)$$
$$- \frac{(N-1)\sigma}{N+1} \log_2 \left(\frac{\sigma}{N+1}\right). \tag{4.54}$$

The conditional output entropy can be evaluated from Eq. (4.27). First

$$\int_{\tau=0}^{\tau=1} g_Q'(\tau)\tau \log_2 \tau \, d\tau = \frac{-\sigma}{4\ln 2}, \tag{4.55}$$

which is valid for all x, as previously discussed. Secondly

$$-\sum_{n=0}^{N} \log_2 \binom{N}{n} P_y(n) = -\sum_{n=1}^{N-1} \log_2 \binom{N}{n} P_y(n)$$

$$= -\frac{\sigma}{N+1} \sum_{n=1}^{N-1} \log_2 \binom{N}{n} - P_y(0) \log_2 1 - P_y(N) \log_2 1$$

$$= -\frac{\sigma}{N+1} \sum_{n=0}^{N} \log_2 \binom{N}{n}$$

$$= \frac{\sigma}{N+1} \sum_{n=2}^{N} (N+1-2n) \log_2 n, \qquad (4.56)$$

since, as before, $-\sum_{n=0}^{N} \log_2 \binom{N}{n} = \sum_{n=1}^{N} (N+1-2n) \log_2 n$. Finally, we have

$$H(y|x) = \sigma \left(\frac{N}{2 \ln 2} + \frac{1}{N+1} \sum_{n=2}^{N} (N+1-2n) \log_2 n \right). \qquad (4.57)$$

It is clear that the rhs of this is exactly σ times that of Eq. (4.39), the average conditional entropy for identical signal and noise distributions at $\sigma = 1$, and that therefore $H(y|x)$ is a linearly increasing function of σ for this specific case. This is due to $P_y(n)$ being uniform at $\sigma = 1$, and $P_y(n)$ being uniform – that is, $P_y(n) = \sigma/(N+1)$ for all n except $n = 0$ and $n = N$. The $n = 0$ and $n = N$ terms do not appear in the conditional entropy term due to being multiplied by the logarithm of unity.

Therefore, the complete exact expression for the mutual information for uniform signal and noise and $\sigma \leq 1$ is

$$I(x, y) = -\left(\frac{2\sigma}{N+1} + 1 - \sigma \right) \log_2 \left(\frac{\sigma}{N+1} + \frac{1}{2} - \frac{\sigma}{2} \right)$$

$$- \frac{(N-1)\sigma}{N+1} \log_2 \left(\frac{\sigma}{N+1} \right)$$

$$- \frac{\sigma}{N+1} \sum_{n=2}^{N} (N+1-2n) \log_2 n - \frac{N\sigma}{2 \ln 2}. \qquad (4.58)$$

This expression is that derived in Stocks (2001a) and stated as Eq. (7) in Stocks (2001c). Note that if $\sigma = 0$ in Eq. (4.58), then $I(x, y) = 1$ and if $\sigma = 1$, Eq. (4.58) becomes identical to Eq. (4.40).

This exact expression for the mutual information is plotted against σ for various values of N in Fig. 4.7, as well as the average conditional output entropy and the output entropy Apart from $N = 1$, it is clear that there is a maximum in the mutual information for some value of σ between zero and unity. As N gets larger, the value of σ at which the maximum occurs gets closer to unity. It is also clear that both the output entropy and the average conditional output entropy are strictly increasing with increasing σ. Note that, as indicated by Eq. (4.57), the average conditional output entropy is a linear function of σ, whereas the output entropy increases with a larger slope for smaller σ.

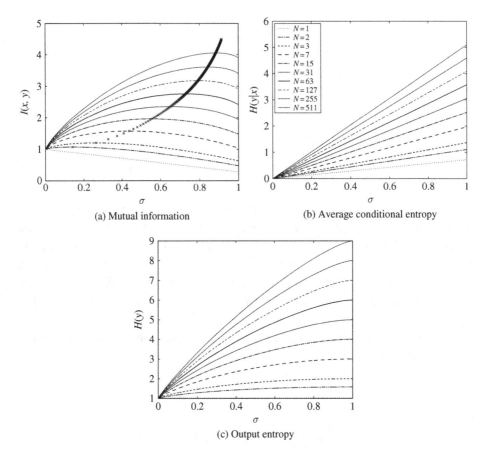

Fig. 4.7. Analytically calculated $I(x, y)$, $H(y|x)$, and $H(y)$ for uniform signal and noise and $\sigma \leq 1$. The exact mutual information is shown in Fig. 4.7(a), while Fig. 4.7(b) shows the average conditional output entropy, and Fig. 4.7(c) shows the output entropy, for the case of uniform signal and noise, and $\sigma \leq 1$, for various values of N. The threshold value is $\theta = 0$. The increasing values of the curves up the y-axis correspond to larger values of N. The average conditional entropy can be seen to increase linearly with increasing σ, as indicated in Eq. (4.57). The output entropy is also an increasing function of σ, but appears to increase faster for smaller σ than for larger σ. The mutual information shows a noise-induced maximum, and the value of σ at which this occurs increases with increasing N. The optimal value of mutual information is shown with a cross in 4.7(a) for $N = 2, \ldots, 1000$.

The maximum in the mutual information occurs for the value of σ for which the slope of the output entropy is equal to the slope of the average conditional output entropy. Label this optimal value as σ_o. This can be found by differentiating $I(x, y)$ with respect to σ and setting the result equal to zero. Carrying this out gives, after some algebra, that

$$\sigma_o = \frac{N+1}{N-1+2^{B+1}},$$ (4.59)

where

$$B = \frac{N(N+1)}{2\ln 2(N-1)} + \frac{1}{N-1}\sum_{n=2}^{N}(N+1-2n)\log_2 n.$$ (4.60)

The mutual information at this optimal value is indicated in Fig. 4.7(a) with crosses, for $N = 2, \ldots, 1000$.

Another exact result

Another exact result for the mutual information of the SSR model is given in Section 5.5 of Chapter 5. This result holds for any signal and noise distributions such that $f_Q(\tau)$ is the PDF of the arcsine distribution

$$f_Q(\tau) = \frac{1}{\pi\sqrt{\tau(1-\tau)}}, \qquad \tau \in [0, 1].$$ (4.61)

In this situation, we show that the output distribution is beta-binomially distributed, and that the mutual information can be written as in Eq. (5.98). Figure 5.13 in Chapter 5 compares this mutual information with that of the uniform case given by Eq. (4.58).

Before discussing these observations further, we next present numerical results for the other signal and noise distributions given in Table 4.1.

Numerical results

In this section we illustrate that SSR is not a phenomenon that occurs only for the specific cases of uniform or Gaussian signal and noise distributions. We begin by considering various cases of 'matched' signal and noise, so that the signal and noise have the same distribution, although with different variances. This is the same situation as studied in Stocks (2000c), Stocks (2001c), and Stocks (2001a). The facts that an exact solution for the mutual information holds in such a case when $\sigma = 1$, and the mutual information is a function of the ratio, σ, and not both the signal and the noise variance independently, are the reasons that this situation is the natural situation to examine first. However, there is no other reason why the signal and noise need have the same distribution.

Therefore, we also look briefly at some examples where the noise is Gaussian, and the signal has some other distribution, to illustrate that the qualitative behaviour of SSR does not depend on the actual signal and noise distribution to any large extent.

The mutual information between the input and output signals of the array of threshold devices can be obtained by numerical integration for any given signal and noise distribution. Some technical details on how the numerical integrations have been carried out are presented in Section A1.5 of Appendix 1.

Matched signal and noise

We now present figures showing the mutual information, average conditional entropy, and output entropy for five different matched signal and noise distributions, for a range of σ and N, and $\theta = 0$. Each of these five distributions has PDFs that are even functions about a mean – or location parameter, in the Cauchy case – of zero.

Gaussian Fig. 4.8 shows the mutual information, average conditional entropy, and output entropy for Gaussian signal and noise.

Uniform Fig. 4.9 shows the mutual information, average conditional entropy, and output entropy for uniform signal and noise. Fig. 4.9 also shows with circles the exact mutual information, average conditional entropy, and output entropy already plotted in Fig. 4.7, for a number of values of σ. Clearly, the numerically calculated results agree with the known exact values in all cases.

Laplacian Fig. 4.10 shows the mutual information, average conditional entropy, and output entropy for Laplacian signal and noise.

Logistic Fig. 4.11 shows the mutual information, average conditional entropy, and output entropy for logistic signal and noise.

Cauchy Fig. 4.12 shows the mutual information, average conditional entropy, and output entropy for Cauchy signal and noise.

In each case, the mutual information shows a noise-induced maximum, and the value of σ at which this occurs increases with increasing N, and corresponds to the point where the slope of the output entropy is equal to the slope of the average conditional entropy. Clearly, the maximum in $I(x, y)$ must occur for $\sigma \leq 1$, since, for $\sigma > 1$, the output entropy decreases, while the average conditional entropy increases. Hence, the slope of each can never be the same, and therefore no stationary point in the mutual information exists for $\sigma > 1$. For $\sigma \leq 1$, both the average conditional entropy and the output entropy can be seen to increase with increasing σ. For $\sigma \geq 1$, the average conditional entropy continues to increase, but

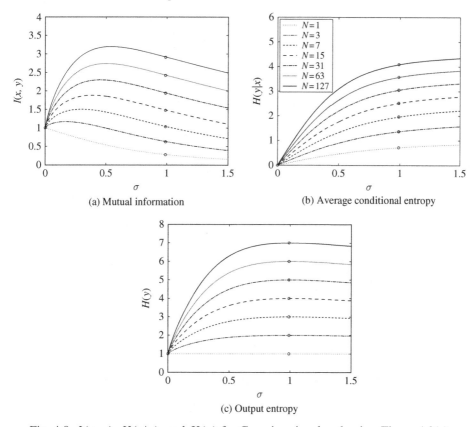

Fig. 4.8. $I(x, y)$, $H(y|x)$, and $H(y)$ for Gaussian signal and noise. Figure 4.8(a) shows the mutual information, Figure 4.8(b) shows the average conditional output entropy, and Figure 4.8(c) shows the output entropy, in the case of Gaussian signal and noise for various values of σ and N. The threshold value is $\theta = 0$. The increasing values of the curves up the y-axis correspond to larger values of N. The circles at $\sigma = 0$ – for which the mutual information and output entropy are always exactly unity – and $\sigma = 1$ indicate exact values for each quantity, calculated using Eqs (4.40) and (4.36). For $\sigma \leq 1$, both the average conditional entropy and the output entropy can be seen to increase with increasing σ. For $\sigma \geq 1$, the average conditional entropy continues to increase, but the output entropy decreases. This makes sense, as the output entropy reaches its maximum value at $\sigma = 1$ of $\log_2 (N + 1)$. The mutual information shows a noise-induced maximum, and the value of σ at which this occurs increases with increasing N, and corresponds to the point where the slope of the output entropy is equal to the slope of the average conditional entropy.

the output entropy decreases. This makes sense, since from Eq. (4.36), the output entropy reaches its maximum value at $\log_2 (N + 1)$ at $\sigma = 1$.

Each figure shows with circles the exact mutual information, average conditional entropy, and output entropy at $\sigma = 1$, calculated from Eqs (4.40) and (4.36). Also

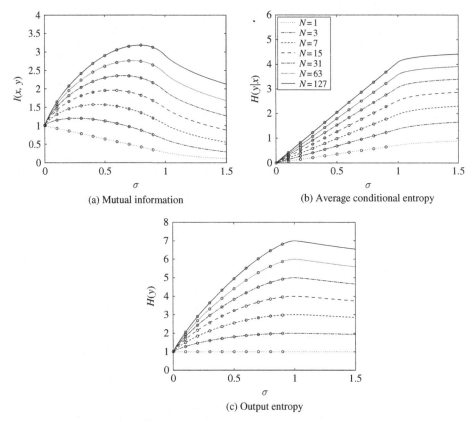

(a) Mutual information

(b) Average conditional entropy

(c) Output entropy

Fig. 4.9. Plots showing (a) the mutual information, (b) average conditional output entropy, and (c) output entropy, against σ for various values of N and uniform signal and noise. The solid line indicates numerical integration and the circles indicate the exact results that appear in Fig. 4.7. The increasing values of the curves up the y-axis correspond to larger values of N.

shown with circles is the fact that the mutual information and output entropy are exactly unity at $\sigma = 0$, and that the average conditional entropy is zero at $\sigma = 0$. Clearly, the numerically calculated results agree with the known exact values in all cases.

Given the very similar qualitative behaviour for these five signal and noise pairs, it is of interest to examine more closely the differences between them. To facilitate this, Fig. 4.13 shows the mutual information, average output entropy, and output entropy for all five cases superimposed, for $N = 127$. As of course should be the case, given the previously presented theory, for $\sigma = 1$ the mutual information, average output entropy, and output entropy are the same for all signal/noise pairs. A close inspection shows that the Gaussian case gives the largest maximum mutual information out of all the plotted cases. A further observation is that the maximum

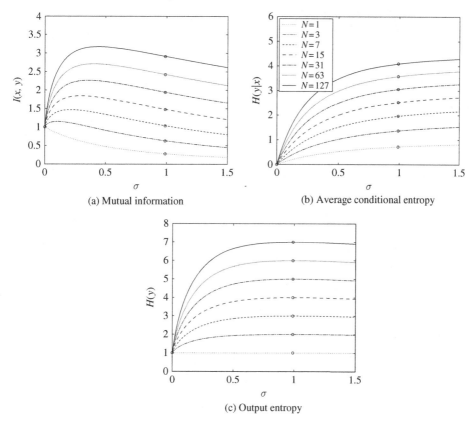

(a) Mutual information

(b) Average conditional entropy

(c) Output entropy

Fig. 4.10. Plots showing (a) the mutual information, (b) average conditional output entropy, and (c) output entropy against σ for various values of N and Laplacian signal and noise. The increasing values of the curves up the y-axis correspond to larger values of N. The solid lines were obtained by numerical integration and the circles at $\sigma = 0$ – for which the mutual information and output entropy is always exactly unity – and $\sigma = 1$ indicate exact values for each quantity, calculated using Eqs (4.40) and (4.36).

mutual information for the Gaussian, Laplacian, and logistic cases occurs for nearly the same values of σ – a value that is much smaller than the maximizing σ for the uniform case. The Cauchy case has a slightly smaller maximum, but also occurs for a much smaller σ than the uniform case.

We now consider two cases of distributions with single-sided PDFs, that is, the PDF of random variables which are defined only for positive values.

Fig. 4.14 shows the mutual information, average conditional entropy, and output entropy for Rayleigh signal and noise, for a range of σ and N, and two values of the threshold. The first case, shown with solid lines, is for the threshold set to the signal mean, which is $\theta = \sigma_x \sqrt{\frac{\pi}{2}}$. This ensures that the mutual information at $\sigma = 0$

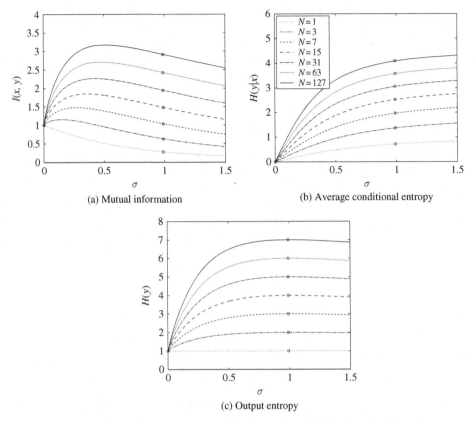

Fig. 4.11. Plots showing (a) the mutual information, (b) average conditional output entropy, and (c) output entropy against σ for various values of N and logistic signal and noise. The increasing values of the curves up the y-axis correspond to larger values of N. The solid lines were obtained by numerical integration and the circles at $\sigma = 0$ – for which the mutual information and output entropy is always exactly unity – and $\sigma = 1$ indicate exact values for each quantity, calculated using Eqs (4.40) and (4.36).

is equal to unity. However, due to the one-sided nature of the noise PDF, this also means that for all values of x greater then the mean, the output will always be N, since the noise can never help these values of the signal cross below any thresholds. This causes the maximum value of the mutual information to be much smaller than for the same N in the Gaussian case. The second case shown in Fig. 4.14 is for the threshold value set to twice the signal mean. Although this decreases the mutual information for small σ, it substantially increases it for large σ, since there are now far fewer values of x for which the output will always be zero.

Fig. 4.15 shows the mutual information, average conditional entropy, and output entropy for exponential signal and noise, for a range of σ and N. The threshold

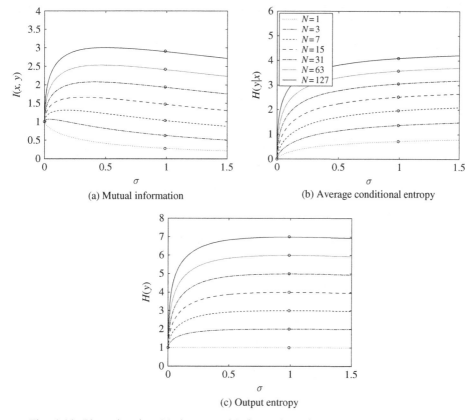

Fig. 4.12. Plots showing (a) the mutual information, (b) average conditional output entropy, and (c) output entropy against σ for various values of N and Cauchy signal and noise. The increasing values of the curves up the y-axis correspond to larger values of N. The solid lines were obtained by numerical integration and the circles at $\sigma = 0$ – for which the mutual information and output entropy is always exactly unity – and $\sigma = 1$ indicate exact values for each quantity, calculated using Eqs (4.40) and (4.36).

value is set to the signal mean, which is $\theta = \sigma_x$. This ensures that the mutual information at $\sigma = 0$ is equal to unity. Although the results shown in Fig. 4.15 show the same qualitative behaviour as in the Gaussian signal and noise case, the maximum value of the mutual information for each N is much smaller. As with the Rayleigh case, this is because with the threshold set to the signal mean, there is a large range of x for which the output is always zero.

To compare the mutual information in the Rayleigh and exponential cases to the even PDF cases, Fig. 4.16 shows the mutual information, average output entropy, and output entropy for both these distributions for $N = 127$, as well as the case of Gaussian signal and noise. The Rayleigh case is shown for both threshold values

Suprathreshold stochastic resonance: encoding

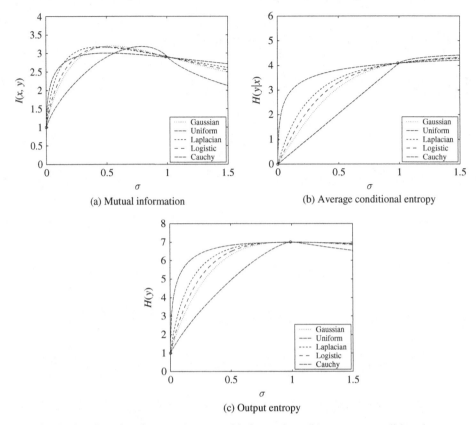

Fig. 4.13. Plots showing (a) the mutual information, (b) average conditional output entropy, and (c) output entropy against σ for $N = 127$ and each of five signal and noise pairs. The solid lines were obtained by numerical integration and the circles at $\sigma = 0$ – for which the mutual information and output entropy are always exactly unity – and $\sigma = 1$ indicate exact values for each quantity, calculated using Eqs (4.40) and (4.36). Notice that in all cases the mutual information, output entropy, and average conditional output entropy have the same value at $\sigma = 1$, as it should be, given the known exact expressions at this point. The maximum mutual information for the Gaussian, Laplacian, and logistic cases occurs for nearly the same values of σ, values which are much smaller than the maximizing σ for the uniform case. For the Cauchy case, the maximum σ is slightly smaller, but also occurs for a much smaller value of σ than the uniform case.

plotted in Fig. 4.14. It is clear that the mutual information is far smaller for the Rayleigh and exponential cases than for the Gaussian case. However in the case of the Rayleigh distribution, where the threshold is set to twice the signal mean, the mutual information for large σ is very close to that obtained for the Gaussian case. Indeed, at $\sigma = 1$, it is only a small fraction less, indicating that even though the exact formula for the mutual information at $\sigma = 1$ given in Eq. (4.40) does not

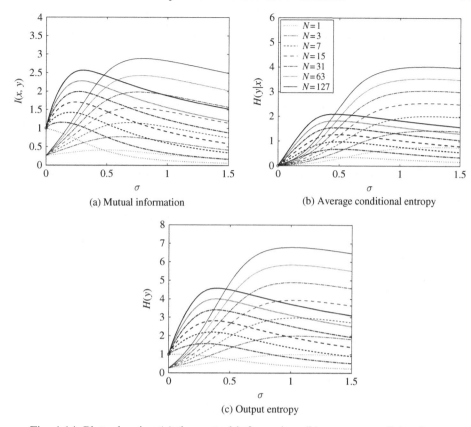

(a) Mutual information   (b) Average conditional entropy

(c) Output entropy

Fig. 4.14. Plots showing (a) the mutual information, (b) average conditional output entropy, and (c) output entropy against σ for various values of N and Rayleigh signal and noise. The solid lines are the results for the threshold set to the signal mean, so that the mutual information is exactly unity in the absence of noise. The dashed lines are the results for the threshold set to twice the signal mean, showing that in this case a much larger maximum mutual information occurs. This happens due to the one-sided nature of the noise PDF; for all signal values above the mean, the noise cannot cause thresholds to be crossed, and the overall output will always be N. For a larger mean, this is less likely to occur, and for sufficiently large σ, the noise is better matched to the signal. The increasing values of the curves up the y-axis correspond to larger values of N.

apply for one-sided signal and noise PDFs, it can give a close upper bound, for an optimally set threshold.

We next briefly consider mixed signal and noise distributions.

Mixed signal and noise

Unlike in the matched cases above, in general the mutual information will be a function of both the signal and noise variances independently. Hence, our results

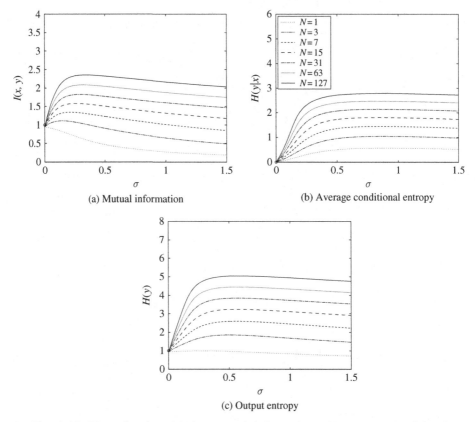

Fig. 4.15. Plots showing (a) the mutual information, (b) average conditional output entropy, and (c) output entropy against σ for various values of N and exponential signal and noise. The threshold is set to the signal mean, $\theta = \sigma_x$, so that the mutual information is exactly unity in the absence of noise. As with the Rayleigh case, the maximum mutual information is far smaller than for the even PDFs, due to the one-sided nature of the noise PDFs. The increasing values of the curves up the y-axis correspond to larger values of N.

for the mutual information in this section are plotted as a function of the ratio of noise standard deviation to signal standard deviation, and the signal standard deviation will be set to unity.

We consider the case of Gaussian noise. This is usually the first assumption to be made when the distribution of a noise source is unknown. Partly this is because Gaussian distributions are ubiquitous in nature, and partly because for all continuously valued distributions with a known variance the Gaussian distribution has the largest entropy.

Figs 4.17, 4.18, 4.19, 4.20, and 4.21 show with dashed lines the numerical results for the cases of uniform, Laplacian, logistic, Cauchy, and Rayleigh signals, with

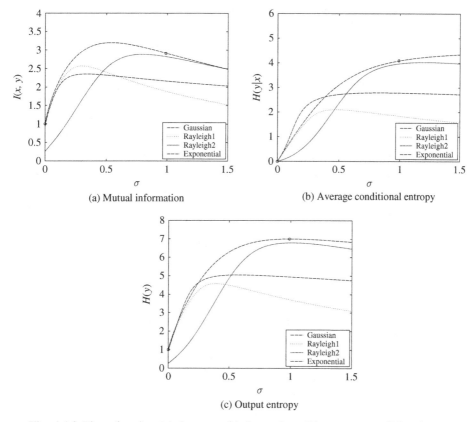

Fig. 4.16. Plots showing (a) the mutual information, (b) average conditional output entropy, and (c) output entropy against σ for $N = 127$ and the cases of Rayleigh signal and noise, and exponential signal and noise. The Rayleigh case is shown for two different threshold values, first the threshold set to the signal mean (Rayleigh1), and second the threshold set to twice the signal mean (Rayleigh2). The Gaussian case is shown for comparison. The circles indicate known exact results at $\sigma = 0$ and $\sigma = 1$ that apply for the Gaussian case. Note that there is no exact result for the Rayleigh or exponential cases, due to the one-sided nature of their PDFs. It is clear that the maximum mutual information is far smaller for the Rayleigh and exponential cases than for the Gaussian case. This is due to the one-sided nature of the noise PDF. However, for a threshold set to allow more threshold crossings, the mutual information in the Rayleigh case approaches that of the Gaussian case for large σ near $\sigma = 1$.

Gaussian noise. The solid lines show the previously plotted results for noise having the same distribution as the signal. In all cases except for a uniform or Rayleigh signal, Gaussian noise provides a slightly larger maximum mutual information than for the matched case. In the Rayleigh case, a much larger mutual information results from Gaussian noise. Fig. 4.22 shows the mutual information for all cases with Gaussian noise for $N = 127$. Very similar qualitative behaviour is seen

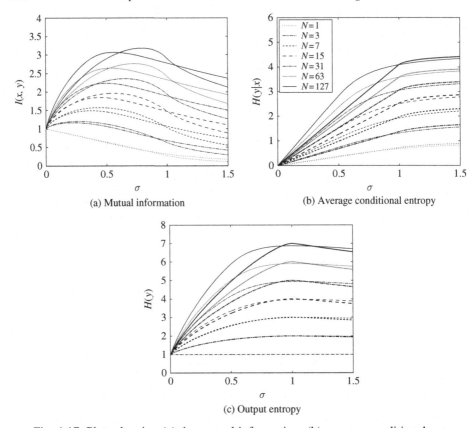

Fig. 4.17. Plots showing (a) the mutual information, (b) average conditional output entropy, and (c) output entropy against noise intensity, σ, for various values of array size, N. The solid lines show uniform signal and uniform noise, and the dashed lines show uniform signal and Gaussian noise, with the signal standard deviation set to $\sigma_x = \sqrt{12}$ in both cases. The increasing values of the curves up the y-axis correspond to larger values of N.

for each. The maximum mutual information is between 3 and 3.2 bits per sample for all cases, and occurs for a value of σ close to that for matched Gaussian signal and noise. Fig. 4.22 also shows that the mutual information for a Rayleigh signal and Gaussian noise is very close to that of Gaussian signal and Gaussian noise, indicating that Gaussian noise is a far better match for a Rayleigh signal than Rayleigh noise.

Such results lead into the focus of the next section, that of *channel capacity*. Channel capacity is defined as the *maximum* possible mutual information, subject to certain constraints. The results presented here lead us to ask questions such as 'what noise distribution provides the maximum mutual information in the SSR

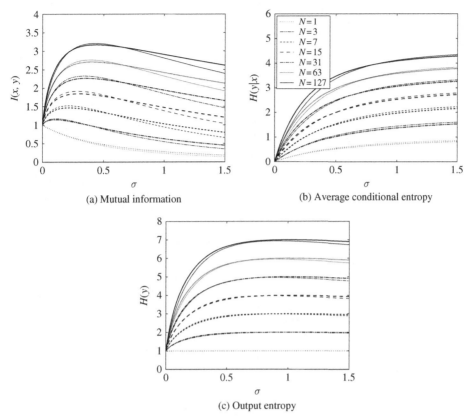

Fig. 4.18. Plots showing (a) the mutual information, (b) average conditional output entropy, and (c) output entropy against σ for various values of N. The solid lines show Laplacian signal and Laplacian noise, and the dashed lines show Laplacian signal and Gaussian noise, with $\sigma_x = 1$ in both cases. The increasing values of the curves up the y-axis correspond to larger values of N.

model for a given signal?' and 'what signal distribution provides the maximum mutual information for a given noise distribution?'

4.4 Channel capacity for SSR

In information theory, the term channel capacity is usually defined as being the maximum possible (on average) mutual information, per sample, that can be transmitted through a given channel. Usually, the channel is fixed and the maximization performed over all possible source probability distributions. This means that the

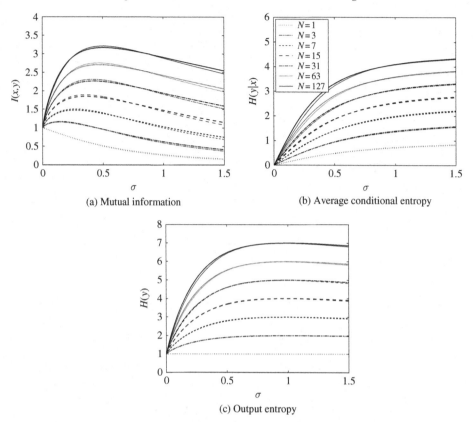

(a) Mutual information

(b) Average conditional entropy

(c) Output entropy

Fig. 4.19. Plots showing (a) the mutual information, (b) average conditional output entropy, and (c) output entropy against σ for various values of N. The solid lines show logistic signal and logistic noise, and the dashed lines show logistic signal and Gaussian noise, with $\sigma_x = 1$ in both cases. The increasing values of the curves up the y-axis correspond to larger values of N.

problem of finding channel capacity, $C(x, y)$, can be expressed as the optimization problem

$$\text{Find:} \quad C = \max_{\{f_x(x)\}} I(x, y). \tag{4.62}$$

Often there are prescribed constraints on the source distribution, $f_x(x)$, such as a fixed average power, or a finite alphabet, that is $f_x(x)$ is discretely valued, with a fixed, finite number of states. Relevant attributes of the given channel are the noise power and distribution, and bandwidth. The second most well-known example of a channel capacity formula is that for a binary symmetric channel (Cover and Thomas 1991).

The best-known example of a channel capacity formula is Shannon's famous, but often misused, formula for the channel capacity of a bandlimited, additive white

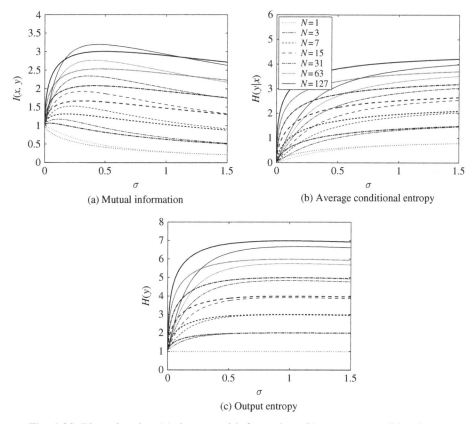

(a) Mutual information

(b) Average conditional entropy

(c) Output entropy

Fig. 4.20. Plots showing (a) the mutual information, (b) average conditional output entropy, and (c) output entropy against noise intensity, σ, for various values of N. The solid lines show Cauchy signal and Cauchy noise, and the dashed lines show Cauchy signal and Gaussian noise. In both cases, for the Cauchy signal, the FWHM for the signal is $\lambda_x = 1$ and the FWHM for the noise is $\lambda_\eta = \sigma/5$. The increasing values of the curves up the y-axis correspond to larger values of N.

Gaussian noise channel (Shannon 1948)

$$C = B \log_2 \left(1 + \frac{P_s}{P_n}\right) \quad \text{bits per second,} \tag{4.63}$$

where B is the channel bandwidth, P_s is the prescribed maximum mean square signal power and P_n is the mean square noise power (within the channel bandwidth). Note that this formula has many names – see Section 2.7 in Chapter 2.

The channel capacity of Eq. (4.63) can be rewritten in its best-known form, in terms of the input SNR, as

$$C = B \log_2 (1 + \text{SNR}) \quad \text{bits per second.} \tag{4.64}$$

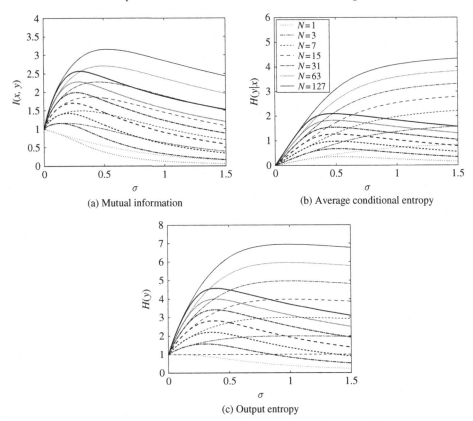

(a) Mutual information

(b) Average conditional entropy

(c) Output entropy

Fig. 4.21. Plots showing (a) the mutual information, (b) average conditional output entropy, and (c) output entropy against σ for various values of N. The solid lines show Rayleigh signal and Rayleigh noise, and the dashed lines show Rayleigh signal and Gaussian noise. In both cases, $\sigma_x = 1$ and the threshold value, θ, is set to the signal mean so that $\theta = \sqrt{2 - 0.5\pi\sigma_x}$. The increasing values of the curves up the y-axis correspond to larger values of N. Clearly, Gaussian noise provides a far larger maximum mutual information than for matched signal and noise in this case.

The main reason that this formula is often misapplied – see Berger and Gibson (1998) for a discussion – is that it applies only when the channel noise has a Gaussian distribution that is independent of the signal and is additive in nature. It also does not apply if the noise is not white within the channel bandwidth – that is, if the power spectral density of the noise is not constant for all frequencies in the passband of the channel – or if there are constraints other than the power constraint on the input signal. Furthermore, capacity can be achieved only if the input signal has a Gaussian distribution.

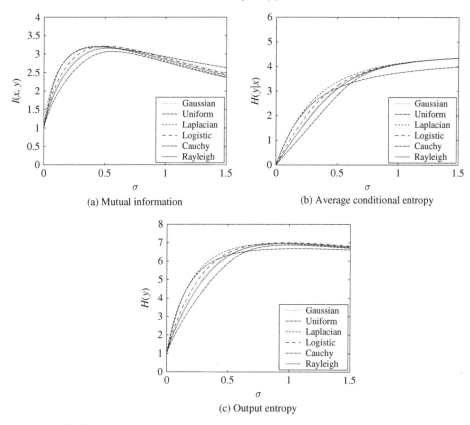

Fig. 4.22. Plots showing (a) the mutual information, (b) average conditional output entropy, and (c) output entropy against σ for $N = 127$ and various signal distributions with Gaussian noise. The maximum mutual information is between 3 and 3.2 bits per sample, and occurs for nearly the same values of σ for all cases.

In channel capacity research, conventionally the channel is prescribed, and subject to certain imposed input signal constraints, the optimal signal distribution found. Previously in this chapter, we have seen that the mutual information in the SSR model is maximized by a nonzero value of noise. If we determine the optimal ratio of noise standard deviation to signal standard deviation, σ, we have effectively found channel capacity. At first glance, when compared to conventional studies of information channels, varying the noise level for a given signal seems as if it is equivalent to modifying the channel, rather than looking for an optimal source distribution.

However, recall that in the case of even PDFs, the mutual information for matched signal and noise is a function of σ. Adjusting σ to find the maximum mutual information for given signal and noise pairs is also equivalent to modifying the source distribution, by increasing or decreasing its variance, for fixed noise.

The end result is that the questions of 'what noise variance maximizes the mutual information, for given matched signal and noise distributions and a fixed signal variance?' and 'what signal variance maximizes the mutual information, for given matched signal and noise distributions and a fixed noise variance?' are equivalent. The following paragraphs look at this new single question, and, in light of the above discussion, this is no different from conventional channel capacity questions, since the channel is fixed – that is, a noise distribution and variance is given – and the source distribution is to be found, subject to the constraint that it has the same PDF as the noise, but can have any variance.

Matched and fixed signal and noise PDFs

In light of the proceeding discussion, consider the case of a channel being prescribed – that is, the number of comparators, N, and the noise PDF, $f_\eta(\eta)$, and variance, σ_η^2, will remain fixed – and the constraint on the signal that its PDF is the same as the noise, other than the variance.

In this situation, finding the SSR channel capacity means finding the optimal signal variance – or, equivalently, power – for a given noise variance. We can express this problem in the following manner

Find: $\qquad C(x, y) = \max_{\text{var}[x]} I(x, y)$

subject to: $\quad f_\eta(\eta)$ and var$[\eta]$ fixed

and $\qquad\quad f_x(x)$ matched with $f_\eta(\eta)$. $\qquad\qquad$ (4.65)

The result of solving this optimization problem is the optimal value of the variance of the signal. Since this value will hold for a specified variance of the noise, the result will give the optimal value of σ.

Although solving this problem analytically is intractable – apart from the uniform signal and noise case – numerical solution is straightforward. The results of such an optimization for the cases of matched Gaussian, Laplacian, and logistic signal and noise are shown in Fig. 4.23, which shows the values of σ at which channel capacity is achieved, for increasing N, and Fig. 4.24(a), which shows the channel capacity. Also shown in these figures is the capacity for uniform signal and noise, where σ_o is calculated from Eq. (4.59) and plotted in Fig. 4.23, and the mutual information at this σ_o, plotted in Fig. 4.24(a). Fig. 4.24(a) also shows the mutual information obtained at $\sigma = 1$ using the exact formula of Eq. (4.40), which holds in all cases.

From Fig. 4.24(a) the difference between the capacity of each of the three infinite tail distributions is very small, and indeed is hard to resolve on the scale shown.

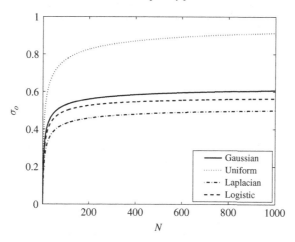

Fig. 4.23. Optimal noise intensity, σ, for Gaussian, Laplacian, and logistic signal and noise. This figure shows σ_o – that is, the value of σ that achieves channel capacity – against increasing N. The Gaussian, Laplacian, and logistic cases were calculated numerically, and the uniform case was calculated from the exact formula of Eq. (4.59). Although it is not completely clear from this figure with the scale limited to $N \leq 1000$, for larger N, σ_o asymptotically approaches a fixed value.

Hence, the difference is plotted in Fig. 4.24(b). This plot shows that the difference in channel capacity appears to be decreasing as N increases, apart from an initial increase in difference for small N.

Fig. 4.24(c) shows the difference between channel capacity, and the mutual information at $\sigma = 1$. The capacity is, of course, always greater than the exact mutual information at $\sigma = 1$, although the difference between the capacity and this value appears to also approach a constant slightly greater than 0.2 bits per sample for large N, apart from the uniform case. In the uniform case, the optimizing σ approaches unity for large N. As previously noted here, and in Stocks (2001a), the maximum in the mutual information appears always to be between $\sigma = 0$ and $\sigma = 1$, since at $\sigma = 0$, the mutual information is always one bit per sample.

It is clear from Fig. 4.24(a) that as N increases, the channel capacity also increases. However, from Fig. 4.23, the optimizing value of σ appears to converge asymptotically to a constant value, σ_o, as N increases. For Gaussian signal and noise, this large N value appears to be $\sigma_o \simeq 0.607$, for Laplacian signal and noise, $\sigma_o \simeq 0.5$, and for logistic signal and noise $\sigma_o \simeq 0.564$. It is not visible on the axes of Fig. 4.23, but for uniform signal and noise, the large N value of $\sigma_o \simeq 1$. This is shown to be the case in Chapter 5.

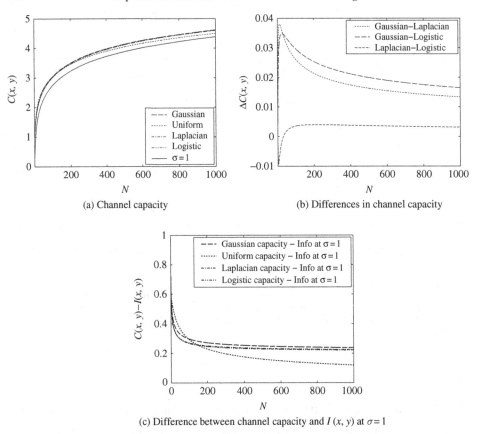

(a) Channel capacity

(b) Differences in channel capacity

(c) Difference between channel capacity and $I(x, y)$ at $\sigma = 1$

Fig. 4.24. Channel capacity for matched signal and noise. Figure 4.24(a) shows the channel capacity calculated numerically for Gaussian, Laplacian, and logistic signal and noise pairs against increasing N. It also shows the exact channel capacity for uniform signal and noise, calculated from Eqs (4.59) and (4.58), and the mutual information that holds for all cases where $f_X(x) = f_\eta(\theta - x)$ so that $\sigma = 1$, calculated from Eq. (4.40). Clearly, the channel capacity is almost identical for the Gaussian, Laplacian, and logistic cases, and is larger than the mutual information at $\sigma = 1$. Figure 4.24(b) shows that the difference between the channel capacity for each of these three pairs is very small, and the difference gets smaller with larger N. Figure 4.24(c) shows the difference between the channel capacity and the mutual information at $\sigma = 1$ for each case. Apart from the uniform case, this difference appears to approach a constant value, again indicating that the maximum mutual information occurs for $\sigma < 1$ for these cases'for all N.

Furthermore, the difference between channel capacities for the various distributions appears as if it might approach zero for very large N. These results indicate the possibility that proofs of this behaviour may be possible in a large N limit. Chapter 5 is devoted to the investigation of expressions for the mutual information and channel capacity that hold for large N.

Channel capacity in general

We saw for the case of mixed signal and noise distributions that Gaussian noise can provide a larger maximum mutual information than noise matched with the signal. Hence, the natural question to ask is 'what signal distribution maximizes the mutual information for a given noise distribution?' In contrast to the previous paragraphs, where we imposed the constraint that the signal distribution must be matched with the source distribution, and therefore the shape of the signal PDF is known, to solve this problem the optimal source PDF must be found. This problem can be expressed as

$$\text{Find:} \quad C(x, y) = \max_{f_x(x)} I(x, y)$$

$$\text{subject to:} \quad f_\eta(\eta) \text{ and } \text{var}[\eta] \text{ fixed.} \tag{4.66}$$

The constraint that specifies that the noise distribution is fixed can be dropped from the expression of the problem, as it simply specifies the nature of the channel. The optimization problem to be solved is

$$\text{Find:} \quad C(x, y) = \max_{f_x(x)} I(x, y). \tag{4.67}$$

For channels with continuously valued inputs and outputs, usually a constraint, such as a peak power or fixed average power constraint, is specified. This is to give some realism to a signal's distribution, since otherwise the optimal $f_x(x)$ could have infinite power. Such channel capacity problems can be solved by use of the Arimoto–Blahut algorithm (Arimoto 1972, Blahut 1972, Cover and Thomas 1991). This has been carried out for the SSR model with Gaussian noise in Hoch *et al.* (2003b) – see Section 5.5 in Chapter 5.

This iterative algorithm begins with an initial guess for $f_x(x)$. It then uses this $f_x(x)$ along with the channel characteristics to calculate the joint probabilities of all pairs of input and output values, $f_{xy}(x, n)$. The current mutual information can be calculated from these distributions. The algorithm then calculates a new $f_x(x)$, one that is optimal for the current joint probabilities. This process is repeated, until the algorithm converges to a solution such that the difference between successive values of the mutual information is smaller than a specified error tolerance, and therefore the channel capacity has been found.

The Arimoto–Blahut algorithm is fairly general, and can be easily extended to incorporate various constraints on the input signal, or specified costs for using various values of the input (Blahut 1972, Hoch *et al.* 2003b). It can also be used to calculate rate-distortion functions – see Chapter 9.

Applying the Arimoto–Blahut algorithm to calculating channel capacity for the SSR model, as specified in Problem (4.67) does indeed find solutions. Specific values for channel capacity for each N are slightly larger than in the case of matched signal and noise. In fact, although not once thought of as being a case of a noise optimized signal, such a calculation has been previously performed in Chang and Davisson (1988). The main purpose of Chang and Davisson is to specify two algorithms based on the Arimoto–Blahut algorithm for finding channel capacity in an infinite-input, finite-output channel. The SSR model is exactly such a channel.

Chang and Davisson (1988) use an example situation to test each of these algorithms, where the input signal is a continuously valued variable, x', between zero and unity, and the output signal is integer valued between zero and L. The channel transition probabilities are specified to be given by the binomial distribution as a function of x' as

$$P(k|x') = \binom{L}{k}(x')^k(1 - x')^{(L-k)} \quad k = 0, \ldots, L. \tag{4.68}$$

If $P_{1|x}$ is substituted for x', Eq. (4.68) is recognizable as being equivalent to Eq. (4.9). This substitution is allowable, since $P_{1|x} \in [0, 1]$ and $x' \in [0, 1]$. With reference to Table 4.1, a situation where $P_{1|x}$ increases linearly with x for all valid x occurs when the noise is uniform, and the signal PDF is limited to the support of the uniform noise. Therefore, the channel of Chang and Davisson (1988) is exactly the same as the SSR model when subjected to uniform noise, provided the signal and noise have the same support.

Replicating the results in Chang and Davisson (1988) – which does not give the actual optimal $f_x(x)$, only the capacity – finds that the input PDF that achieves the channel capacity, $f_x^o(x)$, can be multimodal with small regions of highly probable values located symmetrically a long distance from its mean. The capacity found, although larger than for matched signal and noise, is larger by a few percent than 'matched' signal and noise – see also Section 5.5 in Chapter 5.

For the case of SSR, a more relevant channel capacity question might be one that specifies a set of input PDF constraints such as a maximum variance, combined with a minimum entropy. Furthermore, constraints on the number of local maxima of the PDF might be included. Such problems are interesting open questions.

Alternatively, constraints can be set on the output distribution. Such a situation can arise naturally in a neural coding context, where coding of information might be subject to strict energy constraints. This scenario has been studied in Hoch *et al.* (2003b), which uses the Arimoto–Blahut algorithm to derive optimal input PDFs under an output energy constraint, such as might occur in real neurons.

Coding efficiency

Since the SSR model has a discretely valued output signal, it also has a well-defined limit to its channel capacity. This is $\log_2 (N + 1)$ bits per sample, since the output has $N + 1$ states. This contrasts with the additive Gaussian white noise channel considered as the channel for Shannon's channel capacity formula, for which the output is a continuously valued signal. For such systems, there is no upper bound on the channel capacity, which can theoretically be infinite.

Note that it is possible to achieve this maximum channel capacity in a situation where not all thresholds are equal to the signal mean. For example, in the absence of noise, this maximum channel capacity is achieved by setting the thresholds such that all output states are equally likely. Such a scenario however is beyond the scope of this chapter, and will be considered in Chapter 8. Here, we are considering only the case of all thresholds identical.

The existence of such an upper bound on channel capacity allows the definition of a *coding efficiency* measure as

$$\Gamma(N) = \frac{C(x, y)}{\log_2 (N + 1)}. \tag{4.69}$$

This measure gives the fraction of the maximum possible mutual information provided by given signal and noise distributions in the SSR model. If Γ is multiplied by 100, then the coding efficiency can be measured as a percentage. This idea is expressed in Stocks (2001b). An alternative way of measuring the same thing is to measure the difference between $C(x, y)$ and $\log_2 (N + 1)$. However, since this is an absolute measure, and Γ is a relative measure, we shall focus on Γ.

The results for channel capacity in the previous parts of this section can be re-plotted in terms of coding efficiency. Figure 4.25 shows the coding efficiency for matched signal and noise for the Gaussian, uniform, Laplacian, and logistic cases, as well as the coding efficiency when $f_x(x) = f_\eta(\theta - x)$, so that $\sigma = 1$. For large N, the coding efficiency appears to asymptotically approach a value slightly less than 0.5. As noted in Stocks (2001b), this means that the mutual information for the SSR model for matched signal and noise is less than 50 percent of the maximum noiseless channel capacity. However, given that in most systems noise is unavoidable, it may be the case that a coding efficiency of about 50 percent is quite acceptable.

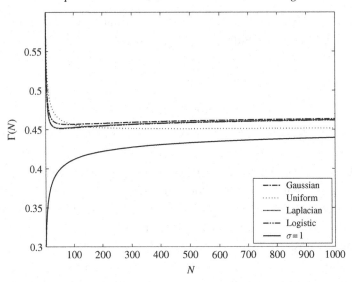

Fig. 4.25. Coding efficiency for matched signal and noise. This figure shows the coding efficiency, Γ, as a function of array size, N, for matched signal and noise for the Gaussian, uniform, Laplacian, and logistic cases. Also shown is the coding efficiency when $f_x(x) = f_\eta(\theta - x)$, so that $\sigma = 1$. For large N, the coding efficiency appears to asymptotically approach a value slightly less than 0.5.

4.5 SSR as stochastic quantization

We have seen in this chapter how the mutual information in the SSR model has a maximum for a nonzero noise intensity. We have also seen how the maximum mutual information in the absence of noise is exactly one bit per sample, since all thresholds have the same value and the output can only ever be zero or N. Why is it that the presence of noise allows a larger information rate per sample? The explanation is that the presence of *independent* noise on each threshold has the effect of distributing all N thresholds to different values, rather than the same value. Instead of modelling the noise as being added to the signal, it is equivalent to model the noise as being threshold noise, so that the threshold is a random variable. Hence, if the threshold is θ, then the signal, x, is thresholded N times by N independent samples taken from the random variable, η, described by the PDF, $f_\eta(\theta - \eta)$.

When noise is present, effectively all threshold values will be unique. This allows the output to take values other than zero and N, and to become a $\log_2(N + 1)$ bit *stochastic quantization* of the input signal. For noise with a small variance compared to the signal, most of the output states will occur with a very small – or zero, for noise with finite support – probability. Hence, very little extra information can be gained, since only a fraction of the $N + 1$ output states are utilized. However, when the noise variance is such that each output state is occupied with a probability that reflects the shape of the input PDF, far more information can be gained, although, as we have seen, this appears to be at most about 50 percent of the maximum possible channel capacity. For a noise variance larger than optimal, the dependence between the input and output signals decreases, and the mutual information starts to decrease again.

Indeed, a description of SSR as stochastic quantization is given in some of the first work on SSR:

At any instant of time, finite noise results in a distribution of thresholds that, in turn, leads to the signal being 'sampled' at N randomly spaced points across the signal space. (Stocks 2000a)

As discussed, the realization that nonzero noise causes a distribution of thresholds is the key to understanding why a nonzero noise intensity gives best performance, the end result being a non-deterministic – or stochastic – quantization of the signal.

Example random distributions of thresholds for Gaussian signal and noise are shown in Fig. 4.26. In the subfigures, Fig. 4.26(a) shows the optimal noiseless threshold values for this Gaussian source, while Figs 4.26(b), 4.26(c), and 4.26(d) show examples of the random threshold distribution in the SSR model, for three different values of Gaussian noise variance. Clearly, although the example of the random threshold distribution does not give the optimal threshold distribution, as we have seen in this chapter it does give a distribution that allows the *average* mutual information of the SSR model to approach half the maximum noiseless channel capacity.

It is now clear why channel capacity occurs at $\sigma_o \to 1$ in the case of uniform signal and noise, but approaches a value much less than unity in the cases of infinite tail PDFs. In the uniform case, when $\sigma > 1$ the random distribution of thresholds for any given input sample sometimes places thresholds outside the dynamic range of the signal. This can only reduce the mutual information. Furthermore, since for the uniform PDF, all values of x are equally probable, it is desirable that the random threshold distribution is just as likely to consist of a threshold near the mean of the signal as it is close to its maximum and minimum value. For large N, this can occur only if the noise PDF has a variance close to that of the signal.

By contrast, for the infinite tail signal and noise cases, the most probable values of the signal occur within several standard deviations of the mean. Therefore, the noise distribution does not need to be as wide as the signal distribution for the optimal random threshold distribution to occur. Hence, channel capacity occurs for $\sigma_o \ll 1$ for these cases.

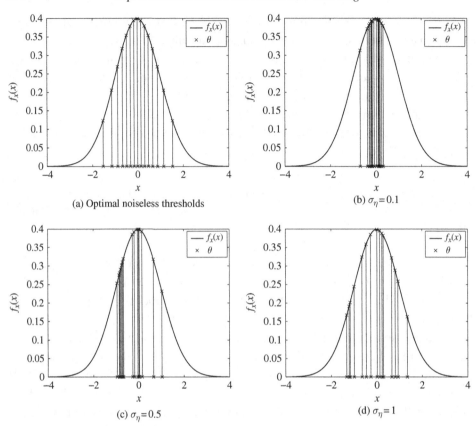

Fig. 4.26. Random threshold distribution for Gaussian signal and noise. Fig. 4.26(a) shows the optimal – that is, maximum mutual information – threshold values, θ, for $N = 15$ and a unity variance Gaussian source. The stem plot indicates with crosses the values of x for each threshold, and the value of $f_X(x)$ at which each threshold occurs. Figs 4.26(b), 4.26(c), and 4.26(d) show particular instances of the random SSR threshold distribution for three different values of Gaussian noise variance, compared with a unity variance Gaussian source. Notice that for small noise variance, most thresholds are likely to end up close to the signal mean. For intermediate and larger variances, more thresholds are likely to be located closer to the tails of the source.

Encoder transfer function

Given this stochastic quantization interpretation of the SSR model, it is natural to wonder how the SSR output compares with a standard quantization of the input. We shall examine this question more closely in Chapter 6; however, here we illustrate the effect of independent threshold noise in this context by comparing the transfer function of a conventional deterministic quantizer's encoder, with the mean transfer function of the SSR model's encoding. When conventional quantization is introduced in standard textbooks, it is usual to graphically illustrate the transfer function by plotting the output values that correspond to certain input values.

For example, suppose a quantizer's output has three bits, so that $N = 7$ thresholds are required, and there are $N + 1 = 8$ output states. For the case where the thresholds uniformly quantize a signal between ± 1, the transfer function is shown with a thick black line in Fig. 4.27(a). Note that this transfer function is deterministic, so therefore there is no uncertainty about which output value is attained for every input value.

In contrast, the average transfer function for SSR is simply the *expected* value of the output, y, given the input, x. Recall from Eq. (4.9) that the conditional probability distribution of the output given the input is given by the binomial distribution. Here, the expected value of the binomial distribution is given by

$$E[y|x] = N P_{1|x}, \tag{4.70}$$

which gives the average transfer function for SSR. However, unlike conventional quantizers, the output is not deterministic. One way of characterizing this uncertainty is with the conditional variance, $\text{var}[y|x]$, which for the binomial distribution is given by

$$\text{var}[y|x] = N P_{1|x}(1 - P_{1|x}). \tag{4.71}$$

Fig. 4.27(a) shows the *average* transfer function for the SSR model, with $N = 7$, for various values of noise standard deviation, σ_η, and Gaussian noise, as a function of x. Fig. 4.27(b) shows the variance corresponding to each value of σ_η as a function of x. We see that for the intermediate values of σ_η, the average transfer function appears to more closely approximate the deterministic transfer function, whereas for $\sigma_\eta = 0.3$ and $\sigma_\eta = 1.2$, the average transfer function appears to have too sharp, or too broad, a slope to give a good approximation. The relationship between the ideal and the average transfer function is broadly measured by what is known as *bias*.

However, the performance of the SSR model as a quantizer cannot be measured only by how well its average transfer function compares with an ideal transfer function. From Fig. 4.27(b), we see that the variance of the output for a given value of the input always increases with increasing σ_η. Hence, there is a tradeoff required between the bias and the variance for a given x. We shall see quantitatively in Chapter 6 how the performance of the SSR model depends on both of these quantities, and also on the input distribution.

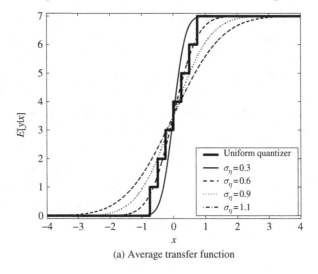

(a) Average transfer function

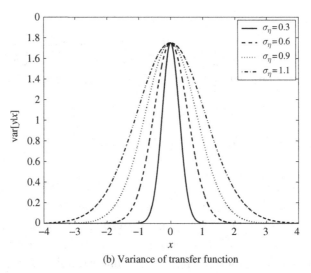

(b) Variance of transfer function

Fig. 4.27. Average transfer function, and its variance, for SSR. Fig. 4.27(a) shows with a thick black line the output of a conventional deterministic three-bit uniform quantizer as a function of the input, x. It shows also the *average* transfer function for the SSR model, with $N = 7$, for various values of noise standard deviation, σ_η, and Gaussian noise. Fig. 4.27(b) shows the variance corresponding to each value of σ_η as a function of x.

Decoding the SSR output

This description of SSR as a stochastic quantization of the signal leads us naturally to ask whether SSR can be described in terms of conventional quantization. The answer is yes. Such theory usually considers a quantizer as having two stages – the encoding stage and the decoding stage. The encoding stage can be described in terms of mutual information, for randomly noisy encoding. As well as describing the interpretation of SSR as an information channel, the current chapter serves as a description for the *encoding* of a signal in a quantization scheme. Analysis of SSR in terms of the *decoder* component of a quantizer is left for Chapter 6.

Furthermore, the realization that SSR can be described in terms of stochastic quantization of a signal also leads to the subject of optimal quantization. Usually a quantization scheme is considered to be deterministic, and optimizing the quantization scheme requires the optimal selection of threshold values. There is no reason why such optimal threshold selection cannot be considered for the case of noisy thresholds. This is the focus of Chapter 8 and is also considered under constraints in 9.

4.6 Chapter summary

This chapter commences with a historical literature review of the initial discovery of suprathreshold stochastic resonance, and all subsequent substantial work published on the topic. We also briefly compare and contrast SSR with other similar work, and discuss why the SSR model can be analyzed from an information-theoretic perspective.

Next, Section 4.3 comprehensively defines and discusses the SSR model, and replicates the approach used by Stocks to measure the mutual information between its input and output signals. We also introduce a generic change of variable and show how this approach can be used to calculate the mutual information. This change of variable results in a new PDF, $f_Q(\tau)$, which we interpret as being the PDF of the average transfer function of the SSR model, and is used to show that the mutual information is a function of σ for specific matched signal and noise. Exact results for identical signal and noise distributions, and the case of uniform signal and noise, as derived by Stocks are expounded. Finally, Section 4.3 presents numerical results for seven different matched signal and noise distributions, as well as for various signal distributions subject to Gaussian noise.

Section 4.4 then discusses finding channel capacity for the SSR model. We first present numerical results showing the capacity obtained, and the value of σ which attains capacity, for four specified matched signal and noise cases. We then discuss briefly how the Arimoto–Blahut algorithm can be applied to find an input PDF that achieves capacity, when the input PDF is not specified. The last part of Section 4.4 gives a short discussion of how a coding efficiency measure can be defined for SSR, since there is a known finite upper bound on the mutual information.

Finally, Section 4.5 gives an interpretation of the SSR model as being a *stochastic quantizer*, since the independent noise on each threshold acts to randomize the effective threshold values. This point is illustrated by plots showing examples of the values of such random thresholds, and plots showing the average transfer function for the SSR model, when compared to a conventional quantizer.

Chapter 4 in a nutshell

This chapter includes the following highlights:

- An overview of all previous work on the SSR model.
- The introduction of a generic change of variable in the equations used to determine the mutual information through the SSR model, and the realization that this change of variable results in a PDF that describes the average transfer function of the SSR model. This PDF is derived for several cases, for which it is proved that the mutual information is a function of the ratio, σ, rather than a function of the noise variance and signal variance independently.
- The entropy of $f_Q(\tau)$ is shown to be equal to the negative of the relative entropy between the signal and noise PDFs.
- Numerical calculations of the output entropy, average conditional entropy, and mutual information, for a number of signal and noise distributions are plotted. These include the Laplacian, logistic, Cauchy, Rayleigh, and exponential distributions. SSR is shown to occur in all cases. The fact that SSR occurs for Cauchy distributed signal and noise illustrates that the effect does not require any finite variance assumptions.
- Numerical calculations of the mutual information for mixed signal and noise distributions are plotted. SSR is shown to occur in all cases.
- Channel capacity is found as a function of N for the Gaussian, Laplacian, uniform, and logistic cases, as is the value of σ that achieves capacity. The difference between channel capacity and the mutual information at $\sigma = 1$ is calculated, and found to converge towards a constant value with increasing N.
- A comparison of coding efficiency for each signal and noise pair.
- Illustrations of why the signal encoding provided by the SSR model can be interpreted as *stochastic quantization*.

Open questions

Interesting open questions arising from this chapter might include:

- The use of the Arimoto–Blahut algorithm to find input PDFs that attain channel capacity in the SSR model, subject to the specification of constraints. Such constraints might include a specified variance, combined with a unimodal constraint.
- Investigation of multi-modal or discretely valued signal and noise distributions.

- More thorough investigation of cases of mixed signal and noise distributions, and impulsive (infinite variance) signal and noise distributions.
- Consideration of how SSR effects may vary with PDF 'tail thickness', when the signal or noise is from the family of α-stable distributions (Kosko and Mitaim 2001). Here, we considered only the Gaussian ($\alpha = 2$) and Cauchy ($\alpha = 1$) cases.

This concludes Chapter 4, which introduces the SSR model, and presents results that apply for any value of N. Chapter 5 now examines approximations for the mutual information in the SSR model, under the assumption of a large number of threshold devices.

5

Suprathreshold stochastic resonance: large N encoding

This chapter discusses the behaviour of the mutual information and channel capacity in the suprathreshold stochastic resonance model as the number of threshold elements becomes large or approaches infinity. The results in Chapter 4 indicate that the mutual information and channel capacity might converge to simple expressions of N in the case of large N. The current chapter finds that accurate approximations do indeed exist in the large N limit. Using a relationship between mutual information and Fisher information, it is shown that capacity is achieved either (i) when the signal distribution is *Jeffrey's prior*, a distribution which is entirely dependent on the noise distribution, or (ii) when the noise distribution depends on the signal distribution via a cosine relationship. These results provide theoretical verification and justification for previous work in both computational neuroscience and electronics.

5.1 Introduction

Section 4.4 of Chapter 4 presents results for the mutual information and channel capacity through the suprathreshold stochastic resonance (SSR) model shown in Fig. 4.1. Recall that σ is the ratio of the noise standard deviation to the signal standard deviation. For the case of matched signal and noise distributions and a large number of threshold devices, N, the optimal value of σ – that is, the value of σ that maximizes the mutual information and achieves channel capacity – appears to asymptotically approach a constant value with increasing N. This indicates that analytical expressions might exist in the case of large N for the optimal noise intensity and channel capacity. We also saw that the channel capacity appeared as if it might become independent of the signal and noise distribution for large N. We now explore these questions and find accurate large N approximations in some specific cases.

In particular, we significantly extend previous results on this subject (Stocks 2001c, Stocks 2001a, Hoch *et al.* 2003b), by showing that the mutual information and output entropy can both be written in terms of simple relative entropy expressions – see Eqs (5.77) and (5.78). This leads to a very general sufficient condition, Eq. (5.80), for achieving capacity in the large N regime that can be achieved either by optimizing the signal distribution for a given noise distribution, or by optimizing the noise for a given signal. Given the neuroscience motivation for studying the SSR model, this result is potentially highly significant in computational neuroscience, where both optimal stimulus distributions, and optimal tuning curves are often considered (Hoch *et al.* 2003b, Brunel and Nadal 1998).

As a means of verification of our theory, we also in this chapter compare our general results with all prior specific results (Stocks 2001c, Stocks 2001a, Hoch *et al.* 2003b). This leads us to find and justify improvements to these previous results.

The mathematical treatment in this chapter follows along the lines of McDonnell *et al.* (2006a) and McDonnell *et al.* (2007). First, however, we briefly summarize relevant results given in Chapter 4 and then describe work already completed on this topic by other authors.

Mutual information

As described in Chapter 4, for N threshold devices, the output of the SSR model is a discretely valued signal, y, with possible values being the integers between zero and N. As explained, y can be considered to be a stochastic encoding of the continuously valued input signal x.

Regardless of the fact that N is allowed to approach infinity in the present chapter, the output can never be considered to be a continuously valued signal, even if it is decoded to a finite range. This fact has its roots in the notion of *countable* and *uncountable* infinities – the set of integers is called *countably infinite*, whereas the set of real numbers is called *uncountably infinite* (Courant and Robbins 1996). While there is a countable infinity of integers, there is an uncountable infinity of real numbers between any two distinct real numbers, no matter how small the separation between them.

This fact is also the basis for the difference between the entropy of a discrete random variable, which is always nonnegative, and the differential entropy of a continuously valued random variable, which may be positive, negative, or zero.

Thus, theoretically, quantization of a continuously valued signal always gives a lossy encoding, even if the *quantization noise* – that is, the noise introduced into an encoding of a number by quantization or truncation – can be made arbitrarily small, or to asymptotically approach zero, by allowing N to approach infinity. This is manifested by a term in the mutual information of the order of $\log_2 N$. This term approaches infinity for large N, and accounts for the fact that the quantization noise can decrease inversely with N. We shall return to this fact in Chapters 6 and 7.

In light of this discussion, it is important that large N approximations to the mutual information between the input and the output of the SSR model do not assume that the output is continuously valued, without rigorous justification.

Before proceeding to examining in detail previous results on this topic in Stocks (2001c) and Hoch *et al.* (2003a), we first summarize the relevant results from Chapter 4. As in Chapter 4, this chapter considers only the case of all threshold values being identical, and equal to θ, say.

Review of key results from Chapter 4

We define $P_y(n), n = 0, \ldots, N$ to be the probability mass function of the output signal, y, and $P_{y|x}(n|x), n = 0, \ldots, N$ to be the set of transition probabilities giving the probability that the output is $y = n$ given input signal value, x.

The mutual information between the input signal, x, and the output, y, of the SSR model can be written as

$$I(x, y) = H(y) - H(y|x)$$

$$= -\sum_{n=0}^{N} P_y(n) \log_2 P_y(n)$$

$$- \int_{-\infty}^{\infty} f_x(x) \sum_{n=0}^{N} P_{y|x}(n|x) \log_2 P_{y|x}(n|x) dx, \qquad (5.1)$$

where $f_x(x)$ is the PDF of the input signal, x. To progress further we use the notation introduced in Stocks (2000c) – see Chapter 4. Let $P_{1|x}$ be the probability of the ith threshold device giving output $y_i = 1$ in response to input signal value, x. If the noise CDF is $F_\eta(\cdot)$, then

$$P_{1|x} = 1 - F_\eta(\theta - x). \qquad (5.2)$$

The entropy of the output for a given value of the input can be expressed in terms of $P_{1|x}$ as

$$\hat{H}(y|x) = N \left(P_{1|x} \log_2 P_{1|x} + (1 - P_{1|x}) \log_2 (1 - P_{1|x}) \right)$$

$$+ \sum_{n=0}^{N} P_{y|x}(n|x) \log_2 \binom{N}{n}, \qquad (5.3)$$

where $P_{y|x}(n|x)$ is given by the binomial distribution as

$$P_{y|x}(n|x) = \binom{N}{n} P_{1|x}^n (1 - P_{1|x})^{N-n} \quad n \in 0, \ldots, N. \tag{5.4}$$

Eq. (5.1) reduces to

$$I(x, y) = -\sum_{n=0}^{N} P_y(n) \log_2 \left(\frac{P_y(n)}{\binom{N}{n}} \right) + N \int_x f_x(x) P_{1|x} \log_2 P_{1|x} dx$$

$$+ N \int_x f_x(x)(1 - P_{1|x}) \log_2 (1 - P_{1|x}) dx. \tag{5.5}$$

Recall also that integrations over the signal PDF's support variable, x, can be simplified to expressions defined in terms of integrations over the variable, τ, which exists only between zero and unity, via the transformation $\tau = P_{1|x} = 1 - F_\eta(\theta - x)$. The resultant expressions are functions of the PDF, $f_Q(\tau)$, where

$$f_Q(\tau) = \left. \frac{f_x(x)}{f_\eta(\theta - x)} \right|_{x=\theta - F_\eta^{-1}(1-\tau)}. \tag{5.6}$$

Making a change of variable in Eq. (5.5) gives

$$I(x, y) = -\sum_{n=0}^{N} P_y(n) \log_2 \left(\frac{P_y(n)}{\binom{N}{n}} \right) + N \int_{\tau=0}^{\tau=1} f_Q(\tau) \tau \log_2 \tau d\tau$$

$$+ N \int_{\tau=0}^{\tau=1} f_Q(\tau)(1 - \tau) \log_2 (1 - \tau) d\tau, \tag{5.7}$$

where

$$P_y(n) = \binom{N}{n} \int_{\tau=0}^{\tau=1} f_Q(\tau) \tau^n (1 - \tau)^{N-n} d\tau. \tag{5.8}$$

If we make a change of variable from τ to x, and note that $f_x(x)dx = f_Q(\tau)d\tau$, the entropy of the new variable can be written as

$$H(Q) = -\int_0^1 f_Q(\tau) \log_2 (f_Q(\tau)) d\tau$$

$$= -\int_x f_x(x) \log_2 \left(\frac{f_x(x)}{f_\eta(\theta - x)} \right) dx$$

$$= -D(f_x(x) \| f_\eta(\theta - x)). \tag{5.9}$$

Literature review

Stocks' seminal paper (Stocks 2000c) on SSR briefly discusses the scaling of SSR with N, for large N. It states that an exact expression for mutual information can be found at $\sigma = 1$ for identical signal and noise distributions. From this expression, it can be shown that the mutual information scales with $0.5 \log_2 (N)$ for large N. Stocks notes that this means that the channel capacity for large N is about half the maximum noiseless channel capacity, since in the absence of noise, the maximum mutual information is the maximum entropy of the output signal, which is $\log_2 (N + 1)$. The mathematical details of this result are not presented in Stocks (2000c), but left for Stocks (2001c) and Stocks (2001a).

Stocks (2001c) presents an exact result for the mutual information in the SSR model for the case of uniform signal and noise and $\sigma \leq 1$. He also derives an approximation to this exact expression that holds in the case of large N, and shows that the mutual information scales with $0.5\sigma \log_2 (N)$. Using this expression, Stocks (2001c) derives formulas for the channel capacity for a given N, and the value of σ at which capacity occurs, σ_o, finding that as $N \to \infty$, $\sigma_o \to 1$. Thus the main result is that for uniform signal and noise and large N, capacity occurs at $\sigma_o = 1$ and scales with $0.5 \log_2 (N)$.

It is shown in Section 4.3 of Chapter 4 that the mutual information at $\sigma = 1$ is independent of the signal and noise distribution, provided $f_x(x) = f_\eta(\theta - x)$ for all valid x. This means that the result of Stocks (2001c) showing that the mutual information at $\sigma = 1$ scales with $0.5 \log_2 (N)$ holds for any matched signal and noise distributions, where $f_x(x) = f_\eta(\theta - x)$. This is also demonstrated in Stocks (2001a).

The only other authors to consider SSR in the large N regime find that, in contrast to uniform signal and noise – as studied in Stocks (2001c) – the channel capacity for *Gaussian* signal and noise occurs for a value of $\sigma_o \simeq \sqrt{1 - 2/\pi} \simeq$ 0.603 (Hoch *et al.* 2003a, Hoch *et al.* 2003b). We have already seen in Section 4.4 that for Gaussian signal and noise, the channel capacity appears to occur for σ_o approaching this value for $N = 1000$.

In contrast to Stocks (2001c) – which makes use of an exact expression for the mutual information, and derives a large N approximation by approximating a summation with an integral – Hoch *et al.* (2003a) begin by using a *Fisher information*-based approximation to mutual information. The optimal value of σ is found by an analytical approximation to the stationary point of the resultant expression. Hoch *et al.* (2003b) and Wenning (2004) also give the results of Hoch *et al.* (2003a), but provide more detail.

In a neural coding context, the question of 'what is the optimal stimulus distribution?' for a given noise distribution is also discussed numerically for the SSR model in Hoch *et al.* (2003b). In addition, Hoch *et al.* (2003b) give an extensive investigation into quantifying SSR in populations of spiking neuron models, including a section on SSR and energy constraints. We also study the behaviour of SSR under constraints on the energy available in the model in Chapter 9.

Chapter structure

The remainder of this chapter is organized as follows. Section 5.2 discusses Stocks'
large N approximation to the mutual information at $\sigma = 1$ and Section 5.3 discusses his approximation to the mutual information for uniform signal and noise
and $\sigma \leq 1$. Hoch *et al.*'s Fisher information approach to approximating the mutual
information for large N and all σ is discussed in Section 5.4, which also provides
an alternative and more accurate derivation of their results.

The central result of this chapter is a general sufficient condition for achieving
channel capacity in the SSR model, for any arbitrary specified signal or noise distribution. This is discussed in Section 5.5. These new general results are compared
with the specific results of Stocks (2001c), Stocks (2001a), and Hoch *et al.* (2003b)
in Section 5.6. Several results relevant to this chapter are given in more detail in
Appendix 2.

5.2 Mutual information when $f_x(x) = f_\eta(\theta - x)$

Recall from Section 4.3 in Chapter 4 that for the specific case of signal and noise
PDFs, and θ, such that $f_x(x) = f_\eta(\theta - x)$, the following exact results are obtained

$$H(y) = \log_2 (N + 1), \tag{5.10}$$

$$H(y|x) = \frac{N}{2 \ln 2} + \frac{1}{N + 1} \sum_{n=1}^{N} (N + 1 - 2n) \log_2 n, \tag{5.11}$$

and

$$I(x, y) = \log_2 (N + 1) - \frac{N}{2 \ln 2} - \frac{1}{N + 1} \sum_{n=2}^{N} (N + 1 - 2n) \log_2 n. \tag{5.12}$$

If $f_x(x)$ and $f_\eta(x)$ have the same distribution, then we have the noise parameter
being $\sigma = 1$. Although we consider only this situation in this section, the above
results hold for any case of θ, $f_x(x)$, and $f_\eta(\eta)$ such that $f_x(x) = f_\eta(\theta - x)$ for all
valid x.

It is shown in Stocks (2001a) that as N approaches infinity, Eq. (5.12) reduces
to the approximation, $I(x, y) \simeq 0.5 \log_2 (N + 1)$, and that this implies that a
maximum must occur in the mutual information for a nonzero noise distribution (Stocks 2001a). This section examines the derivation of this result in detail,
and finds a slightly more accurate expression for the large N mutual information
at $\sigma = 1$.

Average conditional entropy for large N and σ = 1

From Eq. (4.22), the average conditional output entropy is

$$H(y|x) = -\sum_{n=0}^{N} P_y(n) \log_2 \binom{N}{n}$$

$$- N \int_{\tau=0}^{\tau=1} f_Q(\tau) \left(\tau \log_2 \tau \, d\tau + (1-\tau) \log_2 (1-\tau)\right) d\tau. \qquad (5.13)$$

We begin by finding a simplified version of Eq. (5.11) that holds for large N. Rather than beginning with Eq. (5.11) though, we start with Eq. (5.13). It is convenient to express the first term in Eq. (5.13) without the combinatorial term. Note that for $\sigma = 1$, $P_y(n) = \frac{1}{N+1} \, \forall \, n$. This is proven in Section 4.3 in Chapter 4 – see Eq. (4.35). After some algebra – see Section A2.1 of Appendix 2 – we derive the identity

$$-\sum_{n=0}^{N} P_y(n) \log_2 \binom{N}{n} = \log_2 (N!) - \frac{2}{N+1} \sum_{n=1}^{N} n \log_2 n. \qquad (5.14)$$

Note that Eq. (5.14) can also be expressed as in the third term of Eq. (5.12) as

$$-\sum_{n=0}^{N} P_y(n) \log_2 \binom{N}{n} = \frac{1}{N+1} \sum_{n=2}^{N} (N+1-2n) \log_2 n, \qquad (5.15)$$

which is the expression used in Stocks (2001a). However, here it will prove more convenient to use Eq. (5.14). Note that this expression still holds for any value of N, not just large N.

We shall now see that both terms of Eq. (5.14) can be simplified by approximations that hold for large N. First, for the $\log_2 (N!)$ term, we can make use of Stirling's formula (Spiegel and Liu 1999), which is valid for large N

$$N! \sim \sqrt{(2\pi N)} N^N \exp(-N). \qquad (5.16)$$

This approximation is particularly accurate if the log is taken of both sides

$$\log_2 (N!) \sim N \log_2 N + 0.5 \log_2 N - \frac{N}{\ln 2} + 0.5 \log_2 (2\pi), \qquad (5.17)$$

where the absolute error is $O\left(N^{-1}\right)$. This is to be expected, since Stirling's formula results from an asymptotic expansion of the Gamma function, with a second term of $1/(12x)$. The percentage errors are of course far smaller. Hence, this approximation to $\log_2 (N!)$ can be used for moderately small N.

Secondly, the sum in the second term of Eq. (5.14) can be simplified by way of the Euler–Maclaurin summation formula (Spiegel and Liu 1999). Section A2.2 of Appendix 2 shows that

$$\frac{2}{N+1} \sum_{n=1}^{N} n \log_2 n \simeq N \log_2 (N+1) - \frac{N(N+2)}{2 \ln 2(N+1)} + O\left(\frac{\log N}{N}\right). \qquad (5.18)$$

Even though Eq. (5.18) on its own can be reduced further for large N, we are looking for a large N approximation to Eq. (5.14). Thus, it is safer to first subtract the two sides of Eq. (5.18) from the corresponding sides of Eq. (5.17), before letting N become large. Carrying this out gives

$$-\sum_{n=0}^{N} P_y(n) \log_2 \binom{N}{n} \simeq 0.5 \log_2 N - N \log_2 \left(1 + \frac{1}{N}\right) - \frac{N}{2 \ln 2} \left(2 - \frac{N+2}{N+1}\right)$$

$$+ 0.5 \log_2 (2\pi) - O\left(\frac{\log N}{N}\right). \tag{5.19}$$

Noting that for $|x| < 1$, $\ln(1 + x) = x - x^2/2 + x^3/3 - x^4/4 \ldots$ (Spiegel and Liu 1999), we have

$$N \log_2 \left(1 + \frac{1}{N}\right) = \frac{1}{\ln 2} \left(1 - 1/(2N) + 1/(3N^2) - 1/(4N^3) \ldots\right)$$

$$= \frac{1}{\ln 2} + O\left(\frac{1}{N}\right). \tag{5.20}$$

Using Eq. (5.20) allows a simplification of Eq. (5.19), which when combined with Eq. (4.37), and substituted into Eq. (5.13), gives the average conditional entropy as

$$H(y|x) = \frac{N}{2 \ln 2} - \sum_{n=0}^{N} P_y(n) \log_2 \binom{N}{n}$$

$$\simeq 0.5 \log_2 N - \frac{1}{\ln 2} + \frac{1}{2 \ln 2} \frac{N}{N+1} + 0.5 \log_2 (2\pi) - O\left(\frac{\log N}{N}\right)$$

$$= 0.5 \log_2 N + 0.5 \left(\frac{N}{N+1} - 2\right) \log_2 (e)$$

$$+ 0.5 \log_2 (2\pi) - O\left(\frac{\log N}{N}\right). \tag{5.21}$$

For large N, the average conditional output entropy of Eq. (5.21) can be approximated as

$$H(y|x) \simeq 0.5 \log_2 \left(\frac{2\pi N}{e}\right), \tag{5.22}$$

which scales with $0.5 \log_2 N$.

Mutual information for large N and $\sigma = 1$

It is demonstrated in Section 4.3 of Chapter 4 that for the conditions of this section, $H(y) = \log_2 (N + 1)$, as given in Eq. (5.10). Thus, subtracting the two sides of Eq. (5.21) from the corresponding sides of Eq. (5.10) gives the mutual information as

$$
\begin{aligned}
I(x, y) \simeq{}& \log_2(N + 1) - 0.5 \log_2 N - 0.5 \left(\frac{N}{N+1} - 2 \right) \log_2(e) \\
&- 0.5 \log_2(2\pi) + O\left(\frac{\log N}{N} \right) \\
={}& 0.5 \log_2\left(N + 2 + \frac{1}{N} \right) - 0.5 \left(\frac{N}{N+1} - 2 \right) \log_2(e) \\
&- 0.5 \log_2(2\pi) + O\left(\frac{\log N}{N} \right).
\end{aligned} \tag{5.23}
$$

Letting N approach infinity in Eq. (5.23) gives an approximation to the large N mutual information for $\sigma = 1$ as

$$
I(x, y) \simeq 0.5 \log_2\left(\frac{(N+2)e}{2\pi} \right). \tag{5.24}
$$

Eq. (5.24) differs slightly from that stated in Eq. (7) in Stocks (2001a), which can be written as

$$
I(x, y) = 0.5 \log_2\left(\frac{N+1}{e} \right). \tag{5.25}
$$

The explanation of the discrepancy is that Stocks (2001a) uses the Euler–Maclaurin summation formula to implicitly calculate $\log_2(N!)$ in the large N approximation to the average conditional entropy, under the assumption that the remainder of terms are not of consequence. It turns out that these terms are of consequence, the reason being that the Bernoulli numbers, B_p, do not decrease with p, other than for the first few terms – see Section A2.2 of Appendix 2. After the third term, the Bernoulli numbers increase with p. Stirling's approximation for $N!$ used here takes this into account (Abramowitz and Stegun 1972), and therefore gives a more accurate approximation than Stocks (2001a).

The increased accuracy of Eq. (5.24) can be confirmed by comparing both Eq. (5.24) and Eq. (5.25) with the exact expression for $I(x, y)$ of Eq. (5.12), as N increases. Fig. 5.1 shows the mutual information for $\sigma = 1$ obtained by the exact expression given by Eq. (5.12) compared with the approximations of Eq. (5.23) – with the $O\left(\frac{\log N}{N} \right)$ term ignored – Eq. (5.24) and Eq. (5.25). Fig. 5.2 shows that the error between the exact expression and the first two approximations approaches zero as N increases, whereas the error between Eq. (5.12) and Eq. (5.25) approaches a nonzero constant for large N, of about $0.5 \log_2\left(\frac{e^2}{2\pi} \right) \simeq 0.117$.

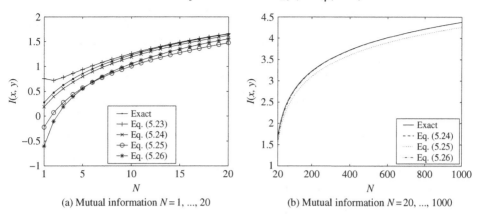

(a) Mutual information $N = 1, ..., 20$ (b) Mutual information $N = 20, ..., 1000$

Fig. 5.1. Mutual information for $\sigma = 1$ and large N. This figure shows the exact mutual information for noise intensity, $\sigma = 1$, given by Eq. (5.12), against increasing array size, N, as well as the four approximations given by Eqs. (5.23), (5.24), (5.25), and (5.26). The plot has been divided into two scales to indicate how well the approximations work for small N and large N. In Fig. 5.1(a), the line plots are used as an aid to the eye, with the marker showing the mutual information at the integer values of N between 1 and 20. It appears from this plot that Eq. (5.24) gives the best approximation for small N. Figure 5.1(b) shows that for larger N, the approximations of Eqs. (5.24) and (5.26) are indistinguishable by eye from the exact mutual information, whereas the approximation of Eq. (5.25) clearly gives a larger error. The actual errors in each approximation are shown in Fig. 5.2.

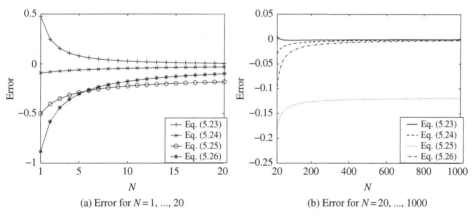

(a) Error for $N = 1, ..., 20$ (b) Error for $N = 20, ..., 1000$

Fig. 5.2. Error in large N approximations to $I(x, y)$ for $\sigma = 1$. Plot showing the error between the exact expression for the mutual information at noise intensity, $\sigma = 1$, given by Eq. (5.12), and the various large N approximations to the mutual information at $\sigma = 1$, given by Eqs. (5.23), (5.24), (5.25), and (5.26). Equations (5.23) and (5.24) give the smallest error for small N, but once N gets close to 1000, all approximations except Eq. (5.25) converge towards an error of zero. By contrast, Eq. (5.25) converges to a nonzero error of $-0.5 \log_2 (0.5e^2/\pi) \simeq -0.117$.

If one was, however, truly interested in a very large N approximation, the percentage error resulting from this term of course becomes very small. This can be seen as follows. For very large N, Eq. (5.24) can be simplified to

$$I(x, y) \simeq 0.5 \log_2 \left(\frac{Ne}{2\pi} \right). \tag{5.26}$$

This expression – also shown in Fig. 5.1 – is obtained in Section 5.4 by an alternative derivation of an approximation to the mutual information at $\sigma = 1$ for large N – see Eq. (5.70).

Since Eq. (5.26) can be written as $I \simeq 0.5 \log_2 (N) - 0.6044$, for very large N the constant term becomes insignificant. However, due to the very slowly increasing nature of the logarithm function, N needs to be very large for 0.6 bits per sample to be considered very small. For example, if $N = 10^{10}$, the relative error is 3.77 percent. It takes $N \simeq 10^{37}$ for the error to fall beneath 1 percent!

Part of our interest in a large N approximation to the mutual information is due to the underlying motivation of studying SSR, which is its possible relationship to neural coding. However, although there are approximately $N = 10^{11}$ neurons in the whole human brain, it is unlikely that the number of neurons devoted to a specific task is any more than 1000 (Hoch *et al.* 2003b), in which case the error incurred by dropping the constant term from Eq. (5.26) is about 14 percent. Since this is quite significant, we suggest that it is appropriate to consider the constant term in Eq. (5.26) if a quantitative approximation to the mutual information in the SSR model is required. However, the main conclusion from this analysis, as first noted in Stocks (2001a), is that the mutual information scales with $0.5 \log_2 (N)$ as N increases – this conclusion remains unchanged whether the constant term is included or not.

From Fig. 5.2 it appears that surprisingly – given that Eq. (5.24) was derived from Eq. (5.23) – Eq. (5.24) gives a better approximation than Eq. (5.23) for small N. This can be seen more clearly in Fig. 5.3, which plots the difference between the absolute errors of these two equations. This discrepancy is due to the discarded terms that make these equations valid for large N. Two of the discarded terms from Eq. (5.23) cancel each other out for small N, which makes Eq. (5.24) misleadingly more accurate. For $N > 6$, the true nature of the approximations starts to show, with Eq. (5.23) being more accurate than Eq. (5.24), as indicated by the negative value of the plot. As expected though, Fig. 5.3 shows that the difference between each equation approaches zero as N gets larger.

In summary, this section has derived a large N expression for the mutual information at $\sigma = 1$, given by Eq. (5.24), which is very accurate – an error of less than 0.01 bits per sample, even for N as low as 20. As N gets larger, the error converges towards zero. This is an improvement on the approximation given in Stocks

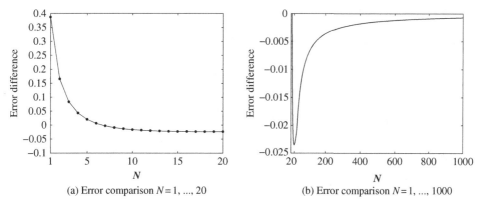

(a) Error comparison $N = 1, ..., 20$ (b) Error comparison $N = 1, ..., 1000$

Fig. 5.3. Error comparison for large N approximations to $I(x, y)$ for $\sigma = 1$. The two large N approximations that give the smallest error are Eqs (5.23) and (5.24). This figure shows how for very small array size ($1 \leq N \leq 6$), Eq. (5.24) gives the smallest absolute error, and for $N > 6$, Eq. (5.23) gives the smallest absolute error. This discrepancy is due to the discarded terms that make these equations valid for large N. In Eq. (5.24) two discarded terms cancel each other out for small N. For $6 \leq N \leq 20$, the difference in the errors gets larger, but, for $N > 20$, it gets smaller again, until as N gets larger than about 250, the difference between the two equations starts to converge asymptotically towards zero. From Fig. 5.2, the actual error in both these approximations approaches zero for large N.

(2001a), for which the error converges towards $0.5 \log_2 \left(\frac{e^2}{2\pi} \right) \simeq 0.117$ bits per sample for large N. For larger N, Eq. (5.24) can be simplified to Eq. (5.26).

Having now derived the specific case of an accurate asymptotic expression for the mutual information for $\sigma = 1$, as carried out in Stocks (2001a), the next logical step is to try to obtain an asymptotic expression for arbitrary σ. This is the focus of the next two sections. First, in Section 5.3 a result is given for the specific case of uniform signal and noise, and then Section 5.4 gives results for arbitrary matched signal and noise distributions.

5.3 Mutual information for uniform signal and noise

Section 4.3 in Chapter 4 included the derivation given in Stocks (2001c) of an exact expression for the mutual information for uniform signal and noise and $\sigma \leq 1$. The relevant expressions, repeated here, are

$$P_y(n) = \begin{cases} \frac{\sigma}{N+1} + \frac{1}{2} - \frac{\sigma}{2} & \text{for} \quad n = 0, N \\ \frac{\sigma}{N+1} & \text{for} \quad 1 \leq n \leq N - 1, \end{cases} \tag{5.27}$$

$$H(y) = -\left(\frac{2\sigma}{N+1} + 1 - \sigma\right)\log_2\left(\frac{\sigma}{N+1} + \frac{1}{2} - \frac{\sigma}{2}\right)$$
$$-\frac{(N-1)\sigma}{N+1}\log_2\left(\frac{\sigma}{N+1}\right), \tag{5.28}$$

$$H(y|x) = \frac{\sigma}{N+1}\sum_{n=2}^{N}(N+1-2n)\log_2 n + \frac{N\sigma}{2\ln 2}, \tag{5.29}$$

$$I(x,y) = -\left(\frac{2\sigma}{N+1} + 1 - \sigma\right)\log_2\left(\frac{\sigma}{N+1} + \frac{1}{2} - \frac{\sigma}{2}\right)$$
$$-\frac{(N-1)\sigma}{N+1}\log_2\left(\frac{\sigma}{N+1}\right)$$
$$-\frac{\sigma}{N+1}\sum_{n=2}^{N}(N+1-2n)\log_2 n - \frac{N\sigma}{2\ln 2}. \tag{5.30}$$

In addition, Stocks (2001c) shows that a large N approximation can be made to these equations. His derivation is repeated here, and as was the case in Section 5.2, we find a similar approximation, with improved accuracy.

Large N mutual information

For large N, Eq. (5.28) reduces to

$$H(y) \simeq \sigma\log_2(N+1) + (1-\sigma)(1-\log_2(1-\sigma)) - \sigma\log_2(\sigma). \tag{5.31}$$

Furthermore, the average conditional entropy is simply σ multiplied by Eq. (5.11). Thus, using the large N approximation to Eq. (5.11) given by Eq. (5.22), we have

$$H(y|x) \simeq \frac{\sigma}{2}\log_2\left(\frac{2\pi N}{e}\right). \tag{5.32}$$

Hence, using the same arguments as for the $\sigma = 1$ case of Section 5.2 the mutual information approximation for large N is

$$I(x,y) \simeq \frac{\sigma}{2}\log_2\left(\frac{(N+2)e}{2\pi}\right) + (1-\sigma)(1-\log_2(1-\sigma)) - \sigma\log_2(\sigma). \tag{5.33}$$

The mutual information is equal to σ times the $\sigma = 1$ mutual information plus the term $(1-\sigma)(1-\log_2(1-\sigma)) - \sigma\log_2(\sigma)$, which reduces to zero for $\sigma = 1$.

Due to the same differences in the derivation as stated in Section 5.2, Eq. (5.33) differs from Eq. (9) given in Stocks (2001c), which can be written as

$$I(x,y) = \frac{\sigma}{2}\log_2\left(\frac{(N+1)}{e}\right) + (1-\sigma)(1-\log_2(1-\sigma)) - \sigma\log_2(\sigma), \tag{5.34}$$

so that the error is exactly that described in the $\sigma = 1$ case, only here it is a function of σ, namely $\frac{\sigma}{2} \log_2 \left(\frac{e^2}{2\pi} \right)$.

This error can be seen in Fig. (3) of Stocks (2001c). To illustrate this, and the accuracy of Eq. (5.33), Fig. (3) of Stocks (2001c) is reproduced here as Fig. 5.4, with the mutual information given by Eq. (5.33) superimposed. It is clear that Eq. (5.33) gives a better approximation. Fig. 5.5 shows that the error between Eq. (5.33) and the exact mutual information of Eq. (5.30) asymptotically approaches zero for all $\sigma \leq 1$ as N increases.

Large N channel capacity

Differentiating the rhs of Eq. (5.33) with respect to σ and setting the result to zero gives the extremum

$$\sigma_o = \frac{\sqrt{(N+2)}}{\sqrt{(N+2)} + \sqrt{\left(\frac{8\pi}{e} \right)}}. \tag{5.35}$$

Taking the second derivative of Eq. (5.33) results in an expression that is always negative and hence the extremum is a maximum. The value of the mutual information at this peak can be found as the very simple formula

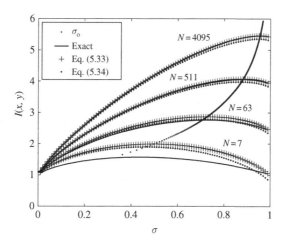

Fig. 5.4. Mutual information for uniform signal and noise and large N against noise intensity, $\sigma \in [0, 1]$. The line plot shows the exact mutual information calculated using Eq. (5.30). Plus signs are the large N approximation of Eq. (5.33) and dots show the slightly less accurate large N approximation of Eq. (5.34), given in Stocks (2001c). The small circles plot the curve of the optimal value of σ obtained from Eq. (5.35) against the corresponding mutual information of Eq. (5.36), for N between 1 and 8000. As N becomes larger, it is clear that Eq. (5.33) gives a more accurate approximation to Eq. (5.30) than Eq. (5.34).

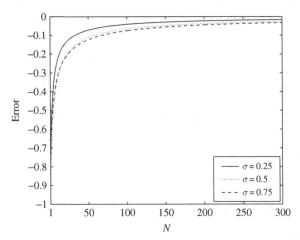

Fig. 5.5. Error in mutual information for uniform signal and noise and large N. Plot showing the difference between the exact mutual information for uniform signal and noise given by Eq. (5.30) and the approximation given by Eq. (5.33), as the array size, N, increases, for various values of noise intensity, σ. The error can be seen to approach zero as N increases, for each value of σ.

$$I_o(x, y) = 1 - \log_2 (1 - \sigma_o) = \log_2 \left(2 + \sqrt{\frac{(N+2)e}{2\pi}}\right). \qquad (5.36)$$

Clearly, as N becomes very large, the optimal value of σ given by Eq. (5.35) approaches unity, and the mutual information at this σ given by Eq. (5.36) approaches $0.5 \log_2 ((N + 2)e/(2\pi))$, which agrees with Eq. (5.33) at $\sigma = 1$, and Eq. (5.22).

Recall that an exact expression for σ_o was given by Eq. (4.59) in Section 4.3 of Chapter 4. Eq. (5.35) and Eq. (4.59) can be seen to be closely related when Eq. (5.35) is rewritten as

$$\sigma_o = \frac{N + 2}{N + 2 + \sqrt{\left(\frac{8\pi(N+2)}{e}\right)}}. \qquad (5.37)$$

The error between Eq. (5.37) and Eq. (4.59) also converges to zero for large N, again validating our approximations.

Stocks (2001c) also gives expressions for the optimal σ and the corresponding mutual information. Again, these are slightly different from Eqs. (5.35) and (5.36), due to the slightly inaccurate terms in the large N approximation to the average conditional entropy. However the important qualitative result remains the same, which is that the maximum mutual information scales with half the logarithm of N, and the value of σ which achieves this asymptotically approaches unity. Figure 5.4 illustrates this behaviour of σ_o and the corresponding $I_o(x, y)$, as N increases from 1 to 8000.

5.4 Mutual information for arbitrary signal and noise

In the previous two sections we gave asymptotic results for (i) $\sigma = 1$ for arbitrary distributions, and (ii) for $\sigma \leq 1$ for uniform signal and noise. In this section we find an expression for the mutual information for arbitrary distributions and all σ. The $\sigma = 1$ result can be used to verify that any new expressions are correct at $\sigma = 1$.

The Gaussian approximation to the binomial distribution

In the SSR model, for each value of the input, x, the probability that the output is $y = n$ is given by the binomial formula, as given by Eq. (4.9), repeated here as

$$P_{y|x}(n|x) = \binom{N}{n} P_{1|x}^n (1 - P_{1|x})^{N-n}. \tag{5.38}$$

Consider only one value of x. For this value, $P_{y|x}(n|x)$ is a function of n. As N becomes large, the binomial distribution is known to approach a Gaussian distribution, with the same mean and variance as the binomial distribution (Kreyszig 1988), provided the mean of the binomial is sufficiently large. The mean is given by $N P_{1|x}$ and the variance is $N P_{1|x}(1 - P_{1|x})$. Thus, provided $0 \ll N P_{1|x} \ll N$

$$P_{y|x}(n|x) \simeq \frac{1}{\sqrt{2\pi N P_{1|x}(1 - P_{1|x})}} \exp\left(-\frac{(n - N P_{1|x})^2}{2N P_{1|x}(1 - P_{1|x})}\right). \tag{5.39}$$

This approximation breaks down when $P_{1|x}$ is close to zero or unity, in which case $P_{y|x}(n|x)$ can be approximated by the Poisson distribution, or the Edgeworth series approximation (Harrington 1955). However, here we shall use Eq. (5.39), and find that it is valid for our needs, since for infinite tail PDFs such as the Gaussian, logistic, and Laplacian distributions, when $P_{1|x}$ is close to zero or unity, $f_x(x)$ is also very close to zero. Note that even though Eq. (5.39) is a Gaussian function, since n is still discretely valued, Eq. (5.39) still represents a discrete probability mass function rather than a PDF.

The large N approximation given by Eq. (5.39) allows us to obtain a general large N approximation for the average conditional entropy. This is the focus of the following section.

Conditional entropy for large N

Using the binomial approximation

Recall in Chapter 4 that we used the notation $\hat{H}(y|x)$ to denote the conditional output entropy for a given value of x. The *average* conditional output entropy is then $H(y|x) = \int_x f_x(x)\hat{H}(y|x)dx$, where

$$\hat{H}(y|x) = -\sum_{n=0}^{N} P_{y|x}(n|x) \log_2 \left(P_{y|x}(n|x) \right). \tag{5.40}$$

Taking the log of the Gaussian approximation to the binomial given by Eq. (5.39), and substituting into Eq. (5.40) gives

$$\hat{H}(y|x) \simeq -\sum_{n=0}^{N} P_{y|x}(n|x)$$

$$* \left(-0.5 \log_2 \left(2\pi N P_{1|x}(1 - P_{1|x}) \right) - \frac{(n - N P_{1|x})^2}{2 \ln(2) N P_{1|x}(1 - P_{1|x})} \right)$$

$$= 0.5 \log_2 \left(2\pi N P_{1|x}(1 - P_{1|x}) \right)$$

$$+ \frac{1}{2 \ln(2) N P_{1|x}(1 - P_{1|x})} \sum_{n=0}^{N} P_{y|x}(n|x)(n - N P_{1|x})^2$$

$$= 0.5 \log_2 \left(2\pi N P_{1|x}(1 - P_{1|x}) \right) + \frac{1}{2 \ln(2) N P_{1|x}(1 - P_{1|x})} \text{var}[y|x]$$

$$= 0.5 \log_2 \left(2\pi N P_{1|x}(1 - P_{1|x}) \right) + \frac{1}{2 \ln(2)}$$

$$= 0.5 \log_2 \left(2\pi e N P_{1|x}(1 - P_{1|x}) \right), \tag{5.41}$$

where we have used the fact that the variance of the binomial distribution, $P_{y|x}(n|x)$, is $N P_{1|x}(1 - P_{1|x})$, and the mean is $N P_{1|x}$.

Multiplying both sides of Eq. (5.41) by $f_x(x)$ and integrating over all x gives

$$H(y|x) \simeq 0.5 \log_2 (2\pi eN) + 0.5 \int_{x=-\infty}^{\infty} f_x(x) \log_2 \left(P_{1|x}(1 - P_{1|x}) \right) dx. \tag{5.42}$$

Equation (5.42) can also be written in terms of $f_Q(\tau)$. Recall that if we let $\tau = P_{1|x}$, then $f_x(x)dx = f_Q(\tau)d\tau$. Making this change of variable in Eq. (5.42) gives

$$H(y|x) = 0.5 \log_2 (2\pi eN) + 0.5 \int_{\tau=0}^{\tau=1} f_Q(\tau) \log_2 (\tau(1 - \tau)) d\tau. \tag{5.43}$$

Verification at $\sigma = 1$

Using the large N approximation to $H(y|x)$ derived in Section 5.2 for the specific case of $\sigma = 1$, we can verify Eq. (5.43) for the case of $\sigma = 1$. At $\sigma = 1$, $f_Q(\tau) = 1$ and therefore $\int_{\tau=0}^{\tau=1} f_Q(\tau) \log_2 (\tau) d\tau = -\log_2 (e)$ and Eq. (5.43) reduces to

$$H(y|x) = 0.5 \log_2 \left(\frac{2\pi N}{e} \right), \tag{5.44}$$

which agrees precisely with Eq. (5.22).

Hoch's approach

An alternative approach, taken in Hoch *et al.* (2003a) and Hoch *et al.* (2003b), which does not explicitly use the Gaussian approximation to the binomial, is to use the Euler–Maclaurin summation formula (Spiegel and Liu 1999), to approximate $\hat{H}(y|x)$ as an integral for large N as

$$\hat{H}(y|x) = -\int_{n=0}^{N} P_{y|x}(n|x) \log_2 (P_{y|x}(n|x)) dn. \tag{5.45}$$

This is the entropy of a continuous random variable defined on the support $n \in [0, N]$. If this random variable has a variance $\sigma_{y|x}^2$, then the entropy is less than or equal to the variance of a Gaussian random variable with the same variance. This is due to the well-known result that the maximum entropy distribution for continuous random variables under a power constraint – that is, a specified variance in this case – is the Gaussian distribution (Cover and Thomas 1991). Hence, we have

$$\hat{H}(y|x) \le 0.5 \log_2 (2\pi e \sigma_{y|x}^2). \tag{5.46}$$

The variance in this case is again the binomial variance of $\mathrm{var}[y|x] = N P_{1|x}(1 - P_{1|x})$. Therefore

$$\hat{H}(y|x) \le 0.5 \log_2 (2\pi e N P_{1|x}(1 - P_{1|x})). \tag{5.47}$$

Multiplying both sides of inequality (5.47) by $f_x(x)$ and integrating over all x leaves

$$H(y|x) \le 0.5 \log_2 (2\pi e N) + 0.5 \int_x f_x(x) \log_2 \left(P_{1|x}(1 - P_{1|x}) \right) dx. \tag{5.48}$$

This is identical to Eq. (5.42) except that (5.48) is an inequality. The equality given by the Gaussian approximation to the binomial of Eq. (5.42) will only hold for values of x for which $P_{y|x}(n|x)$ is exactly Gaussian.

Numerical verification

Consider the exact conditional entropy, $\hat{H}(y|x)$, given by Eq. (5.40) and the approximation given by Eq. (5.41). These are both functions of N and $P_{1|x}$. However, since $P_{1|x}$ is always between zero and unity, we do not need to specify a noise PDF to compare the approximation to the exact formula – we need only plot both as a function of $P_{1|x} \in [0, 1]$. Figure 5.6 shows the absolute error between Eqs (5.40) and (5.41), calculated numerically for $P_{1|x} \in [0.01, 0.99]$, for various values of N. It is clear from Fig. 5.6 that the absolute error decreases with increasing N, but is larger for $P_{1|x}$ near zero and unity. This illustrates the caveat that the Gaussian approximation to the binomial requires $0 \ll N P_{1|x} \ll N$ for high accuracy. Figure 5.6 verifies this, by clearly showing that the absolute error in the approximation decreases as $P_{1|x}$ gets closer to 0.5 and as N gets larger.

We shall see, however, that the inaccuracy for $N P_{1|x} \to 0$ and $P_{1|x} \to 1$ becomes more important for $\sigma \ll 1$; that is, when the variance of the noise is much smaller than the variance of the signal. Assuming that $f_x(x)$ and $f_\eta(x)$ have long tails, such as in the Gaussian distribution, if $\sigma \ll 1$, then $P_{1|x}$ gets very close to zero or unity for quite a large range of highly probable x values. Thus, we can expect that the accuracy of the approximation to $H(y|x)$ decreases for smaller σ.

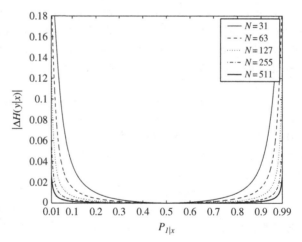

Fig. 5.6. Error in the large N approximation to $\hat{H}(y|x)$. This figure shows the absolute error between the exact conditional entropy – that is, the conditional entropy of y, given a particular value of x – $\hat{H}(y|x)$, of Eq. (5.40), and the large N approximation of Eq. (5.41), and demonstrates that it decreases with increasing N. The data were calculated numerically for various values of N for $P_{1|x} \in [0.01, 0.99]$. The error clearly gets larger for $P_{1|x} \to 0$ and $P_{1|x} \to 1$. As the absolute error gets very large at zero and one, the data were not calculated at these values.

This is indeed the case, as can be seen from Fig. 5.7, which shows plots of the approximation to $H(y|x)$ given by Eq. (5.42), against increasing σ, compared with the exact $H(y|x)$ of Eq. (5.13). These results were calculated numerically for the cases of Gaussian signal and noise, Laplacian signal and noise, and logistic signal and noise, and $\sigma = 0.2, 0.3, \ldots, 1.6$. It is clear from Fig. 5.7 that the approximation to $H(y|x)$ is highly accurate for $\sigma > 0.7$, and although less accurate for smaller σ, the accuracy increases with increasing N.

Having obtained a large N approximation to the average conditional entropy that, unlike the result of Section 5.2, is valid for σ other than $\sigma = 1$, we now wish to find a similar approximation for the output entropy.

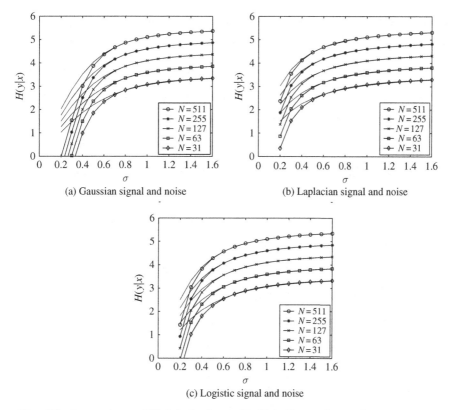

Fig. 5.7. Approximate $H(y|x)$ for large N. This figure shows the exact average conditional output entropy, $H(y|x)$, of Eq. (5.13) and the approximate average conditional entropy of Eq. (5.42), for Gaussian signal and noise, Laplacian signal and noise, and logistic signal and noise. The exact $H(y|x)$ is shown with a thin solid line, while the approximate $H(y|x)$ is shown with circles and a thicker solid line to interpolate between values of noise intensity, σ. Clearly, the approximation is very accurate for $\sigma > 0.7$ in all cases. The accuracy for $\sigma < 0.7$ decreases with decreasing σ, but increases for increasing N.

Output distribution and entropy for large N

This section aims to find a large N approximation to the output entropy. As discussed in the introduction to this chapter, since the output is discretely valued, the output entropy is that of a discrete random variable, regardless of how large N is. Thus, we require a large N approximation to the output probability mass function, $P_y(n)$.

Although the output of the SSR model is a discrete random variable, in the presence of noise the expected value of the output of the SSR array, y, *given* x, is a continuously valued variable, giving an *average* transfer function, $\bar{y}(x)$. Since $E[y|x] = NP_{1|x}$, the average transfer function is $\bar{y}(x) = N - NF_\eta(\theta - x)$.

If we can find the PDF of the average transfer function, $f_{\bar{y}}(\bar{y})$, for the continuous random variable, $\bar{y}$, then for large N we can use this PDF to approximate the actual output discrete probability mass function, $P_y(n)$.

PDF of the average transfer function

The PDF of the average transfer function can be easily derived from the input signal's PDF by a transformation from coordinate x, with PDF $f_x(x)$, to coordinate $\bar{y}$, with PDF $f_{\bar{y}}(\bar{y})$.

In general, a transformation from coordinate a with PDF $f_a(a)$ to coordinate b with PDF $f_b(b)$, where $b = g(a)$ is a differentiable and invertible function, is given by Fry (1928) and Hogg *et al.* (2005)

$$f_b(b) = f_a(g^{-1}(b)) \left| \frac{dg^{-1}(b)}{db} \right|. \tag{5.49}$$

In this case, we use the average transfer function, $\bar{y}(x)$, so that the inverse transfer function is $x = \theta - F_\eta^{-1}\left(1 - \frac{\bar{y}}{N}\right)$, and the result for $f_{\bar{y}}(\bar{y})$ is

$$
\begin{aligned}
f_{\bar{y}}(\bar{y}) &= f_x\left(\theta - F_\eta^{-1}\left(1 - \frac{\bar{y}}{N}\right)\right) \left| \frac{d\left(\theta - F_\eta^{-1}\left(1 - \frac{\bar{y}}{N}\right)\right)}{d\bar{y}} \right| \\
&= f_x\left(\theta - F_\eta^{-1}\left(1 - \frac{\bar{y}}{N}\right)\right) \left(\frac{1}{Nf_\eta(\theta - x)}\right) \\
&= \frac{f_x(x)}{Nf_\eta(\theta - x)}. \tag{5.50}
\end{aligned}
$$

Recall from Section 4.3 in Chapter 4 the definition of the PDF $f_Q(\tau)$, where $\tau = P_{1|x} \in [0, 1]$. It is clear that Eq. (5.50) can be rewritten as

$$f_{\bar{y}}(\bar{y}) = \frac{f_Q\left(\frac{\bar{y}}{N}\right)}{N}, \tag{5.51}$$

which is a PDF for the continuously valued variable, $\bar{y}$. Unlike $f_Q(\tau)$, which has support $[0, 1]$, the support of $f_{\bar{y}}(\bar{y})$ is $[0, N]$.

If we make a change of variable so that $\tau = \frac{\bar{y}}{N}$, Eq. (5.51) becomes

$$f_\tau(\tau) = f_Q(\tau). \tag{5.52}$$

This agrees with the discussion about $f_Q(\tau)$ in Chapter 4, where it was suggested that $f_Q(\tau)$ is the PDF of the continuous random variable given by $E[y|x]$.

Approximating the output distribution

For large N, we can assume that $\bar{y}(x) \to n$ and hence the PDF of Eq. (5.51) satisfies

$$f_{\bar{y}}(\bar{y}) d\bar{y} \simeq P_y(n). \tag{5.53}$$

Thus, the PDF $f_{\bar{y}}(\bar{y})$ approximates the discretely valued probability mass function

$$P_y(n) \simeq \frac{f_Q\left(\frac{n}{N}\right)}{N}. \tag{5.54}$$

An alternative approach is to examine the large N behaviour of the Gaussian approximation to $P_{y|x}(n|x)$ under a change of variable. Upon substituting for $\tau = P_{1|x}$, Eq. (5.39) can be rewritten as

$$P_{y|x}(n|x) \simeq \frac{1}{\sqrt{2\pi s^2}} \exp\left(-\frac{\left(\tau - \frac{n}{N}\right)^2}{2s^2}\right) \frac{1}{N}$$

$$= \frac{P_{y|\tau}(n|\tau)}{N}, \tag{5.55}$$

where $s^2 = \frac{\tau(1-\tau)}{N}$. The term $P_{y|\tau}(n|\tau)$ is a function defined on $\tau \in [0, 1]$, and, hence, it cannot be Gaussian. However, consider $0 \ll n \ll N$. In this region, $P_{y|\tau}(n|\tau)$ approaches zero for τ near zero or unity and hence $P_{y|\tau}(n|\tau)$ is approximately a Gaussian PDF with a variance that decreases with $\frac{1}{N}$. Such a Gaussian with a variance that approaches zero can be approximated by a delta function located at the mean of the Gaussian

$$P_{y|\tau}(n|\tau) \simeq \delta\left(\tau - \left(\frac{n}{N}\right)\right). \tag{5.56}$$

Thus, $P_{y|\tau}(n|\tau) = 1$ if and only if $\tau = n/N$. This means $P_{y|x}(n|x)$ approaches $1/N$ at $x = \theta - F^{-1}(1 - \tau) = \theta - F_\eta^{-1}(1 - \frac{n}{N})$. This value of x is the *maximum likelihood* value of the output, given x, and can be easily shown – see Eq. (4.12) – without recourse to large N – to be the mode of $P_{y|x}(n|x)$ for each n.

Using the fact that $P_y(n) = \int_x f_x(x) P_{y|x}(n|x) dx$, changing the variable x to τ gives $P_y(n) \simeq \int_{\tau=0}^{\tau=1} f_Q(\tau) P_{y|\tau}(n|\tau) \frac{1}{N} d\tau$, which can be simplified by substitution of Eq. (5.56) to

$$
\begin{aligned}
P_y(n) &\simeq \frac{1}{N} \int_{\tau=0}^{\tau=1} f_Q(\tau) \delta\left(\tau - \frac{n}{N}\right) d\tau \\
&= \frac{f_Q\left(\frac{n}{N}\right)}{N},
\end{aligned}
\tag{5.57}
$$

which verifies Eq. (5.54). As with the Gaussian approximation to the binomial, we can expect that this approximation gets less accurate for values of n near zero and N.

A second alternative approach due to Amblard *et al.* (2006) follows from noting the output distribution of our model can be expressed in terms of the PDF of a beta distribution (Yates and Goodman 2005), with parameters $n+1$ and $N-n+1$, that is from Eq. (5.8)

$$
\begin{aligned}
P_y(n) &= \binom{N}{n} \int_0^1 \tau^n (1-\tau)^{(N-n)} f_Q(\tau) d\tau \\
&= \frac{1}{N+1} \int_0^1 \left(\frac{\tau^n (1-\tau)^{(N-n)}}{\int_0^1 \phi^n (1-\phi)^{(N-n)} d\phi} \right) f_Q(\tau) d\tau \\
&= \frac{1}{N+1} \int_0^1 f_T(\tau) f_Q(\tau) d\tau,
\end{aligned}
\tag{5.58}
$$

where $f_T(\tau)$ is the PDF of the beta distribution with parameters $n+1$, $N-n+1$, and we have used the fact that the beta function

$$
\begin{aligned}
\beta(n+1, N-n+1) &= \int_0^1 \phi^n (1-\phi)^{(N-n)} d\phi \\
&= \frac{(n)!(N-n)!}{(N+1)!}
\end{aligned}
\tag{5.59}
$$

cancels out most of the combinatorial term. For the case where $f_x(x) = f_n(\theta - x)$, $f_Q(\tau)=1$, and since $f_T(\tau)$ is a PDF, $P_y(n) = 1/(N+1)$, which is a result derived previously (Stocks 2001c). The PDF of a beta distribution is unimodal, and can be shown to behave as a delta function located at $\tau = n/N$, when N is large, that is $f_T(\tau) \to \delta\left(\frac{n}{N}\right)$ (Amblard *et al.* 2006). Substituting this into Eq. (5.58) gives the result

$$
P_y(n) = \frac{f_Q\left(\frac{n}{N}\right)}{N+1} \qquad n = 1, \ldots, N-1.
\tag{5.60}
$$

We note that this result for the large N output distribution can be derived more rigorously using a saddlepoint method known as *Laplace's method* (Brunel and Nadal 1998, Amblard *et al.* 2006, McDonnell *et al.* 2006a).

Numerical verification of large N approximation to $P_y(n)$

We have for the Gaussian, Laplacian, and logistic cases that when the noise intensity, $\sigma > 1$, $f_Q(0) = f_Q(1) = 0$, whereas for $\sigma < 1$, we have $f_Q(0) = f_Q(1) = \infty$. From Eq. (5.54), this would mean that $P_y(0)$ and $P_y(N)$ are either zero or infinite. However, for finite N, there is some finite nonzero probability that all output states are on or off. Indeed, at $\sigma = 1$, we know that $P_y(n) = \frac{1}{N+1}$ $\forall$ n, and at $\sigma = 0$, $P_y(0) = P_y(N) = 0.5$. Furthermore, for finite N, the approximation of Eq. (5.57) does not guarantee that $\sum_{n=0}^{N} P_y(n) = 1$. A little trial and error finds that two simple adjustments to the approximation deal with these problems. The two different cases of $\sigma < 1$ and $\sigma \geq 1$ require different adjustments to achieve a high accuracy that increases with increasing N.

First, for $\sigma \geq 1$, we adjust Eq. (5.54) to become

$$P'_y(n) = \frac{f_Q\left(\frac{n+1}{N+2}\right)}{\sum_{m=0}^{N} f_Q\left(\frac{m+1}{N+2}\right)} \quad n = 0, \ldots, N. \tag{5.61}$$

Eq. (5.61) ensures $\sum_{n=0}^{N} P'_y(n) = 1$ and also ensures that $P_y(0)$ and $P_y(N)$ are nonzero for finite N as required, while remaining identical to Eq. (5.54) in the large N limit.

Secondly, for $\sigma < 1$ we adjust Eq. (5.54) to become

$$P'_y(n) = \frac{f_Q\left(\frac{n}{N}\right)}{N} \quad n = 1, \ldots, N - 1. \tag{5.62}$$

Recall that the approximation is expected to be less valid for n near zero and N for smaller σ. Hence, to increase the accuracy of our approximation, and to ensure $P_y(n)$ forms a valid probability mass function, we define our approximation as

$$P'_y(n) = \begin{cases} \frac{f_Q\left(\frac{n}{N}\right)}{N} & \text{for} \quad n = 1, \ldots, N - 1 \\ 0.5\left(1 - \sum_{m=1}^{N-1} \frac{f_Q\left(\frac{m}{N}\right)}{N}\right) & \text{for} \quad n = 0, n = N. \end{cases} \tag{5.63}$$

For $\sigma < 1$, $P_y(0)$ and $P_y(N)$ are the most likely values of y. Eq. (5.61) ensures that $P_y(0)$ and $P_y(N)$ are reasonably accurate.

Fig. 5.8 shows that the approximation to $P_y(n)$ is highly accurate for N as small as 63, for Gaussian signal and noise, and three values of σ. Similar results can be obtained for the Laplacian or logistic cases. Note from Fig. 5.8(a), where

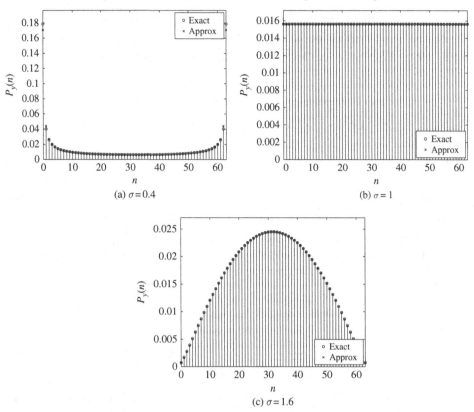

Fig. 5.8. Large N approximation to $P_y(n)$. This figure shows for various values of noise intensity, σ, that for $N = 63$ and Gaussian signal and noise, the approximation to the output probability mass function, $P_y(n)$, given by Eqs. (5.61) and (5.63), gives a highly accurate approximation to the output probability mass function. The circles indicate the exact $P_y(n)$ obtained by numerical integration and the crosses show the approximations. Similar results can be obtained for the Laplacian or logistic cases. Figure 5.4 shows that $P_y(0)$ and $P_y(N)$ are the most inaccurate approximations for $\sigma = 0.4$. This is not surprising, given the approximation is expected to decrease in accuracy for small σ, and n close to zero or N. Figure adapted with permission from McDonnell *et al.* (2007). Copyright 2007 by the American Physical Society.

$\sigma = 0.4$, how $P_y(0)$ and $P_y(N)$ are the most inaccurate approximations. This is not surprising, given that we expected the approximation to decrease in accuracy for small σ, and n close to zero or N.

The approximations to the output probability mass function given by Eqs (5.61) and (5.63) can also be used to approximate the output entropy as

$$H(y) \simeq -\sum_{n=0}^{N} P_y'(n) \log_2 \left(P_y'(n) \right). \tag{5.64}$$

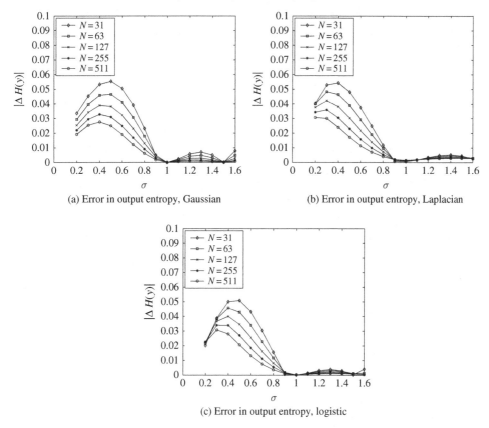

(a) Error in output entropy, Gaussian

(b) Error in output entropy, Laplacian

(c) Error in output entropy, logistic

Fig. 5.9. Absolute error in output entropy using large N approximation to $P_y(n)$. This figure shows that the approximation to the output probability mass function, $P_y(n)$, given by Eqs. (5.61) and (5.63), gives a highly accurate approximation to the output entropy, as expressed by Eq. (5.64). The accuracy increases as the array size, N, increases. Notice that the approximation is more accurate near a noise intensity of $\sigma = 1$.

The result of this is shown in Fig. 5.9, where the absolute error between Eq. (5.64) and the exact output entropy, calculated numerically, is shown for the cases of Gaussian signal and noise, Laplacian signal and noise, and logistic signal and noise. The accuracy can be seen to increase as N increases for all σ. The approximation is more accurate near $\sigma = 1$.

Approximating the output entropy

Consider the entropy of the discrete random variable y. Making use of Eq. (5.57), we have

$$H(y) = -\sum_{n=0}^{N} P_y(n) \log_2 (P_y(n))$$

$$= -\sum_{n=0}^{N} \frac{f_Q\left(\frac{n}{N}\right)}{N} \log_2 \left(\frac{f_Q\left(\frac{n}{N}\right)}{N}\right)$$

$$= -\frac{1}{N} \sum_{n=0}^{N} f_Q\left(\frac{n}{N}\right) \log_2 \left(f_Q\left(\frac{n}{N}\right)\right) + \frac{\log_2 (N)}{N} \sum_{n=0}^{N} f_Q\left(\frac{n}{N}\right). \quad (5.65)$$

Suppose that the summations above can be approximated by integrals, without any remainder terms. Carrying this out and then making the change of variable $\tau = n/N$ gives

$$H(y) \simeq -\frac{1}{N} \int_{n=0}^{N} f_Q\left(\frac{n}{N}\right) \log_2 \left(f_Q\left(\frac{n}{N}\right)\right) dn + \frac{\log_2 (N)}{N} \int_{n=0}^{N} f_Q\left(\frac{n}{N}\right) dn$$

$$= -\int_{\tau=0}^{\tau=1} f_Q(\tau) \log_2 \left(f_Q(\tau)\right) d\tau + \log_2 N \int_{\tau=0}^{\tau=1} f_Q(\tau) d\tau$$

$$= \log_2 N - \int_{\tau=0}^{\tau=1} f_Q(\tau) \log_2 \left(f_Q(\tau)\right) d\tau$$

$$= \log_2 N + H(Q), \quad (5.66)$$

where $H(Q)$ is the differential entropy corresponding to the PDF $f_Q(\tau)$. This analysis agrees with Theorem 9.3.1 of Cover and Thomas (1991), which states that the entropy of an M bit quantization of a continuous random variable Z is approximately the sum of M and the entropy of Z. Here, we have $Q = \bar{y}/N$ as the continuous random variable that approximates the $N + 1$ state discrete output distribution. Hence the discrete output distribution is a $\log_2 (N + 1)$ bit quantization of Q, and has entropy approximately equal to the differential entropy of Q plus $\log_2 (N + 1)$, which for large N agrees with Eq. (5.66).

Performing a change of variable in Eq. (5.66) of $\tau = 1 - F_\eta(\theta - x)$ gives

$$H(y) \simeq \log_2 N - \int_{x=-\infty}^{x=\infty} f_x(x) \log_2 \left(\frac{f_x(x)}{f_\eta(\theta - x)}\right) dx$$

$$= \log_2 (N) - D(f_x(x)\|f_\eta(\theta - x)), \quad (5.67)$$

where $D(f_x(x)\|f_\eta(\theta - x))$ is the relative entropy between $f_x(x)$ and $f_\eta(\theta - x)$.

Thus, $H(y)$ for large N is approximately the sum of the number of output bits and the negative of the relative entropy between $f_x(x)$ and $f_\eta(\theta - x)$. Therefore, since relative entropy is always non-negative, the approximation to $H(y)$ given by

Eq. (5.67) is always less than or equal to $\log_2(N)$. This agrees with the known expression for $H(y)$ in the specific case of $\sigma = 1$ of $\log_2(N+1)$, which holds for any N.

Figure 5.10 shows the approximation of Eq. (5.67), as well as the exact output entropy, for the Gaussian, Laplacian, and logistic cases, for a range of σ and increasing N. The approximation is quite good for $\sigma > 0.7$, but worsens for smaller σ. However, as N increases the approximation improves. This indicates that approximating the summation in Eq. (5.65) with the relative entropy between $f_x(x)$ and $f_\eta(x)$ gets more accurate for small σ with increasing N.

We are now in a position to combine the results of this section and the previous to obtain a result for the mutual information.

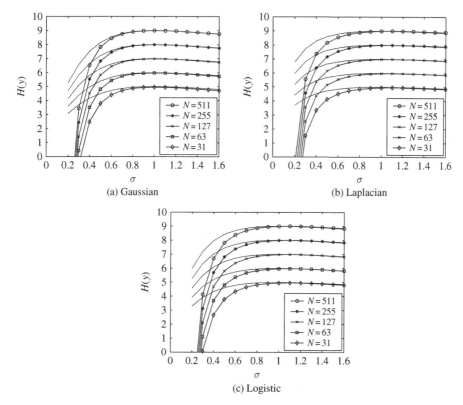

Fig. 5.10. Large N approximation to output entropy. This figure shows the approximate output entropy given by Eq. (5.67), as well as the exact output entropy calculated numerically. Clearly the approximation is quite good for noise intensities, $\sigma > 0.7$. The exact expression is shown by thin solid lines, and the approximation by circles, with a thicker solid line interpolating between values of σ as an aid to the eye. The approximation can be seen to always be a lower bound on the exact entropy.

Mutual information for large N

We have obtained a large N approximation to the average conditional output entropy, given by Eq. (5.43), and a large N approximation to the output entropy, given by Eq. (5.66). These two equations can be combined to give a large N approximation to the mutual information as

$$I(x, y) \simeq H(y) - H(y|x)$$

$$= 0.5 \log_2 \left(\frac{N}{2\pi e}\right) - \int_{\tau=0}^{\tau=1} f_Q(\tau) \log_2 \left(\sqrt{\tau(1-\tau)} f_Q(\tau)\right) d\tau \quad (5.68)$$

$$= 0.5 \log_2 \left(\frac{N}{2\pi e}\right) - 0.5 \int_{\tau=0}^{\tau=1} f_Q(\tau) \log_2 (\tau(1-\tau)) d\tau + H(Q).$$

$$(5.69)$$

The integral on the rhs of Eq. (5.68) is independent of N and therefore, for large N, the mutual information scales with $0.5 \log_2 (N)$. As noted previously in this chapter, and in Stocks (2001a), this is half the maximum channel capacity for an N comparator system. The integral on the rhs of Eq. (5.68) is insignificant when compared to $\log_2 (N)$, but its importance is that it determines how the mutual information varies from $0.5 \log_2 \left(\frac{N}{2\pi e}\right)$ as σ varies.

For the specific case of $\sigma = 1$, that is when $f_x(x) = f_\eta(\theta - x)$, $f_Q(\tau) = 1$ so that $-\int_{\tau=0}^{\tau=1} \log_2 (\tau) d\tau = \log_2 (e)$. Therefore

$$I(x, y) = 0.5 \log_2 \left(\frac{Ne}{2\pi}\right), \quad (5.70)$$

which for large N agrees precisely with Eq. (5.24), validating the new formula in this specific case.

Note that Eq. (5.68) can be rewritten in terms of x as

$$I(x, y) = 0.5 \log_2 \left(\frac{N}{2\pi e}\right)$$

$$- 0.5 \int_{x=-\infty}^{x=\infty} f_x(x) \log_2 (P_{1|x}(1 - P_{1|x}) dx - D(f_x||f_\eta). \quad (5.71)$$

Figure 5.11 shows the approximation of Eq. (5.68), as well as the exact mutual information, for the Gaussian, Laplacian, and logistic cases, for a range of σ and increasing N. As with the output entropy, and conditional output entropy approximations, the mutual information approximation is quite good for $\sigma > 0.7$, but worsens for smaller σ. However, as N increases the approximation improves.

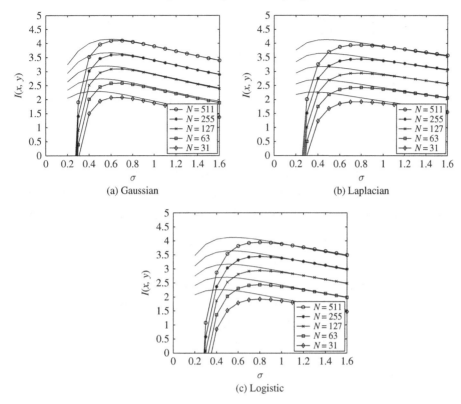

Fig. 5.11. Large N approximation to mutual information. This figure shows the approximate output entropy given by Eq. (5.68), as well as the exact mutual information calculated numerically. Clearly the approximation is quite good for noise intensities, $\sigma > 0.7$. The exact expression is shown by thin solid lines, and the approximation by circles, with a thicker solid line interpolating between values of σ as an aid to the eye. The approximation can be seen to always be a lower bound on the exact mutual information. Figure adapted with permission from McDonnell *et al.* (2007). Copyright 2007 by the American Physical Society.

Relationship to Fisher information

Chapters 6 and 7 in this book make extensive use of Fisher information (Cover and Thomas 1991) in relation to the decoding of the SSR model. Hoch *et al.* (2003a) shows that the Fisher information for SSR is given by

$$J(x) = \left(\frac{d P_{1|x}}{dx}\right)^2 \frac{N}{P_{1|x}(1 - P_{1|x})}. \tag{5.72}$$

Equation (5.72) is also derived in Section A3.6 of Appendix 3. Note that the Fisher information is always non-negative and has the same support as $f_\eta(\theta - x)$, since the derivative of $P_{1|x}$ is $f_\eta(\theta - x)$.

Rewriting the large N expression for mutual information of Eq. (5.71) gives

$$I(x, y) = H(x) - 0.5 \log_2 (2\pi e) + 0.5 \int_{x=-\infty}^{x=\infty} f_x(x) \log_2 \left(\frac{N f_\eta(x)^2}{P_{1|x}(1 - P_{1|x})} \right) dx$$

$$= H(x) - 0.5 \log_2 (2\pi e) + 0.5 \int_{x=-\infty}^{x=\infty} f_x(x) \log_2 (J(x)) dx$$

$$= H(x) - 0.5 \int_{x=-\infty}^{x=\infty} f_x(x) \log_2 \left(\frac{2\pi e}{J(x)} \right) dx. \tag{5.73}$$

Eq. (5.73) is precisely the same as that derived in Hoch *et al.* (2003a) as an asymptotic large N expression for the mutual information. Hoch *et al.* (2003a) derived this expression by considering a result from Brunel and Nadal (1998), which shows that in the limit of large N, the mutual information in a system becomes equal to the mutual information between the input signal and an *efficient* Gaussian estimator for that signal (Clarke and Barron 1990, Stemmler 1996, Brunel and Nadal 1998, Kang and Sompolinsky 2001, Hoch *et al.* 2003b). *Efficient* has a precise technical meaning, which we shall explore in Chapter 6. Furthermore, discussion of this result will also be examined later, since understanding it depends on the concept of optimally decoding the SSR system. Decoding is the focus of Chapter 6 for small N, and Chapter 7 for large N.

5.5 A general expression for large N channel capacity

We now show that a channel capacity achieving input PDF can be found for any given noise PDF, whenever the approximation of Eq. (5.68) holds. This is a very general result, and one that may be of significance for both computational neuroscience and electronics.

To begin, Eq. (5.73) can be written as

$$I(x, y) = -0.5 \log_2 (2\pi e) - \int_x f_x(x) \log_2 \left(\frac{f_x(x)}{\sqrt{J(x)}} \right) dx. \tag{5.74}$$

The integral in Eq. (5.74) can be recognized as being very similar to the relative entropy between $f_x(x)$ and $\sqrt{J(x)}$. However, in general, the Fisher information as a function of x is not a PDF and the integral cannot be relative entropy. In the case of SSR though, remarkably – as shown in Section A2.3 of Appendix 2 – the function,

$$f_S(x) = \frac{\sqrt{J(x)}}{\pi \sqrt{N}} \tag{5.75}$$

is indeed a PDF, one which has the same support as $f_\eta(\theta - x)$, and, for the noise distributions considered in this chapter, is a function of the single parameter, σ_η.

Note that the PDF, $f_S(x)$, can be written in terms of the Fisher information as

$$f_S(x) = \frac{\sqrt{J(x)}}{\int_\phi \sqrt{J(\phi)}d\phi}. \tag{5.76}$$

Such a PDF is therefore simply a normalization of the square root of the Fisher information. A PDF with this property is known as *Jeffrey's prior* (Rissanen 1996, Jaynes 2003).

Using Eq. (5.75), Eq. (5.74) can be written as

$$I(x, y) = 0.5 \log_2 \left(\frac{N\pi}{2e} \right) - D(f_x || f_S). \tag{5.77}$$

Furthermore, our approach in the previous two sections led us to find a large N approximation to the output entropy, that is from Eq. (5.67)

$$H(y) = \log_2 (N) - D(f_x(x) || f_\eta(\theta - x)). \tag{5.78}$$

By inspection of Eq. (5.75), $f_S(x)$ can be derived from knowledge of the noise PDF, $f_\eta(\eta)$, since we can write

$$f_S(x) = \frac{f_\eta(\theta - x)}{\pi \sqrt{F_\eta(\theta - x)(1 - F_\eta(\theta - x))}}. \tag{5.79}$$

The fact that Jeffrey's prior can be written in the form of Eq. (5.75) and Eq. (5.79) for the SSR model is quite significant, as in general Jeffrey's prior has no such simple form.

A sufficient condition for optimality

Since relative entropy is always non-negative, from Eq. (5.77) a sufficient condition for achieving the large N channel capacity is that

$$f_x(x) = f_S(x) \quad \forall\, x, \tag{5.80}$$

with the resultant capacity as

$$C(x, y) = 0.5 \log_2 \left(\frac{N\pi}{2e} \right) \simeq 0.5 \log_2 N - 0.3956. \tag{5.81}$$

Equation (5.81) holds provided the conditions for the approximation given by Eq. (5.73) hold. Otherwise, the rhss of Eqs (5.78) and (5.77) give lower bounds. This means that for the situations considered previously in (Hoch *et al.* 2003b, Stocks 2001c), where the signal and noise both have the same distribution (but different variances), we can expect to find channel capacity that is less than or equal to that of Eq. (5.81). This is discussed in Section 5.6.

The derived sufficient condition of Eq. (5.80) leads to two ways in which capacity can be achieved: (i) an optimal signal PDF for a given noise PDF, and (ii) an optimal noise PDF for a given signal PDF.

Optimizing the signal distribution

Assuming Eq. (5.73) holds, the channel capacity achieving input PDF, $f_x^o(x)$, can be found for any given noise PDF from Eqs (5.79) and (5.80) as

$$f_x^o(x) = \frac{f_\eta(\theta - x)}{\pi \sqrt{F_\eta(\theta - x)(1 - F_\eta(\theta - x))}}. \tag{5.82}$$

Example: uniform noise

Suppose the *iid* noise at the input to each threshold device in the SSR model is uniformly distributed on the interval $[-\sigma_\eta/2, \sigma_\eta/2]$ so that it has PDF

$$f_\eta(\eta) = \frac{1}{\sigma_\eta}, \quad \eta \in [-\sigma_\eta/2, \sigma_\eta/2]. \tag{5.83}$$

Substituting Eq. (5.83) and its associated CDF into Eq. (5.82), we find that the optimal signal PDF is

$$f_x^o(x) = \frac{1}{\pi \sqrt{\frac{\sigma_\eta^2}{4} - (x - \theta)^2}}, \quad x \in [\theta - \sigma_\eta/2, \theta + \sigma_\eta/2]. \tag{5.84}$$

This PDF is in fact the PDF of a sine-wave with uniformly random phase, amplitude $\sigma_\eta/2$, and mean θ. A change of variable to the interval $\tau \in [0, 1]$ via the substitution $\tau = (x - \theta)/\sigma_\eta + 0.5$ results in the PDF of the beta distribution with parameters 0.5 and 0.5, also known as the arcsine distribution.

This beta distribution is bimodal, with the most probable values of the signal those near zero and unity. It also has a relatively large variance. This result provides theoretical justification for a proposed heuristic method for analogue-to-digital conversion based on the SSR model (Nguyen 2007). In this method, the input signal is transformed so that it has a large variance and is bimodal.

Similar results for an optimal input distribution in an information theoretic optimization of a neural system have been found in Schreiber *et al.* (2002). These results were achieved numerically with the Blahut–Arimoto algorithm often used in information theory to find channel capacity achieving source distributions or rate-distortion functions (Arimoto 1972, Blahut 1972, Cover and Thomas 1991).

Gaussian noise

Suppose the *iid* noise at the input to each threshold device has a zero mean Gaussian distribution with variance σ_η^2, with PDF

$$f_\eta(\eta) = \frac{1}{\sqrt{2\pi\sigma_\eta^2}} \exp\left(-\frac{\eta^2}{2\sigma_\eta^2}\right). \tag{5.85}$$

Substituting from Eq. (5.85) and its associated CDF into Eq. (5.82) gives the optimal signal PDF. The resultant expression for $f_x^o(x)$ does not simplify much, and contains the standard error function, erf($\cdot$) (Spiegel and Liu 1999).

We are able to verify that the resultant PDF has the correct shape via Fig. 8 in Hoch *et al.* (2003b), which presents the result of *numerically* optimizing the signal PDF, $f_x(x)$, for unity variance zero mean Gaussian noise, $\theta = 0$, and $N = 10\,000$. As with the work in Schreiber *et al.* (2002), the numerical optimization is achieved using the Blahut–Arimoto algorithm. It is remarked in Hoch *et al.* (2003b) that the optimal $f_x(x)$ is close to being Gaussian. This is illustrated by plotting both $f_x(x)$ and a Gaussian PDF with nearly the same peak value as $f_x(x)$. It is straightforward to show that a Gaussian with the same peak value as our analytical $f_x^o(x)$ has variance $0.25\pi^2$. If the signal was indeed Gaussian, then we would have $\sigma = 2/\pi \simeq 0.6366$, which is very close to the value calculated for actual Gaussian signal and noise in Section 5.6.

Our analytical $f_x^o(x)$ from Eqs (5.85) and (5.82), with $\theta = 0$, is plotted on the interval $x \in [-3, 3]$ in Fig. 5.12, along with a Gaussian PDF with variance $0.25\pi^2$. Clearly, the optimal signal PDF is very close to the Gaussian PDF. Our Fig. 5.12 is virtually identical to Fig. 8 in Hoch *et al.* (2003b). It is emphasized that the results in Hoch *et al.* (2003b) were obtained using an entirely different method that involves numerical iterations, and therefore provides excellent validation of our theoretical results.

Optimizing the noise distribution

We now assume that the signal distribution is known and fixed. We wish to achieve channel capacity by finding the optimal noise distribution. It is easy to show by integrating Eq. (5.79) that the CDF corresponding to the PDF, $f_S(\cdot)$, evaluated at x, can be written in terms of the CDF of the noise distribution as

$$F_S(x) = 1 - \frac{2}{\pi} \arcsin\left(\sqrt{F_\eta(\theta - x)}\right). \tag{5.86}$$

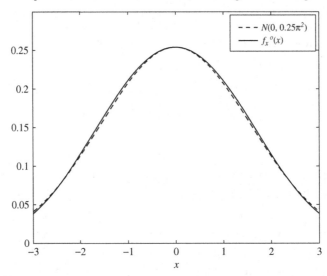

Fig. 5.12. The optimal signal PDF, $f_x^o(x)$, for zero mean, unity variance Gaussian noise, and threshold value $\theta = 0$, as obtained from Eq. (5.82). Superimposed is a Gaussian PDF with the same peak value as $f_x^o(x)$, so that it has variance $0.25\pi^2$. This figure uses our new theoretical results to analytically replicate Fig. 8 in Hoch *et al.* (2003b), which was calculated numerically. Reprinted figure with permission after McDonnell *et al.* (2007). Copyright 2007 by the American Physical Society.

If we now let $f_x(x) = f_S(x)$, then $F_x(s) = F_S(x)$, and rearranging Eq. (5.86) gives the optimal noise CDF in terms of the signal CDF as

$$F_\eta^o(x) = \sin^2\left(\frac{\pi}{2}(1 - F_x(\theta - x))\right) = 0.5 + 0.5\cos\left(\pi F_x(\theta - x)\right). \quad (5.87)$$

Differentiating $F_\eta^o(x)$ gives the optimal noise PDF as a function of the signal PDF and CDF

$$f_\eta^o(x) = \frac{\pi}{2}\sin\left(\pi(1 - F_x(\theta - x))\right)f_x(\theta - x). \quad (5.88)$$

Unlike optimizing the signal distribution, which is the standard way for achieving channel capacity in information theory (Cover and Thomas 1991), we have assumed a signal distribution, and found the 'best' noise distribution, which is equivalent to optimizing the channel, rather than the signal.

Example: uniform signal

Suppose the signal is uniformly distributed on the interval $x \in [-\sigma_x/2, \sigma_x/2]$. From Eqs. (5.87) and (5.88), the capacity achieving noise distribution has CDF

$$F_\eta^o(x) = 0.5 + 0.5\sin\left(\frac{\pi(x - \theta)}{\sigma_x}\right), \quad x \in [\theta - \sigma_x/2, \theta + \sigma_x/2] \quad (5.89)$$

and PDF

$$f_\eta^o(x) = \frac{\pi}{2\sigma_x} \cos\left(\frac{\pi(x-\theta)}{\sigma_x}\right), \qquad x \in [\theta - \sigma_x/2, \theta + \sigma_x/2]. \qquad (5.90)$$

Substitution of $F_\eta^o(x)$ and $f_\eta^o(x)$ into Eq. (5.72) finds the interesting result that the Fisher information is constant for all x

$$J(x) = N\frac{\pi^2}{\sigma_x^2}. \qquad (5.91)$$

This is verified in Eq. (5.92) below.

Consequences of optimizing the large N channel capacity

Optimal Fisher information

Regardless of whether we optimize the signal for given noise, or optimize the noise for a given signal, it is straightforward to show that the Fisher information can be written as a function of the signal PDF

$$J^o(x) = N\pi^2(f_x(x))^2. \qquad (5.92)$$

Therefore, the Fisher information at large N channel capacity is constant for the support of the signal if and only if the signal is uniformly distributed. The optimality of constant Fisher information in a neural coding context is studied in Bethge *et al.* (2002).

The optimal PDF $f_Q(\tau)$

A further consequence that holds in both cases is that the ratio of the signal PDF to the noise PDF is

$$\frac{f_x(x)}{f_\eta(\theta - x)} = \frac{2}{\pi \sin(\pi(1 - F_x(x)))}. \qquad (5.93)$$

This is not a PDF. However, if we make a change of variable via $\tau = 1 - F_\eta(\theta - x)$ we get the PDF $f_Q(\tau)$, which for channel capacity is

$$f_Q^o(\tau) = \frac{1}{\pi\sqrt{\tau(1-\tau)}}, \qquad \tau \in [0, 1]. \qquad (5.94)$$

This optimal $f_Q(\tau)$ is in fact the PDF of the beta distribution with parameters 0.5 and 0.5, that is the arcsine distribution. It is emphasized that this result holds *regardless* of whether the signal PDF is optimized for a given noise PDF or *vice versa*.

Output entropy at channel capacity

From Eq. (5.9), the entropy of Q is equal to the negative of the relative entropy between $f_x(x)$ and $f_\eta(\theta - x)$. The entropy of Q when capacity is achieved can be calculated from Eq. (5.94) using direct integration as

$$H(Q) = \log_2 (\pi) - 2. \tag{5.95}$$

From Eqs (5.78) and (5.9), the large N output entropy at channel capacity in the SSR model is

$$H(y) = \log_2 \left(\frac{N\pi}{4} \right). \tag{5.96}$$

The output PMF at large N capacity is beta-binomial

Suppose we have signal and noise such that $f_Q(\tau) = f_Q^o(\tau)$ – that is, the signal and noise satisfy the sufficient condition, Eq. (5.80) – but that N is not necessarily large. We can derive the output probability mass function (PMF) for this situation, by substituting Eq. (5.94) into Eq. (5.8) to get

$$\begin{aligned} P_y(n) &= \binom{N}{n} \frac{1}{\pi} \int_0^1 \tau^{(n-0.5)}(1 - \tau)^{(N-n-0.5)} d\tau \\ &= \binom{N}{n} \frac{\beta(n + 0.5, N - n + 0.5)}{\beta(0.5, 0.5)}, \end{aligned} \tag{5.97}$$

where $\beta(a, b)$ is a beta function. This PMF can be recognized as that of the beta-binomial – or negative hypergeometric – distribution with parameters N, 0.5, 0.5 (Evans *et al.* 2000). It is emphasized that Eq. (5.97) holds as an exact analytical result for *any* N.

Analytical expression for the mutual information

The exact expression for the output PMF of Eq. (5.97) allows exact calculation of both the output entropy and the mutual information without need for numerical integration, using Eq. (5.7). This is because when $f_Q(\tau) = f_Q^o(\tau)$, the integrals in Eq. (5.7) can be evaluated exactly to get

$$I_o(x, y) = -\sum_{n=0}^{N} P_y(n) \log_2 \left(\frac{P_y(n)}{\binom{N}{n}} \right) + N \log_2 \left(\frac{e}{4} \right). \tag{5.98}$$

The exact values of $I_o(x, y)$ and the corresponding output entropy, $H_o(y)$, are plotted in Fig. 5.13(a) for $N = 1, \ldots, 1000$. For comparison, the exact $I(x, y)$ of Eq. (5.12), which holds for $f_x(x) = f_\eta(\theta - x)$, is also plotted, as well as the corresponding entropy, $H(y) = \log_2 (N + 1)$. It is clear that $I_o(x, y)$ is always larger than the mutual information of the $f_x(x) = f_\eta(\theta - x)$ case,

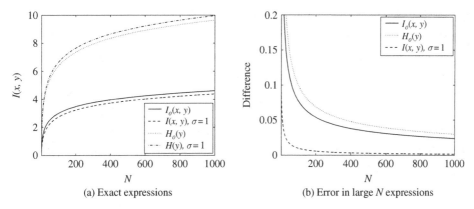

(a) Exact expressions   (b) Error in large *N* expressions

Fig. 5.13. Exact expressions for $I(x, y)$ and $H(y)$ at channel capacity. (a) Exact expressions obtained using $f_Q^o(\tau)$, for $I_o(x, y)$, and $H_o(y)$, as well as the exact mutual information and output entropy when $f_x(x) = f_\eta(\theta - x)$ (denoted as $\sigma = 1$), as a function of N. (b) The difference between the exact expressions for $I_o(x, y)$, $H_o(y)$, and $I(x, y)$ for $f_x(x) = f_\eta(\theta - x)$, and the corresponding large N expressions given by Eqs. (5.77), (5.96), and (5.24). Reprinted figure with permission after McDonnell *et al.* (2007). Copyright 2007 by the American Physical Society.

and that $H_o(y)$ is always less than its entropy, which is the maximum output entropy.

To illustrate that the large N expressions derived are lower bounds to the exact formula plotted in Fig. 5.13(a), and that the error between them decreases with N, Fig. 5.13(b) shows the difference between the exact and the large N mutual information and output entropy. This difference clearly decreases with increasing N.

A note on the output entropy

The SSR model is described in terms of signal quantization theory in Chapters 6 and 7. It has also been compared with the related process of companding in Amblard *et al.* (2006). For a deterministic scalar quantizer with $N + 1$ output states, N threshold values are required. In quantization theory, there is a concept of high resolution quantizers, in which the distribution of $N \to \infty$ threshold values can be described by a point density function, $\lambda(x)$. For such quantizers, it can be shown that the quantizer output, y, in response to a random variable, x, has entropy $H(y) \simeq \log_2 N - D(f_x||\lambda)$ (Gersho and Gray 1992). This is strikingly similar to our Eq. (5.78) for the large N output entropy of the SSR model. In fact, since the noise that perturbs the fixed threshold value, θ, is additive, each threshold acts as an *iid* random variable with PDF $f_\eta(\theta - x)$, and therefore for large N, $f_\eta(\theta - x)$ acts as a density function describing the relative frequency of threshold values as a function of x, just as $\lambda(x)$ does for a high-resolution deterministic quantizer.

For deterministic quantizers, the point density function can be used to approximate the high resolution distortion incurred by the quantization process. For the SSR model, however, since the quantization has a random aspect, the distortion has a component due to randomness as well as lossy compression, and cannot be simply calculated from $f_\eta(\cdot)$. Instead, we can use the Fisher information to calculate the asymptotic mean square error distortion, which is not possible for deterministic high-resolution quantizers.

5.6 Channel capacity for 'matched' signal and noise

Unlike the previous section, we now consider channel capacity under the constraint of 'matched' signal and noise distributions – that is, where both the signal and noise, while still independent, have the same distribution, except for a scale factor. The mean of both signal and noise is zero and the threshold value is also $\theta = 0$. In this situation the mutual information depends solely on the ratio $\sigma = \sigma_\eta/\sigma_x$, which is the only free variable. Finding channel capacity is therefore equivalent to finding the optimal value of noise intensity, σ. Such an analysis provides verification of the more general capacity expression of Eq. (5.81), which cannot be exceeded.

From Eq. (5.77), channel capacity for large N occurs for the value of σ that minimizes the relative entropy between $f_x(x)$ and $f_S(x)$. If we let

$$f(\sigma) = \int_{x=-\infty}^{x=\infty} f_X(x) \ln\left(\frac{1}{J(x)}\right) dx, \tag{5.99}$$

then from Eq. (5.73), it is also clear that this minimization is equivalent to solving the following problem

$$\sigma_o = \min_\sigma f(\sigma). \tag{5.100}$$

This is exactly the formulation stated in Hoch *et al.* (2003b). For simplicity, we assume that both the signal and noise PDFs are even functions. We use a change of variable of $\tau = P_{1|x} = F_\eta(x)$, so that Problem (5.100) can be equivalently expressed as

$$\lim_{N\to\infty} \sigma_o = \min_\sigma f(\sigma) = \int_{\tau=0}^{\tau=1} f_\varrho(\tau) \ln(\tau f_\varrho(\tau)) d\tau, \tag{5.101}$$

or as

$$\lim_{N\to\infty} \sigma_o = \min_\sigma f(\sigma) = D(f_x\|f_\eta) + \int_{x=-\infty}^{x=\infty} f_x(x) \log_2(P_{1|x}) dx. \tag{5.102}$$

Thus, the channel capacity depends on a term that comes from the large N approximation to the output entropy, and a term – the relative entropy between the signal and noise PDFs – that comes from the large N approximation to the average conditional output entropy. For 'matched' signal and noise, this latter term depends only on σ. This shows that for large N the channel capacity occurs for the same value of σ – which we denote as σ_o – for all N. We saw that this was the case for uniform signal and noise in Section 5.3 – see also Stocks (2001c). This fact is also recognized in Hoch *et al.* (2003a) and Hoch *et al.* (2003b), who derive an analytical approximation for σ_o for the case of Gaussian signal and noise. The derivation is given in Hoch *et al.* (2003b), and depends on a Taylor expansion of the Fisher information inside the integral in Eq. (5.73).

Here, we investigate the value of σ_o and the mutual information at σ_o for other signal and noise distributions, and compare the channel capacity obtained with the case where $f_X(x) = f_S(x)$. This comparison finds that the results of Hoch *et al.* (2003b) overstate the true capacity. We begin by repeating and discussing the derivation for Gaussian signal and noise of Hoch *et al.* (2003b).

Gaussian signal and noise

We give a slightly different derivation of the optimal value of σ for large N and Gaussian signal and noise from that of Hoch *et al.* (2003b) that begins with Problem (5.102). Solving this problem requires differentiating $f(\sigma)$ with respect to σ and equating to zero. This means

$$\frac{d}{d\sigma} D(f_x || f_\eta) + \frac{d}{d\sigma} \int_{x=-\infty}^{x=\infty} f_x(x) \log_2 (P_{1|x}) dx = 0. \tag{5.103}$$

Recall from Table 4.2 in Chapter 4 that the entropy between $f_x(x)$ and $f_\eta(x)$ is

$$D(f_x || f_\eta) = \log_2 (\sigma) + \frac{1}{2 \ln 2} \left(\frac{1}{\sigma^2} - 1 \right). \tag{5.104}$$

Therefore

$$\frac{d}{d\sigma} D(f_x || f_\eta) = \frac{1}{\ln 2} \left(\sigma^{-1} - \sigma^{-3} \right). \tag{5.105}$$

For the other term in Eq. (5.103), we take a lead from Hoch *et al.* (2003b) and approximate $\ln (P_{1|x})$ by its second-order Taylor series expansion (Spiegel and Liu 1999) about the mean of the noise. Hoch *et al.* (2003b) use an arbitrary mean, however here, for simplicity, we set the signal and noise means to be zero. The result of this, after some algebra, and using the fact that for Gaussian noise, $P_{1|x} = 0.5$ at $x = 0$, and $f_\eta(0) = 1/\sqrt{2\pi\sigma_\eta^2}$, where σ_η is the standard deviation of the noise, is

$$- \ln (P_{1|x}) = \ln 2 - \sqrt{\frac{2}{\pi}} \frac{x}{\sigma_\eta} + \frac{x^2}{\pi \sigma_\eta^2} - O\left(\frac{x^4}{\sigma_\eta^4}\right). \tag{5.106}$$

Multiplying Eq. (5.106) by $f_x(x)$ and integrating over all x gives

$$- \int_{x=-\infty}^{x=\infty} f_x(x) \ln (P_{1|x}) dx \simeq \ln 2 - \sqrt{\frac{2}{\pi}} \frac{1}{\sigma_\eta} E[x] + \frac{1}{\pi \sigma_\eta^2} E[x^2]. \tag{5.107}$$

Since for a Gaussian signal $E[x] = 0$ and $E[x^2] = \sigma_x^2$ Eq. (5.107) becomes a function of σ,

$$g(\sigma) = - \int_{x=-\infty}^{x=\infty} f_x(x) \log_2 (P_{1|x}) dx \simeq 1 + \frac{1}{\pi \sigma^2 \ln 2}, \tag{5.108}$$

where as before, $\sigma = \sigma_\eta / \sigma_x$.

The approximation of Eq. (5.108) appears to be quite accurate for all σ, as can be seen in Fig. 5.14. Numerical experiments show that the relative error is no more than about 10 percent for $\sigma > 0.2$. However, as we shall see, this is inaccurate enough to cause the end result for the approximate channel capacity to significantly overstate the true channel capacity.

Differentiation of Eq. (5.108) with respect to σ gives

$$- \frac{d}{d\sigma} \int_{x=-\infty}^{x=\infty} f_x(x) \log_2 (P_{1|x}) dx \simeq - \frac{2}{\pi \sigma^3 \ln 2}. \tag{5.109}$$

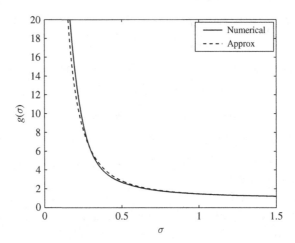

Fig. 5.14. The approximation of Eq. (5.108), compared with the corresponding exact result, calculated numerically (for Gaussian signal and noise). Clearly, the approximation is quite accurate for most values of the noise intensity, σ; however, as shown in the text, it is inaccurate enough to cause the approximation to significantly overstate the true channel capacity.

Combining Eq. (5.109) with Eq. (5.105) and substituting into Eq. (5.103) gives

$$\frac{2}{\pi \sigma^3 \ln 2} + \frac{1}{\ln 2}\left(\sigma^{-1} - \sigma^{-3}\right) = 0. \tag{5.110}$$

Solving Eq. (5.110) gives the optimal value of σ as

$$\sigma_o \simeq \sqrt{1 - \frac{2}{\pi}} \simeq 0.6028, \tag{5.111}$$

as found in Hoch *et al.* (2003b).

An expression for the mutual information at σ_o can be found by substituting Eq. (5.111) into Eqs. (5.108) and (5.104), and substituting the results into Eq. (5.71). Carrying this out gives the large N channel capacity for Gaussian signal and noise as

$$C(x, y) \simeq 0.5 \log_2 \left(\frac{2N}{e(\pi - 2)}\right), \tag{5.112}$$

which can be written as $C(x, y) \simeq 0.5 \log_2 N - 0.3169$.

Although this expression for the channel capacity is close to correct, recall that we showed above that the channel capacity must be less than $0.5 \log_2 N - 0.3956$. Hence, Eq. (5.112) gives a value for channel capacity that is too large. The reason for this is that the Taylor expansion approximation we used requires consideration of more terms. An expression including such higher-order terms is given in Hoch *et al.* (2003b); however, this also appears not to give quite the right answer. Hence, the next section solves Problem (5.102) numerically.

Numerical verification and other distributions

Due to the slight inaccuracy found above, and the fact that no analytical expression could be found for any other signal and noise distributions, we now solve Problem (5.102) numerically. The term $g(\sigma) = \int_x f_x(x) \log_2 (P_{1|x}) dx$ can be found for any specified signal and noise distribution by numerical integration, just as carried out for the relative entropy plotted in Fig. 4.5 in Chapter 4 for cases other than Gaussian. Thus, the function of Problem (5.102) can be easily found as a function of x, and minimized. This function is plotted in Fig. 5.15.

Figure 5.15 also shows the Gaussian case obtained using the approximation of Eq. (5.106). It is clear that although this approximation is close to being correct, it does understate the true value of σ_o, and gives a more negative value of the function, which means overstating the channel capacity.

Table 5.1 gives the result of numerically calculating the value of σ_o and the corresponding large N channel capacity. Note that, in each case, $C(x, y) - 0.5 \log_2 (N) < -0.3956$, as required by Eq. (5.81). Note that the difference between

Table 5.1. *Large N channel capacity (less* $0.5 \log_2 (N)$*) and the optimal noise intensity,* σ_o*, for matched signal and noise. Also shown is the difference between channel capacity and the mutual information at* $\sigma = 1$*. Table adapted with permission from McDonnell et al (2007). Copyright 2007 by the American Physical Society.*

Distribution	$C(x, y) - 0.5 \log_2 (N)$	σ_o	$C(x, y) - I(x, y)_{\sigma=1}$
Gaussian	−0.3964	0.6563	0.208
Logistic	−0.3996	0.5943	0.205
Laplacian	−0.3990	0.5384	0.205

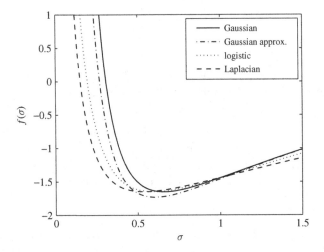

Fig. 5.15. Finding channel capacity for large N. This figure shows the function that requires minimization to find the channel capacity, that is $f(\sigma) = D(f_x \| f_\eta) + \int_{x=-\infty}^{x=\infty} f_x(x) \log_2 (P_{1|x}) dx$, as a function of noise intensity, σ, for the Gaussian, logistic, and Laplacian cases. The Gaussian case obtained using the approximation of Eq. (5.106) is also shown. Clearly, although the approximation gives σ_o close to the right answer, it underestimates the true value slightly, and also provides a minimum smaller than the true minimum, and hence significantly overstates the channel capacity.

capacity and $0.5 \log_2 (N)$ is about 0.4 bits per sample in each case. Thus, in the limit of very large N, this shows that capacity is almost identical, regardless of the distribution. The value of σ_o at which this capacity occurs though is different in each case. The results of Table 5.1 compare favourably with the results presented in Section 4.4 in Chapter 4, where capacity was found for N up to 1000. At $N = 1000$, for Gaussian signal and noise, $\sigma_o \simeq 0.607$, for Laplacian signal and noise, $\sigma_o \simeq 0.5$, and for logistic signal and noise $\sigma_o \simeq 0.564$. Since we also saw that σ_o increases with increasing N, these results are consistent with Table 5.1.

Difference between channel capacity and $I(x, y)$ at $\sigma = 1$

Section 4.4 in Chapter 4 shows in Fig. 4.24(c) the difference between the channel capacity and the mutual information at $\sigma = 1$. This figure shows that for $N = 1000$, this difference appears to be asymptotically converging towards a constant value for each matched signal and noise case. The value of this difference appears to be of the order of 0.22–0.24 bits per sample. Recall from Section 5.2 that at $\sigma = 1$, the mutual information is identical for each signal and noise pair considered here, and is approximately $I(x, y) = 0.5 \log_2(N) - 0.6444$. Thus, given that the channel capacity is slightly larger than this, as indicated by Table 5.1, for each case there is a constant difference between the channel capacity and the mutual information at $\sigma = 1$, for large N. This value is also listed in Table 5.1, and compares favourably with the results of Fig. 4.24(c) at $N = 1000$, which is simply not large enough to show the final asymptotic value of this difference.

The main conclusion to be drawn from this analysis, though, is that the channel capacity is only of the order of 0.2 bits per sample larger at σ_o than it is at $\sigma = 1$. As N gets larger, this difference becomes more and more insignificant.

5.7 Chapter summary

The initial section of this chapter gives a brief discussion of previous work on large N limit results for the SSR model.

Next, Section 5.2 re-derives and improves a large N expression for the mutual information first found in Stocks (2001a). This expression holds for the case of matched signal and noise; that is, for the case where $f_x(x) = f_\eta(\theta - x)$, so that the mutual information is independent of the actual signal and noise distributions.

Section 5.3 re-derives, and improves, a large N expression first found in Stocks (2001c) for the mutual information for the specific case of uniform signal and noise and $\sigma \le 1$. It also gives an expression for the value of σ at which channel capacity occurs in this case. The mutual information is found to scale with $0.5 \log_2 N$.

Section 5.4 then looks at the more general case of any σ, and any matched signal and noise distributions. Work on this topic was first published in Hoch *et al.* (2003a), who found an expression for the large N mutual information that is more accurate for large N and σ greater than about 0.7. We obtain an alternative derivation of this expression, by finding separate expressions for the output entropy, conditional output entropy, and output probability mass function. Furthermore, it is shown that all expressions can be written either in terms of the support of the signal distribution, x, and its associated PDF, $f_x(x)$, or in terms of the transformed random variable, τ, and its associated PDF, $f_Q(\tau)$, with support $\tau \in [0, 1]$.

Section 5.5 uses the large N expression for mutual information to find a sufficient condition for achieving large N channel capacity. This condition is related to the Fisher information, and holds for distributions for which the large N mutual information approximation holds. We discuss how this condition leads to two ways in which to achieve large N capacity; an optimal noise distribution for a given signal distribution, or an optimal signal distribution for a given signal distribution.

Finally Section 5.6 numerically finds the large N channel capacity for the cases of matched Gaussian signal and noise, Laplacian signal and noise, and logistic signal and noise. It is found in each case that the large N channel capacity is only about 0.2 bits per sample greater than the channel capacity at $\sigma = 1$, although the optimal noise intensity, σ, ranges between about 0.5 and 0.65. Furthermore, for the Gaussian signal and noise case, the optimal value of σ reported in Hoch *et al.* (2003a) is found to underestimate the true value by about 0.05.

Chapter 5 in a nutshell

This chapter includes the following highlights:

- An improved approximation to the large N average conditional output entropy, and mutual information at $\sigma = 1$, is obtained by making use of Stirling's formula. The improved mutual information approximation is shown to have an error that approaches zero as N increases, whereas the formula given in Stocks (2001a) is shown to have an error approaching about 0.117 bits per sample.
- An improved approximation to the large N average conditional output entropy, and mutual information for uniform signal and noise, and $\sigma \leq 1$ is derived. The improvement is due to the same reasoning as the improvement at $\sigma = 1$. The improved formula for the mutual information also leads to an improved formula for the optimal value of σ, and the channel capacity. As in Stocks (2001c), these new formulas show that the channel capacity occurs for $\sigma = 1$ at $N = \infty$.
- A new approximation to the conditional entropy of y given x, $\hat{H}(y|x)$, is given, based on the fact that the binomial distribution approaches a Gaussian for large N. This approximation is shown to give the same result as the approximation used in Hoch *et al.* (2003b), and is used to approximate the *average* conditional entropy, $H(y|x)$, for large N. The approximation to $\hat{H}(y|x)$ is numerically verified to become more accurate for increasing N, and the approximation to $H(y|x)$ is verified at $\sigma = 1$ by the results of Section 5.2.
- A derivation of the PDF of the average transfer function, $\bar{y}$, of the SSR model is given. This PDF is shown to be $f_Q\left(\frac{\bar{y}}{N}\right)/N$.

- A large N approximation of the output probability mass function, $P_y(n)$, that is highly accurate for all σ is derived. This approximation is obtained in two different ways, and numerically verified to be accurate.
- A large N approximation to the output entropy, $H(y)$, is found, and is shown to be the difference between $\log_2(N)$ and the relative entropy between $f_x(x)$ and $f_\eta(\theta - x)$, or equivalently, the sum of $\log_2(N)$ and the entropy, $H(Q)$.
- The new methods of approximating $H(y|x)$ and $H(y)$ are used to obtain the same large N approximation to $I(x, y)$ as expressed in Hoch *et al.* (2003b).
- An expression for a channel capacity achieving input PDF for any given noise PDF or noise PDF is found. The channel capacity for these conditions is derived as $C(x, y) = 0.5 \log_2\left(\frac{N\pi}{2e}\right)$, thus giving an upper bound for the achievable channel capacity for SSR, under the conditions for which the large N mutual information approximation formula holds.
- Asymptotic large N expressions are found for the channel capacity, and optimal noise intensity, σ, for the Gaussian, logistic and Laplacian cases. It is shown that the difference between capacity and the mutual information at $\sigma = 1$ is of the order of 0.2 bits per sample for large N. It is also shown that the optimal σ for Gaussian signal and noise given in Hoch *et al.* (2003a) understates the true value by about 0.05.

Open questions

Possible future work and open questions arising from this chapter might include:

- Analytical proofs that the large N approximations converge to zero error as N approaches infinity.
- Analytical proof that the large N approximation to the output probability mass function converges to the exact output probability mass function.
- More rigorous justification of the use of the Fisher information approximation to the mutual information, and its relationship to the concept of stochastic complexity (Rissanen 1996) and minimum description length (MDL) (Grunwald *et al.* 2005). Such work could begin by building on the work of Davisson and Leon-Garcia (1980), Clarke and Barron (1990), Barron and Cover (1991), Nadal and Parga (1994), Rissanen (1996), and Brunel and Nadal (1998), and may find an alternative derivation of the SSR channel capacity and large N mutual information in the MDL framework.
- The only other work on SR we are aware of that has focused on finding optimal noise distributions is that of Chen *et al.* (2007), who have tackled this problem from the signal detection perspective, alongside previous work on controlling

SR (Mitaim and Kosko 1998, Gammaitoni *et al.* 1999, Löcher *et al.* 2000, Kosko and Mitaim 2001). Such an approach may be fruitful in understanding the capabilities of SR in other systems, and for other signal processing goals.

This concludes Chapter 5. Chapters 4 and 5 studied information transmission and encoding in the SSR model, by analytical and numerical studies of the mutual information and channel capacity. The following chapter studies the SSR model as a quantization or lossy source coding model, by examining decoding of the SSR model output, to achieve low distortion between the input and output signals.

6

Suprathreshold stochastic resonance: decoding

The initial research into suprathreshold stochastic resonance described in Chapter 4 considers the viewpoint of information transmission. As discussed briefly in Chapter 4, the suprathreshold stochastic resonance effect can also be modelled as stochastic quantization, and therefore results in nondeterministic lossy compression of a signal. The reason for this is that the effect of independently adding noise to a common signal before thresholding the result a number of times, with the same static threshold value, is equivalent to quantizing a signal with random thresholds. This observation leads naturally to measuring and describing the performance of suprathreshold stochastic resonance with standard quantization theory. In a context where a signal is to be reconstructed from its quantized version, this requires a *reproduction value* or *reproduction point* to be assigned to each possible state of a quantized signal. The quantizing operation is often known as the *encoding* of a signal, and the assignment of reproduction values as the *decoding* of a signal. This chapter examines various methods for decoding the output of the suprathreshold stochastic resonance model, and evaluates the performance of each technique as the input noise intensity and array size change. As it is the performance criterion most often used in conventional quantization theory, the measure used is the *mean square error distortion* between the original input signal and the decoded output signal.

6.1 Introduction

We begin this chapter by very briefly reviewing the SSR model introduced in Chapter 4. We then introduce the concept of *decoding* the output of a quantizer's encoding to reconstruct the input signal, and consider measuring the performance of such a reconstruction. Three measures are considered, the *mean square error* (MSE) distortion, the *signal-to-quantization-noise ratio* (SQNR), and the *correlation coefficient*. The latter two measures are, however, shown to be very closely related to the MSE distortion. The remainder of this chapter is outlined in Section 6.1.

The mathematical treatment in this chapter follows along the lines of McDonnell *et al.* (2002a), McDonnell *et al.* (2002b), McDonnell *et al.* (2003a), McDonnell *et al.* (2005c), and McDonnell *et al.* (2005d).

Summary of relevant results from Chapters 4 and 5

Chapter 4 introduced the array of threshold devices in which SSR occurs. The model is shown in Fig. 4.1, and consists of N binary threshold devices. As in Chapters 4 and 5, we consider each threshold device to have an identical value, θ, and to receive the same input signal, x, which is a sequence of independent samples from the random signal with PDF given by $f_x(x)$ and a standard deviation being a function of σ_x. However, *iid* additive noise, with PDF $f_\eta(\eta)$, and a standard deviation being a function of σ_η, is present at the input to each threshold, so that the output of each threshold device is unity if $x+\eta > \theta$ or zero otherwise. The overall output of the SSR model is the sum of all the threshold outputs, $y \in [0, N]$, which is an integer encoding of the input signal.

We saw in Chapter 4 that the performance of the SSR model can be measured by the mutual information between x and y, and that the maximum mutual information occurs for a nonzero noise intensity. Calculating the mutual information depends on knowing the *transition probabilities*, $P_{y|x}(y = n|x)$, which we abbreviate to $P_{y|x}(n|x)$. For the SSR model, this conditional probability mass function is given by the binomial distribution, in terms of the probability that any given threshold device is on, given x, denoted as $P_{1|x}$. Furthermore, the output probability mass function can be found by integrating the joint input–output probability density function, $f_{xy}(x, y) = f_x(x)P_{y|x}(n|x)$ to get $P_y(n) = \int_x f_x(x)P_{y|x}(n|x)dx$.

A convenient parameterization of the noise intensity is given by the ratio of the noise standard deviation to the signal standard deviation, which for 'matched' signal and noise distributions is $\sigma = \sigma_\eta/\sigma_x$. For such a case, the mutual information can be expressed as a function of σ, and is therefore independent of the signal's variance for a given σ. In this chapter, however, we shall see that sometimes the performance of the decoded SSR array depends on both σ_η and σ_x.

Furthermore, under certain conditions on the support of $f_x(x)$ and $f_\eta(x)$, the mutual information can be expressed, by a transformation using the inverse cumulative distribution function (ICDF) of the noise, $F_\eta^{-1}(\cdot)$, in terms of a PDF with support between zero and unity

$$f_Q(\tau) = \frac{f_x(x)}{f_\eta(\theta - x)}\bigg|_{x=\theta - F_\eta^{-1}(1-\tau)}. \tag{6.1}$$

Chapter 5 shows that $f_Q(\tau)$ is the PDF of the *average* transfer function of the SSR model divided by N – that is, the PDF of the random variable, $E[y|x]/N$ – and how, for large N, the mutual information is highly dependent on the entropy of $f_Q(\tau)$.

In this chapter, we shall consider only cases where $f_x(x)$ and $f_\eta(x)$ are even functions about a mean of zero and the threshold value is $\theta = 0$.

We next introduce the concept of decoding a quantized signal to reconstruct an approximation to the input signal, and show how the performance of such a reconstruction can be measured by the MSE distortion.

Decoding a quantized signal

The output of the SSR model, y, is a non-deterministic integer encoding of the input signal, x. The set of possible values of y are the integers from zero to N, and is therefore a quantization of x. We consider now any scalar quantization scheme. To enable a reconstruction of the input signal from such a quantization, each value of y must be assigned a *reconstruction point*. Label the nth reconstruction point, that is the point corresponding to $y = n$, as $\hat{y}_n$. Therefore, the decoded output is the discretely valued signal $\hat{y}$, which has possible values $\hat{y}_0, \ldots, \hat{y}_n, \ldots, \hat{y}_N$.

Assuming suitable values of $\hat{y} = \{\hat{y}_n\}$ have been set, then it is possible to define an error signal between the input x, and the reconstructed input, $\hat{y}$. Let this error be

$$\epsilon = x - \hat{y}. \tag{6.2}$$

Clearly, this error is a function of x. Since x is a continuously valued variable, some measure of the *average* error – or *distortion* – is required.

Mean square error distortion

There are many possible ways to define such an average error, the most common of which are the absolute error and the MSE distortion. The MSE distortion is also sometimes known as the quantization noise. We shall only focus on the MSE distortion, as this is very commonly used in estimation and quantization theories (Gray and Neuhoff 1998).

It is sometimes convenient to consider the conditional MSE distortion for a given value of the input, x, which we label as

$$D(x) = E[\epsilon^2|x]$$
$$= E[(x - \hat{y})^2|x]$$
$$= \sum_{n=0}^{N} P_{y|x}(n|x)(x - \hat{y}_n)^2. \tag{6.3}$$

Note also that

$$D(x) = var[\epsilon|x] + E[\epsilon|x]^2$$
$$= var[\epsilon|x] + E[\hat{y} - x|x]^2$$
$$= var[\epsilon|x] + (E[\hat{y}|x] - x)^2$$
$$= var[\epsilon|x] + b_{\hat{y}}(x)^2, \tag{6.4}$$

where $b_{\hat{y}}(x)$ is the *bias* of the decoded output, $\hat{y}$, as a function of x. The concept of bias is defined and discussed in Section 6.7, but for now we simply comment that the above equation shows that the conditional MSE distortion is the sum of the conditional variance of the error and the square of the bias. This illustrates that the distortion is strongly dependent on both the variance of the error and the bias.

The MSE distortion is defined as the average value of $D(x)$ over all x, that is

$$MSE = E[\epsilon^2]$$
$$= E[D(x)]$$
$$= \sum_{n=0}^{N} \int_{x=-\infty}^{\infty} f_{xy}(x, y)(x - \hat{y}_n)^2 dx$$
$$= \int_{x=-\infty}^{x=\infty} f_x(x) \left(\sum_{n=0}^{N} P_{y|x}(n|x)(x - \hat{y}_n)^2 \right) dx. \tag{6.5}$$

This definition can be used to calculate the MSE numerically, provided the decoding and the transition probabilities are known.

Also, the MSE can be expressed as

$$MSE = \int_x f_x(x) D(x) dx$$
$$= \int_x f_x(x) \left(var[\epsilon|x] + E[\epsilon|x]^2 \right) dx$$
$$= E[var[\epsilon|x]] + E[b_{\hat{y}}(x)^2]. \tag{6.6}$$

We shall call the first term in Eq. (6.6) the *average error variance* and the second term the *mean square bias*.

The expression given by Eq. (6.5) can also be simplified to give

$$\begin{aligned} \text{MSE} &= E[\epsilon^2] \\ &= E[\hat{y}^2] - 2E[x\hat{y}] + E[x^2], \end{aligned} \tag{6.7}$$

where $E[\hat{y}^2]$ is the mean square value of the decoded output, $\hat{y}$, $E[x^2]$ is the mean square value of the input, x, and $E[x\hat{y}]$ is the correlation between x and $\hat{y}$. In the SSR model, since the input signal has a mean of zero, the mean square value of x is also its variance. Without exception, the decoded output mean will also always be zero in this chapter and, therefore, the mean square value of the decoded output is also always the variance of the decoded output. In contrast, the *encoded* output signal for SSR has a nonzero mean – which is shown in Section A3.2 of Appendix 6 to be $N/2$ – and therefore has a variance

$$\text{var}[y] = E[y^2] - N^2/4. \tag{6.8}$$

Output signal-to-quantization-noise ratio

The lossy source coding and quantization literature often defines an output signal-to-noise ratio measure as the ratio of the input signal power to the MSE distortion – or quantization noise – power (Proakis and Manolakis 1996, Gray and Neuhoff 1998, Lathi 1998). Since we are analyzing only signals that are stationary random variables with zero means, the input signal's power is simply its mean square value, and the output SQNR is

$$\text{SQNR} = \frac{E[x^2]}{\text{MSE}} = \frac{E[x^2]}{E[(x - \hat{y})^2]}. \tag{6.9}$$

Note that it is conventional to express SNR measures in terms of decibels (dB), which for the SQNR is $10\log_{10}(\text{SQNR})$.

Correlation coefficient

The correlation coefficient between two random variables, x and $\hat{y}$, is defined as

$$\rho_{x\hat{y}} = \frac{\text{cov}[x\hat{y}]}{\sqrt{\text{var}[x]\text{var}[\hat{y}]}}, \tag{6.10}$$

where $\text{cov}[x\hat{y}] = E[x\hat{y}] - E[x]E[\hat{y}]$ is the covariance between x and $\hat{y}$. Since, here, $E[x] = 0$ and $E[\hat{y}] = 0$

$$\rho_{x\hat{y}} = \frac{E[x\hat{y}]}{\sqrt{E[x^2]E[\hat{y}^2]}}. \tag{6.11}$$

The correlation coefficient is a measure of the linearity between two random variables. It is equal to unity if an exact linear relationship can be found (Yates and Goodman 2005). It is zero if the two random variables are independent. It will be shown that simple relationships exist between the correlation coefficient and SQNR for optimal linear and nonlinear decoding, and that therefore the correlation coefficient and SQNR are equivalent measures for such decoding schemes.

Chapter structure

Having introduced the concept of decoding a quantized signal, and measures of the performance of such decodings, we are now ready to discuss in detail their application to the SSR model. However, first in Section 6.2, we shall use this theory to discuss noise removal via averaging for analogue signals. This is for later comparison with the performance of the SSR effect – which can be seen as an averaging of N single bit digital signals – for noise removal.

Next, Sections 6.3 and 6.4 consider *linear* decoding, while Section 6.5 considers *nonlinear* decoding. It is shown in Section 6.6 that, although nonlinear decoding can provide a smaller MSE, linear decoding is simpler to specify and calculate.

Section 6.7 shows how the output of the decoded SSR model can be considered as an estimate of the input, x, and how the performance of its decoding can be understood using estimation theory. The main result is to derive an expression for the *Fisher information* of the SSR model, and to use the Fisher information to derive theoretical lower bounds on the MSE performance of the system. Section 6.7 also demonstrates how the performance of the decoded SSR model depends strongly both on *bias* and on *variance*.

Finally, Section 6.8 discusses an approach proposed in Stocks (2001a) to measure the SSR model's output in terms of an SNR measure that is different from the SQNR.

Some results relevant to this chapter are given in more detail in Appendix 3.

6.2 Averaging without thresholding

It is well known that averaging N independently noisy versions of a signal reduces the noise by a factor of $1/\sqrt{N}$. We begin by proving this result.

Suppose x is a random signal consisting of a sequence of samples drawn from some continuously valued probability distribution with PDF $f_x(x)$, zero mean, and variance σ_x^2. Suppose $i = 1, \ldots, N$ and η_i is a set of uncorrelated random signals drawn from a distribution with PDF $f_\eta(\eta)$, zero mean, and variance σ_η^2. Let $y_i = x + \eta_i$ be the sum of the signal and the ith noise signal. The error between y_i and x is $\epsilon_i = y_i - x = \eta_i$, and the MSE distortion between the raw signal, x, and the noisy signal is the noise variance, $D(x) = \sigma_\eta^2 \; \forall \, x$.

Suppose now that the N independently noisy signals are averaged so that

$$y(x) = \frac{1}{N} \sum_{i=1}^{N} y_i$$

$$= \frac{1}{N} \sum_{i=1}^{N} (x + \eta_i)$$

$$= x + \frac{1}{N} \sum_{i=1}^{N} \eta_i. \tag{6.12}$$

The error between $y(x)$ and x has a mean of zero and is given by

$$\epsilon(x) = y(x) - x = \frac{1}{N} \sum_{i=1}^{N} \eta_i \quad \forall x. \tag{6.13}$$

The mean square error for a given x is

$$D(x) = \mathrm{E}[\epsilon(x)^2] = \frac{1}{N^2} \mathrm{E}\left[\left(\sum_{i=1}^{N} \eta_i\right)^2\right]. \tag{6.14}$$

Since each noise signal is uncorrelated with all other noise signals, Eq. (6.14) simplifies to

$$D(x) = \frac{1}{N^2} \mathrm{E}\left[\sum_{i=1}^{N} \eta_i^2\right]$$

$$= \frac{1}{N^2} \sum_{i=1}^{N} \mathrm{E}[\eta^2]$$

$$= \frac{1}{N^2} N \sigma_\eta^2$$

$$= \frac{\sigma_\eta^2}{N} \quad \forall x. \tag{6.15}$$

Thus, the conditional MSE distortion is reduced by a factor of N, and is constant for all values of x. This means the MSE distortion is also the conditional mean square distortion, and is independent of the signal variance. This fact also means that the signal, x, need not be a random signal at all, but can be a deterministic signal of any form. Note that each y_i is an *iid* random variable, with a mean of x, and variance σ_η^2.

The square root of the MSE distortion is often referred to as the root mean square (rms) error. The well-known result of a $\frac{1}{\sqrt{N}}$ reduction in noise due to averaging refers to using the rms error as the noise measure.

In terms of SNR, the SNR after averaging is

$$\text{SNR} = \frac{\sigma_x^2}{\text{MSE}} = \frac{N}{\sigma^2}, \tag{6.16}$$

where $\sigma^2 = \sigma_\eta^2/\sigma_x^2$ is the reciprocal of the SNR of a single noisy input signal, y_i. This means that the SNR after averaging is N times the SNR of a single noisy version of the signal, y_i.

In Section 6.8 we shall compare this result with measures of the output SNR in the SSR model.

We can envisage situations where thresholding first before averaging may be advantageous, or indeed the only possible way of averaging, most particularly in digital systems. In particular, the advantage of thresholding first is that it provides compression to a discrete number of possible values.

We now proceed by discussing specific decoding methods for quantization schemes.

6.3 Linear decoding theory

Consider linear decoding of a quantizer's encoding stage output, $y \in [0, N]$. Such a decoding operation can be written as

$$\hat{y} = ay + b, \quad y = 0, \ldots, N, \tag{6.17}$$

where a and b are constants for all values of y. Such linear decoding will give a set of outputs that are evenly spaced. We label the nth value of $\hat{y}$ as $\hat{y}_n$.

Assume in general that y is a nondeterministic function of an input signal, x, and that $E[y] = N/2$. If we impose the condition that the mean of the decoded output, $\hat{y}$, is also zero, this requires that $E[\hat{y}] = aE[y] + b = 0$. Therefore setting $a = -2b/N$ is sufficient to provide a zero-meaned decoding. Changing notation by letting $c = -b$ gives

$$\hat{y} = \frac{2c}{N}y - c, \quad y = 0, \ldots, N. \tag{6.18}$$

We note that since the mean is zero, the mean square value of $\hat{y}$ is also its variance.

Conditional linear MSE distortion

As well as the overall MSE distortion, we shall also be interested in the *conditional* moments of the decoding and the conditional MSE distortion for a given value of x; each value of x has its own average distortion.

Unlike the overall output, the conditional mean of $\hat{y}$ given x is, in general, nonzero, since

$$E[\hat{y}|x] = \frac{2c}{N}E[y|x] - c. \qquad (6.19)$$

Consequently, the conditional variance of $\hat{y}$ will not equal the conditional mean square value of $\hat{y}$ for a given x.

The conditional mean square value of the decoding is

$$E[\hat{y}^2|x] = \frac{4c^2}{N^2}E[y^2|x] - \frac{4c^2}{N}E[y|x] + c^2, \qquad (6.20)$$

and the conditional variance is

$$\begin{aligned} \mathrm{var}[\hat{y}|x] &= \frac{4c^2}{N^2}\left(E[y^2|x] - E[y|x]^2\right) \\ &= \frac{4c^2}{N^2}\mathrm{var}[y|x]. \end{aligned} \qquad (6.21)$$

Equations (6.20) and (6.21) show that the conditional variance of the decoded output is proportional to the conditional variance of the 'un-decoded' output, while, since y and $\hat{y}$ have different means, the mean square value is not.

It is straightforward to show that the variance of the error, given x, is equal to the variance of $\hat{y}$ given x, so that

$$\mathrm{var}[\epsilon|x] = \mathrm{var}[\hat{y}|x] = \frac{4c^2}{N^2}\mathrm{var}[y|x]. \qquad (6.22)$$

The conditional MSE distortion is

$$\begin{aligned} D(x) &= \mathrm{var}[\hat{y}|x] + E[\hat{y} - x|x]^2 \\ &= \mathrm{var}[\hat{y}|x] + E[\hat{y}|x]^2 - 2xE[y|x] + x^2 \\ &= E[\hat{y}^2|x] - 2xE[\hat{y}|x] + x^2. \end{aligned} \qquad (6.23)$$

Alternatively

$$\begin{aligned} D(x) &= E[(\hat{y} - x)^2|x] \\ &= E[\hat{y}^2|x] - 2xE[\hat{y}|x] + x^2. \end{aligned} \qquad (6.24)$$

Substituting Eqs. (6.19) and (6.20) into Eq. (6.23) and simplifying gives

$$D(x) = \frac{4c^2}{N^2}E[y^2|x] - \frac{4c}{N}E[y|x](x + c) + (x + c)^2. \qquad (6.25)$$

Linear mean square error distortion

The average MSE distortion is the expected value of the conditional distortion

$$\text{MSE} = \text{E}[D(x)]$$

$$= \int_x f_x(x)D(x)dx$$

$$= \text{E}[(\hat{y} - x)^2]$$

$$= \text{E}[\hat{y}^2] - 2\text{E}[x\hat{y}] + \text{E}[x^2], \tag{6.26}$$

as described by Eq. (6.7) in Section 6.1. Substituting from Eq. (6.18) into Eq. (6.26) and simplifying gives

$$\text{MSE} = \frac{4c^2}{N^2}\text{E}[y^2] - \frac{4c}{N}\text{E}[xy] + \text{E}[x^2] - c^2. \tag{6.27}$$

Correlation coefficient

Consider the linear decoding of Eq. (6.17). It is straightforward to show that the correlation coefficient between x and $\hat{y}$ is identical for any a and b in this equation. Hence, to find an expression for the correlation coefficient, it is sufficient to set $a = 1$ and $b = 0$, and obtain the correlation coefficient between x and the 'un-decoded' output y, which is

$$\rho_{x\hat{y}} = \rho_{xy} = \frac{\text{E}[xy] - \text{E}[x]\text{E}[y]}{\sqrt{\text{var}[x]\text{var}[y]}}. \tag{6.28}$$

Since $\text{E}[x] = 0$ and $\text{E}[y] = N/2$

$$\rho_{x\hat{y}} = \rho_{xy} = \frac{\text{E}[xy]}{\sqrt{\text{E}[x^2]\left(\text{E}[y^2] - \frac{N^2}{4}\right)}}. \tag{6.29}$$

Due to its invariance to linear decoding, the correlation coefficient is not a relevant measure when comparing different specific linear decoding schemes. However, because of this fact, it does provide a means of measuring the effectiveness of the best possible linear decoding. It will also prove useful as a means of comparing the effectiveness of a nonlinear decoding when compared to a linear decoding.

Having now defined a linear decoding scheme, and having derived general expressions for the MSE, SQNR, and the correlation coefficient, we now give three different possible ways for specifying the value of c in Eq. (6.18).

Constant decoding

If either the signal variance or the noise variance are unknown, decoding that is independent of either is necessary. Since the aim is to minimize the MSE distortion, a naive decoding choice is one that has the same range of possible values of x as the input signal, so that

$$\min \hat{y} = \min x \quad \text{and} \quad \max \hat{y} = \max x. \tag{6.30}$$

However, for signal distributions that have infinite tails, such as the Gaussian distribution, it is obvious that there should be no reconstruction point at $\pm\infty$. One solution for such a source is to set the minimum and maximum reconstruction points to a set number of standard deviations.

Matched moments linear decoding

Having considered a simple method of decoding in the event that the signal variance is unknown, we now consider decoding that depends on knowledge of the variance of the input signal, and the linear correlation coefficient between the input and output. The aim is to find the linear decoding scheme that provides the minimum MSE distortion.

Assume the decoding is linear, as given by Eq. (6.17). Under the hypothesis that the output moments should be equal to the input moments, a naive approach is to find a and b in Eq. (6.17) such that $E[\hat{y}] = E[x]$ and $E[\hat{y}^2] = E[x^2]$. For the distributions we are considering, $E[x] = 0$, and $E[y] = N/2$. This immediately imposes $b = -aN/2$. We have

$$E[\hat{y}^2] = a^2 \left(E[y^2] - \frac{N^2}{4} \right) = E[x^2], \tag{6.31}$$

which gives

$$a = \sqrt{\frac{E[x^2]}{E[y^2] - \frac{N^2}{4}}}$$

$$= \sqrt{\frac{E[x^2]}{\text{var}[y]}}. \tag{6.32}$$

Consequently, the decoded output signal is

$$\hat{y} = \sqrt{\frac{E[x^2]}{\text{var}[y]}} \left(y - \frac{N}{2} \right). \tag{6.33}$$

With reference to Eq. (6.18)

$$c = \frac{N}{2}\sqrt{\frac{E[x^2]}{\text{var}[y]}},$$
(6.34)

and substituting from Eq. (6.34) into Eq. (6.27) gives

$$\text{MSE} = 2\left(E[x^2] - \frac{\sqrt{E[x^2]}E[xy]}{\sqrt{\text{var}[y]}}\right)$$

$$= 2E[x^2]\left(1 - \frac{E[xy]}{\sqrt{E[x^2]\text{var}[y]}}\right)$$

$$= 2E[x^2](1 - \rho_{xy}).$$
(6.35)

The result is that the MSE of this decoding scheme can be simply expressed in terms of the mean square value of the input signal, and the correlation coefficient of a linear decoding scheme. Furthermore, rearranging Eq. (6.35) gives an expression for the SQNR that is independent of $E[x^2]$ as

$$\text{SQNR} = \frac{1}{2(1 - \rho_{xy})}.$$
(6.36)

Wiener optimal linear decoding

What is the optimal linear decoding scheme? The only unknown in Eq. (6.18) is c. The optimal value of c can be found by differentiating Eq. (6.27) with respect to c, setting the result to zero, and solving for c, the result being

$$c = \frac{NE[xy]}{2\text{var}[y]}.$$
(6.37)

Such decoding is known as Wiener decoding, and is the optimal linear decoding scheme for the MSE distortion (Yates and Goodman 2005). Note that with this decoding, it is straightforward to show that

$$E[\hat{y}^2] = E[x\hat{y}] = \frac{E[xy]^2}{\text{var}[y]}.$$
(6.38)

Substituting from Eq. (6.37) into Eq. (6.27) and simplifying gives

$$\text{MSE} = E[x^2] - \frac{E[xy]^2}{\text{var}[y]}$$

$$= E[x^2] - E[\hat{y}^2].$$
(6.39)

It can be seen that with this decoding, the MSE distortion is independent of the correlation between x and y. We have also that

$$\text{MSE} = E[x^2]\left(1 - \frac{E[xy]^2}{E[x^2]\text{var}[y]}\right)$$
$$= E[x^2](1 - \rho_{xy}^2). \qquad (6.40)$$

Eq. (6.40) can be found in Yates and Goodman (2005). Just as with matched-moments linear decoding, the MSE distortion given by Eq. (6.40) is entirely dependent on the correlation coefficient and the mean square value of the input. It is also straightforward to show that with this decoding the correlation between the encoded output, y, and the error, ϵ, is zero, that is $E[\epsilon y] = 0$ (Yates and Goodman 2005).

By subtraction, Eqs (6.40) and (6.35) give the difference between the MSE distortions of matched-moments decoding and Wiener decoding as

$$2E[x^2](1 - \rho_{xy}) - E[x^2](1 - \rho_{xy}^2) = E[x^2](\rho_{xy} - 1)^2 > 0. \qquad (6.41)$$

This shows that the MSE distortion obtained by Wiener decoding is always smaller than that obtained with matched-moments decoding. Since Eq. (6.40) was derived to give the smallest possible linear MSE distortion, Inequality (6.41) simply acts as verification of this fact.

The SQNR for Wiener decoding can be obtained from rearrangement of Eq. (6.40) as

$$\text{SQNR} = \frac{1}{1 - \rho_{xy}^2}. \qquad (6.42)$$

In decibels, this can be written as

$$10\log_{10}(\text{SQNR}) = -10\log_{10}(1 - \rho_{xy}^2). \qquad (6.43)$$

6.4 Linear decoding for SSR

Section 6.3 does not say anything specific about the SSR model; it discusses only the MSE, SQNR, and correlation coefficient between the signals x and $\hat{y}$. We assumed the decoded signal, $\hat{y}$, is given by the linear transform of Eq. (6.18), where y is a nondeterministic function of x, with a mean of $N/2$ and that x has a mean of zero and a PDF that is an even function of x.

Recall that we also assume in this chapter that the input of the SSR model has a zero-meaned and even PDF, $f_x(x)$. Section A3.2 in Appendix 3 gives a proof that, for the SSR model, and such an input signal, $E[y] = N/2$. This means the equations derived in Section 6.3 apply to the SSR model. The remainder of this section simplifies the equations of Section 6.3 for the SSR model, and then considers some specific cases of signal and noise distributions.

Decoding measures for SSR

MSE distortion

Expressions for the mean of y given x and the mean square value of y given x, for the SSR model, are given in Section A3.1 of Appendix 3 in terms of $P_{1|x}$, by Eqs (A3.3) and (A3.6). Substituting these into Eq. (6.25) gives the conditional MSE distortion for SSR in terms of $P_{1|x}$ as

$$D(x) = 4c^2 P_{1|x}(1 - P_{1|x})\left(\frac{1-N}{N}\right) - 4c P_{1|x}x + (x+c)^2. \tag{6.44}$$

An expression for the correlation between x and y, for the SSR model, is given in Section A3.3 of Appendix 3 in terms of $P_{1|x}$, by Eq. (A3.25). Substituting this and Eq. (A3.17) into Eq. (6.27) gives the MSE distortion for SSR as

$$\text{MSE} = \frac{4c^2(N-1)}{N}E[P_{1|x}^2] - 4cE[x P_{1|x}] + E[x^2] - \frac{c^2(N-2)}{N}. \tag{6.45}$$

We see from Eq. (6.45) that the MSE distortion of a linear decoding of the SSR model can be obtained for any c, provided $E[P_{1|x}^2]$ and $E[x P_{1|x}]$ are known.

Recall the change of variable transformation used in Chapters 4 and 5, $\tau = F_\eta(x) = P_{1|x}$, where $F_\eta(\cdot)$ is the cumulative distribution function (CDF) of the noise. Using this same change of variable, the quantity $E[P_{1|x}^2]$ can be recognized as the mean square value of the PDF, $f_Q(\tau)$, that is $E[\tau^2]$.

SQNR

Using the definition given by Eq. (6.9), the SQNR for a linear decoding of the SSR model is

$$\text{SQNR} = \frac{E[x^2]}{\frac{4c^2(N-1)}{N}E[P_{1|x}^2] - 4cE[x P_{1|x}] + E[x^2] - \frac{c^2(N-2)}{N}}. \tag{6.46}$$

Correlation coefficient

Substituting from Eqs (A3.17) and (A3.25) into Eq. (6.29), and recalling $E[P_{1|x}] = 0.5$, gives the correlation coefficient for SSR and a linear decoding as

$$\rho_{x\hat{y}} = \rho_{xy} = \frac{\sqrt{N}E[x P_{1|x}]}{\sqrt{E[x^2]\left((N-1)E[P_{1|x}^2] - \frac{(N-2)}{4}\right)}}. \tag{6.47}$$

So as with the MSE distortion, the correlation coefficient can be calculated given knowledge of $E[P_{1|x}^2]$ and $E[x P_{1|x}]$.

Having now obtained expressions for the MSE distortion, SQNR, and correlation coefficient for the SSR model with a linear decoding, it is time to examine these measures for specific signal and noise distributions. We find that exact closed-form analytical expressions are available in certain cases. In other cases, numerical evaluation of the MSE distortion, SQNR, or correlation coefficient is required.

Specific signal and noise distributions

We now derive closed-form analytical expressions for $E[P_{1|x}^2]$ and $E[x P_{1|x}]$ for the cases of Gaussian signal and noise, uniform signal and noise, and Laplacian signal and noise. Using these expressions, formulas for the linear decoding correlation coefficient are derived. No such exact expressions could be found for logistic signal and noise, but results obtained using numerical integration are obtained for this distribution. This section does not explicitly state expressions for the MSE distortion or SQNR, but such analytical expressions can easily be found by use of Eq. (6.45) or (6.46) and the equations stated in this section, once a value of c is specified. Specifying c is left for later.

Gaussian signal and noise

For Gaussian signal and noise, the mean square value of $P_{1|x}$ can be derived as follows

$$
\begin{aligned}
E[P_{1|x}^2] &= \int_{-\infty}^{\infty} P_{1|x}^2 f_x(x) dx \\
&= \int_{-\infty}^{\infty} \left(\frac{1}{2} + \frac{1}{2}\mathrm{erf}\left(\frac{x}{\sqrt{2}\sigma_\eta}\right) \right)^2 f_x(x) dx \\
&= \int_{-\infty}^{\infty} \left(\frac{1}{4} + \frac{1}{2}\mathrm{erf}\left(\frac{x}{\sqrt{2}\sigma_\eta}\right) + \frac{1}{4}\mathrm{erf}\left(\frac{x}{\sqrt{2}\sigma_\eta}\right)^2 \right) f_x(x) dx \\
&= \frac{1}{4}\int_{-\infty}^{\infty} f_x(x) dx + \frac{1}{2}\int_{-\infty}^{\infty} \mathrm{erf}\left(\frac{x}{\sqrt{2}\sigma_\eta}\right) f_x(x) dx \\
&\quad + \frac{1}{4}\int_{-\infty}^{\infty} \mathrm{erf}^2\left(\frac{x}{\sqrt{2}\sigma_\eta}\right) f_x(x) dx \\
&= \frac{1}{4} + 0 + \frac{1}{4}\int_{-\infty}^{\infty} \mathrm{erf}^2\left(\frac{x}{\sqrt{2}\sigma_\eta}\right) f_x(x) dx.
\end{aligned} \tag{6.48}
$$

The second term above is zero, since $f_x(x)$ is even and $\mathrm{erf}(x)$ is odd. We make use of the following result proven in Section A3.4 of Appendix 3

$$\int_{x=-\infty}^{x=\infty} \exp(-a^2 x^2) \mathrm{erf}^2(x) dx = \frac{2}{a\sqrt{\pi}} \arctan\left(\frac{1}{a\sqrt{a^2+2}}\right), \qquad (6.49)$$

which gives

$$
\begin{aligned}
\mathrm{E}[P_{1|x}^2] &= \frac{1}{4} + \frac{1}{4\sqrt{2\pi\sigma_x^2}} \int_{-\infty}^{\infty} \mathrm{erf}^2\left(\frac{x}{\sqrt{2}\sigma_\eta}\right) \exp\left(-\frac{x^2}{2\sigma_x^2}\right) dx \\
&= \frac{1}{4} + \frac{\sigma}{4\sqrt{\pi}} \int_{-\infty}^{\infty} \mathrm{erf}^2(\tau) \exp(-\sigma^2\tau^2) d\tau \\
&= \frac{1}{4} + \frac{1}{2\pi} \arctan\left(\frac{1}{\sigma\sqrt{\sigma^2+2}}\right) \\
&= \frac{1}{4} + \frac{1}{2\pi} \arcsin\left(\frac{1}{\sigma^2+1}\right). \qquad (6.50)
\end{aligned}
$$

The conversion from arctan to arcsin holds provided $\sigma > 0$, which is always true. It can be seen that $\mathrm{E}[P_{1|x}^2]$ is a function of the parameter σ. It is shown in Section A3.3 of Appendix 3 that the correlation between the inputs to any two devices – that is, the correlation between $x + \eta_i$ and $x + \eta_j$ – in the SSR model is $\rho_i = \frac{1}{\sigma^2+1}$. Therefore

$$\mathrm{E}[P_{1|x}^2] = \frac{1}{4} + \frac{1}{2\pi} \arcsin(\rho_i). \qquad (6.51)$$

Note that substituting from Eq. (6.51) into Eq. (A3.13) gives the variance of the output of the SSR model encoding as

$$\mathrm{var}[y] = \frac{N(N-1)}{2\pi} \arcsin(\rho_i) + \frac{N}{4}. \qquad (6.52)$$

For a decoding given by $c = N$, so that $\hat{y}$ can take values between $-N$ and N, with a mean of zero, we have

$$\mathrm{var}[\hat{y}] = \mathrm{E}[\hat{y}^2] = N + N(N-1)\frac{2}{\pi} \arcsin(\rho_i). \qquad (6.53)$$

Equation (6.53) is precisely the same as an equation derived in Remley (1966) for the output mean square signal power for Gaussian signal and Gaussian noise in a DIMUS sonar array – see also Section 4.2 in Chapter 4. This result shows that a DIMUS sonar array model where a signal is assumed present, is equivalent to the decoded SSR model. However, the results presented in this chapter are far more general, as nowhere in the DIMUS literature is the output of the DIMUS array considered to be a quantized approximation to a random input signal, nor is the output response considered as a function of varying input noise.

Note that Remley (1966) derives Eq. (6.53) in a different manner from that given here, by making use of a result known as Van Vleck's relation[1] (Price 1958, Van Vleck and Middleton 1966). Van Vleck's result is a formula for the normalized autocorrelation, ρ_2, of 'clipped' Gaussian noise – that is, a one-bit quantization of the noise – in terms of the normalized autocorrelation of the unclipped noise, ρ_1

$$\rho_2(\tau) = \frac{2}{\pi} \arcsin\left(\rho_1(\tau)\right). \qquad (6.54)$$

When applied to the multi-threshold DIMUS sonar array, with Gaussian signal and noise, ρ_1 is the correlation coefficient between the inputs to any two of the thresholds; that is, the correlation coefficient between $x + \eta_i$ and $x + \eta_j$, where η_i and η_j are the independent Gaussian noise samples on two thresholds. At the output, ρ_2 is the correlation coefficient between the binary variables, $2y_i - 1$ and $2y_j - 1$. The possible values of these variables are -1 and 1.

Section A3.3 in Appendix 3 derives the correlation coefficient between any two thresholds' outputs in the SSR model as Eq. (A3.20). Substituting from Eq. (6.51) into Eq. (A3.20) gives

$$\rho_o = 4\mathrm{E}[P_{1|x}^2] - 1$$
$$= \frac{2}{\pi} \arcsin\left(\rho_i\right), \qquad (6.55)$$

which is exactly Van Vleck's result, thus indicating how Remley (1966) derived Eq. (6.53).

We also have

$$\mathrm{E}[x P_{1|x}] = \int_{-\infty}^{\infty} x f_x(x) P_{1|x} dx$$
$$= \int_{-\infty}^{\infty} x f_x(x) \left(\int_{-\infty}^{x} f_\eta(\eta) d\eta \right) dx$$
$$= \int_{-\infty}^{\infty} f_\eta(\eta) \left(\int_{\eta}^{\infty} x f_x(x) dx \right) d\eta$$
$$= \int_{-\infty}^{\infty} f_\eta(\eta) \left(\int_{\eta}^{\infty} \frac{x}{\sqrt{2\pi}\sigma_x} \exp\left(-\frac{x^2}{2\sigma_x^2}\right) dx \right) d\eta$$
$$= \frac{\sigma_x}{\sqrt{2\pi}} \int_{-\infty}^{\infty} f_\eta(\eta) \exp\left(-\frac{\eta^2}{2\sigma_x^2}\right) d\eta$$

---

[1] In Remley (1966), Van Vleck's relation is referred to as 'Van Fleck's' relation, even though the paper referenced actually uses the correct spelling of 'Van Vleck' – see Van Vleck and Middleton (1966).

$$= \frac{\sigma_x}{\sqrt{2\pi}} \int_{-\infty}^{\infty} \frac{1}{\sqrt{2\pi}\,\sigma_\eta} \exp\left(-\frac{\eta^2}{2\sigma_\eta^2}\right) \exp\left(-\frac{\eta^2}{2\sigma_x^2}\right) d\eta$$

$$= \frac{1}{2\pi\sigma} \int_{-\infty}^{\infty} \exp\left(-\frac{\eta^2}{2}\left(\frac{1+\sigma^2}{\sigma_\eta^2}\right)\right) d\eta, \tag{6.56}$$

since $\sigma = \sigma_\eta/\sigma_x$. The final integrand is a Gaussian PDF, with variance $\sigma_\eta^2/(1+\sigma^2)$. Hence the integral from $-\infty$ to $+\infty$ is unity times the normalizing factor

$$E[x P_{1|x}] = \frac{1}{2\pi\sigma} \sqrt{2\pi\left(\frac{\sigma_\eta^2}{1+\sigma^2}\right)} = \frac{\sigma_x}{\sqrt{2\pi(1+\sigma^2)}}, \tag{6.57}$$

therefore

$$E[x P_{1|x}] = \frac{\sigma_x}{\sqrt{2\pi}} \sqrt{\rho_i}. \tag{6.58}$$

Interestingly, this expression is a function not just of σ but also σ_x. However, referring to the correlation coefficient, as given by Eq. (6.47), since $E[x^2] = \sigma_x^2$ for a Gaussian signal, σ_x will be cancelled out of Eq. (6.47), and therefore the correlation coefficient is a function only of σ. Upon substitution from Eqs (6.50) and (6.57) into Eq. (6.47) and simplification

$$\rho_{xy} = \frac{\sqrt{N}}{\sqrt{\sigma^2+1}\sqrt{(N-1)\arcsin\left(\frac{1}{\sigma^2+1}\right) + \frac{\pi}{2}}}$$

$$= \frac{\sqrt{N\rho_i}}{\sqrt{(N-1)\arcsin(\rho_i) + \frac{\pi}{2}}}$$

$$= \frac{\sqrt{N}\sin(\rho_o)}{\sqrt{(N-1)\rho_o + 1}}. \tag{6.59}$$

However, referring to the MSE distortion given by Eq. (6.45), σ_x will not necessarily be eradicated from the MSE distortion. In fact, we shall see later in this section that, in general, the MSE will be a function of the mean square value of the signal.

Uniform signal and noise

For uniform signal and noise, it is necessary to derive expressions for $E[P_{1|x}^2]$ and $E[x P_{1|x}]$ separately for the cases of $\sigma \le 1$ and $\sigma \ge 1$. The two expressions will be equal at $\sigma = 1$.

For $\sigma \leq 1$

$$E[P_{1|x}^2] = \int_{-\infty}^{\infty} P_{1|x}^2 f_x(x)dx$$

$$= \frac{1}{\sigma_x} \int_{-\sigma_\eta/2}^{\sigma_\eta/2} \left(\frac{1}{2} + \frac{x}{\sigma_\eta} \right)^2 dx + \frac{1}{\sigma_x} \int_{\sigma_\eta/2}^{\sigma_x/2} dx$$

$$= \frac{1}{2} - \frac{\sigma}{6}, \qquad (6.60)$$

and

$$E[x P_{1|x}] = \int_{-\infty}^{\infty} x f_x(x) P_{1|x} dx$$

$$= \int_{-\sigma_\eta/2}^{\sigma_\eta/2} x \left(\frac{1}{2} + \frac{x}{\sigma_\eta} \right) \frac{1}{\sigma_x} dx + \int_{\sigma_\eta/2}^{\sigma_x/2} \frac{x}{\sigma_x} dx$$

$$= \frac{\sigma_x}{24} (3 - \sigma^2). \qquad (6.61)$$

For $\sigma \geq 1$

$$E[P_{1|x}^2] = \int_{-\infty}^{\infty} P_{1|x}^2 f_x(x)dx$$

$$= \frac{1}{\sigma_x} \int_{-\sigma_x/2}^{\sigma_x/2} \left(\frac{1}{2} + \frac{x}{\sigma_\eta} \right)^2 dx$$

$$= \frac{1}{4} + \frac{1}{12\sigma^2}, \qquad (6.62)$$

and

$$E[x P_{1|x}] = \int_{-\infty}^{\infty} x f_x(x) P_{1|x} dx$$

$$= \frac{1}{\sigma_x} \int_{-\sigma_x/2}^{\sigma_x/2} x \left(\frac{1}{2} + \frac{x}{\sigma_\eta} \right) dx$$

$$= \frac{\sigma_x}{12\sigma}. \qquad (6.63)$$

Thus, the correlation coefficient for uniform signal and noise is

$$\rho_{x,y} = \begin{cases} \frac{\sqrt{N}(3-\sigma^2)}{2\sqrt{\sigma}(2-2N)+3N} & (\sigma \leq 1), \\[2mm] \frac{\sqrt{N}}{\sqrt{3\sigma^2+N-1}} & (\sigma \geq 1), \end{cases} \qquad (6.64)$$

which again is dependent only on the ratio σ.

Laplacian signal and noise

For Laplacian signal and noise, the mean square value of $P_{1|x}$ can be derived as

$$E[P_{1|x}^2] = \frac{\sigma^2 + 3\sigma + 4}{4(\sigma + 1)(\sigma + 2)}. \qquad (6.65)$$

Substituting from Eq. (6.65) into Eq. (A3.20) gives the correlation coefficient between the output of any two thresholds as

$$\rho_o = \frac{2}{(\sigma + 1)(\sigma + 2)}. \qquad (6.66)$$

The expected value of $x P_{1|x}$ can be derived as

$$E[x P_{1|x}] = \frac{\sigma_x(2\sigma + 1)}{2\sqrt{2}(\sigma + 1)^2}. \qquad (6.67)$$

Substitution from Eqs (6.65) and (6.67) into Eq. (6.47) gives the linear decoding correlation coefficient for Laplacian signal and noise as

$$\rho_{xy} = \frac{\sqrt{N}(2\sigma + 1)\sqrt{(\sigma + 1)(\sigma + 2)}}{\sqrt{2}(\sigma + 1)^2\sqrt{2N + \sigma(\sigma + 3)}}$$

$$= \frac{\sqrt{N}(2\sigma + 1)}{\sqrt{\rho_o}(\sigma + 1)^2\sqrt{2N + \sigma(\sigma + 3)}}. \qquad (6.68)$$

Illustrations

Figure 6.1 shows the correlation coefficient between the output of any two threshold devices, ρ_o, the quantity $E[P_{1|x}^2]$, and the quantity $E[x P_{1|x}]$ for each of four cases of matched signal and noise distributions. The Gaussian, uniform, and Laplacian cases were plotted from the exact formulas derived in this section, while the logistic case was calculated by numerical integration. Figure 6.4 also shows the correlation coefficient between the inputs of any two comparators, ρ_i, which is the same for all distributions. All quantities are decreasing functions of σ. Note from Fig. 6.4 how ρ_i is always greater than ρ_o, which shows that the one-bit quantization that results from thresholding decreases the correlation between the signals at any two devices. Fig. 6.4 shows that at $\sigma = 1$, $E[P_{1|x}^2]$ – and, consequently, ρ_o – is identical in all cases.

Figure 6.2 shows the linear correlation coefficient for the Gaussian, uniform, Laplacian, and logistic cases. Except for the logistic case, the correlation coefficient was calculated from the exact formulas specified in this section. The logistic case was calculated by numerical integration of $E[P_{1|x}^2]$ and $E[x P_{1|x}]$. It is clear in all cases that the correlation coefficient has a maximum for a nonzero value of σ. The optimal value of σ increases with increasing N. For the Gaussian, Laplacian, and logistic cases, the maximum value of the correlation coefficient occurs for $\sigma > 1$

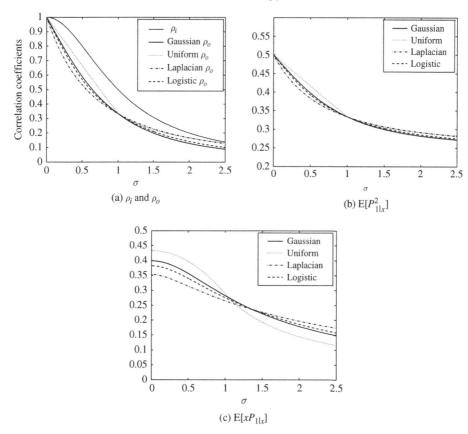

(a) ρ_i and ρ_o

(b) $E[P_{1|x}^2]$

(c) $E[xP_{1|x}]$

Fig. 6.1. Quantities for linear decoding. This figure shows the correlation coefficient between the output of any two threshold devices, ρ_o, the quantity $E[P_{1|x}^2]$, and the quantity $E[xP_{1|x}]$ for four cases of matched signal and noise distributions. The Gaussian, uniform, and Laplacian cases were calculated from the exact formulas, and the logistic case was calculated by numerical integration. Note that Fig. 6.1(a) also shows the correlation coefficient between the inputs of any two comparators, ρ_i. All quantities are decreasing functions of σ. Note how ρ_i is always greater than ρ_o, which shows that thresholding decreases the correlation between the signals at any two devices. At $\sigma = 1$, $E[P_{1|x}^2]$ – and, consequently, ρ_o – is identical in all cases.

for sufficiently large N, while for the uniform case, the optimal value of σ appears to get closer and closer to unity as N increases. The results shown in Fig. 6.2 illustrate that stochastic resonance occurs in the linear decoding correlation coefficient for the SSR model.

To compare the linear correlation coefficient for different distributions, Fig. 6.3 shows ρ_{xy} for the four different matched signal and noise distributions for $N = 127$. The uniform case gives the largest correlation coefficient; however, the same qualitative behaviour can be seen in each case.

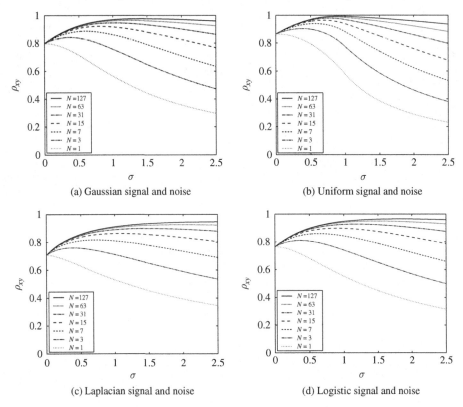

Fig. 6.2. Linear decoding correlation coefficient, ρ_{xy}, against increasing σ for four different matched signal and noise distributions and various values of N. Apart from the logistic case, for which ρ_{xy} was calculated only numerically, each plot was calculated from the exact formulas derived in this section, and also validated numerically. Notice how the correlation coefficient increases with increasing N, and has a maximum for a nonzero value of σ. The optimal value of σ also increases with increasing N, and, except for the uniform case, becomes larger than unity. In the uniform case, the optimal value of σ gets closer to unity with increasing N.

Constant decoding

This section examines the case of suboptimal constant decoding for the cases of uniform signal and noise, and Gaussian signal and noise. Such a decoding remains constant for all σ and a given signal variance.

Uniform signal and noise

Without loss of generality, we set the support of the uniformly distributed signal to be between $\pm\sigma_x/2$. Thus, we set $c = \sigma_x/2$, so that the decoded output varies between $\pm\sigma_x/2$, as does the input signal. Substitution from Eqs (6.60), (6.61), (6.62), and (6.63) into Eq. (6.45) gives the MSE distortion as

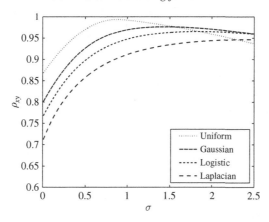

Fig. 6.3. Linear decoding correlation coefficient, ρ_{xy}, for $N = 127$ and four matched signal and noise distributions. The uniform case gives the largest correlation coefficient; however, each case shows the same qualitative behaviour.

$$\text{MSE} = \begin{cases} \sigma_x^2 \left(\frac{\sigma^2+1}{12} - \frac{\sigma(N-1)}{6N} \right) & (\sigma \leq 1), \\[2ex] \sigma_x^2 \left(\frac{\sigma^2(3+N)-2N\sigma+N-1}{12N\sigma^2} \right) & (\sigma \geq 1). \end{cases} \tag{6.69}$$

From Eq. (6.69), for a given nonzero value of σ_x, the MSE distortion for $\sigma \leq 1$ is a quadratic function of σ and has a minimum of

$$\text{MSE}_o = \frac{\sigma_x^2(2N - 1)}{12N^2}, \tag{6.70}$$

at a minimizing σ of $\sigma_o = (N - 1)/N$. It is clear from Eq. (6.70) that the MSE will decrease with increasing N.

For $\sigma \geq 1$, it is straightforward to show that the MSE of Eq. (6.69) is strictly increasing. Hence, the MSE is minimized for a nonzero value of σ; just as in Chapter 4, we showed that the mutual information is maximized for a nonzero value of σ. Here, σ_o is independent of the size of the signal variance; however, the minimum distortion is a function of σ_x^2, and therefore the signal variance.

An alternative constant decoding is one that sets the maximum and minimum reconstruction points slightly smaller than the maximum and minimum possible signal values. Such a decoding provides the same reconstruction points that are optimal for an optimally quantized uniform source, when no input noise is present. This situation will be discussed in more detail in Chapter 8, but for now we simply let $c = \frac{\sigma_x N}{2(N+1)}$, and derive the MSE distortion for this decoding as

$$\text{MSE} = \begin{cases} c^2 \left(\frac{3N - 2\sigma(N-1)}{3N} \right) + c\sigma_x \left(\frac{\sigma^2 - 3}{6} \right) + \frac{\sigma_x^2}{12} & \sigma \leq 1 \\ \frac{4c^2(N-1+3\sigma^2) - 4cN\sigma_x\sigma + \sigma^2\sigma_x^2 N}{12N\sigma^2} & \sigma \geq 1. \end{cases} \tag{6.71}$$

This has a minimum at $\sigma_o = 2(N-1)c/(N\sigma_x) = (N-1)/(N+1)$, which gives a minimum MSE distortion at this value of

$$\text{MSE}_o = \frac{\sigma_x^2}{12(N+1)^2} \left(\frac{2N^2 - N + 1}{N+1} \right). \tag{6.72}$$

Again, the MSE is minimized for a nonzero value of σ, while the MSE will get smaller with increasing N.

Figure 6.4 shows the MSE distortion for the two constant decoding schemes given by Eqs (6.69) and (6.71), with $\sigma_x = 1$. It is clear that the second decoding gives a smaller MSE, although, as N increases, the difference between the MSE distortion for the two decoding schemes becomes smaller. It is also clear from this figure that the optimal value of σ gets closer to unity as N increases, while the MSE distortion becomes closer to zero at this value of σ, as the theory predicted.

Gaussian signal and noise

Choosing an appropriate value of c is more problematical for signals with PDFs with infinite tails, such as a Gaussian signal. One possible method is to select c such that the maximum and minimum output values are equal to a certain number of standard deviations of the input signal. Hence, let $c = k\sigma_x$, where k is the number of standard deviations desired. Substitution from Eqs (6.58) and (6.51) into Eq. (6.45) gives

$$\text{MSE} = \sigma_x^2 \left(\frac{k^2}{N} + \frac{2(N-1)}{\pi N} \arcsin(\rho_i) - \frac{2\sqrt{2}k}{\sqrt{\pi}} \sqrt{\rho_i + 1} \right). \tag{6.73}$$

How to choose k? We might desire the maximum and minimum values of the output to be about those of the input. In this case, the temptation is to set k to 3 for example. However, the variance of the output will not in general be the same as that of the input. For example, for small σ the most probable output states will be those close to zero or N. Hence, if the output is decoded with $c = 3\sigma_x$, the output will very often be near $\pm 3\sigma_x$, whereas the input is not often near that value. If k is to remain fixed for all σ, the best solution appears to be to find the value of k that minimizes the distortion as a function of σ. This value of k can be used to find the value of σ that minimizes the distortion. If that k is used for all σ, the end result will be a function of N.

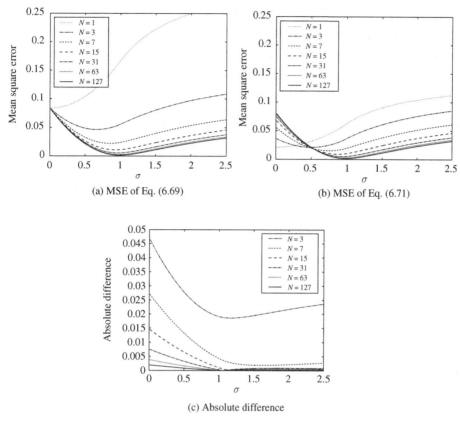

(a) MSE of Eq. (6.69)   (b) MSE of Eq. (6.71)

(c) Absolute difference

Fig. 6.4. MSE distortion for uniform signal and noise, and the two constant decoding schemes given by Eqs (6.69) and (6.71), with $\sigma_x = 1$. The optimal value of σ is nonzero, which means that SR occurs, for $N > 1$, as expected. As N increases, the MSE distortion at the optimal σ gets closer to zero, and the optimal value of σ gets closer to unity. Figure 6.4(c) shows that the absolute difference between the MSE of each decoding becomes smaller as N increases. Note that if $\sigma_x \neq 1$, the actual values of the reconstruction points scale in proportion. This is also the case for all similar plots in this chapter.

However, as we shall see shortly, this procedure is very close to a method for finding the optimal linear decoding scheme as a function of σ. So, instead, we point out only that the behaviour of the MSE distortion, as σ varies for a fixed k, will strongly depend on the actual value of k. Due to this fact, we next concentrate on decoding schemes that are not constant for all σ.

Matched moments linear decoding

The previous paragraphs consider only decoding schemes that have reproduction points that are constant for all σ. Section 6.3 derives an expression for a linear decoding that ensures the first two moments of the decoded output match the first two moments of the input distribution. This decoding is a function of σ, and the value of c is given by Eq. (6.34). The MSE distortion is a function of the linear correlation coefficient, as given by Eq. (6.35). Given these expressions, and the expressions for $E[P_{1|x}^2]$ and $E[x P_{1|x}]$ derived earlier for matched Gaussian, uniform, and Laplacian signal and noise, the MSE distortion and the decoding, $\hat{y}(n)$, can be calculated exactly. For the logistic case, the MSE distortion and $\hat{y}(n)$ can be calculated numerically.

Figure 6.5 shows the MSE distortion for matched-moments, decoding for each of the four matched signal and noise distributions considered in this chapter, with $\sigma_x = 1$. Note that for the uniform case, the MSE distortion is far smaller than for the other cases. This is due to the fact that the variance of a uniform signal that

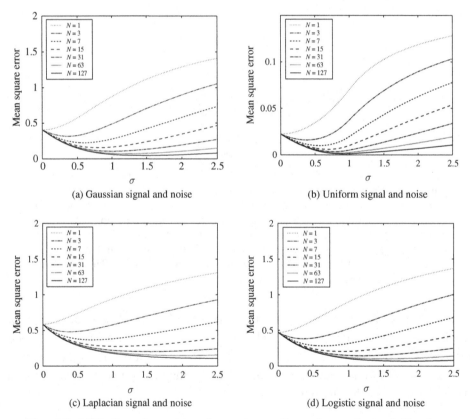

(a) Gaussian signal and noise

(b) Uniform signal and noise

(c) Laplacian signal and noise

(d) Logistic signal and noise

Fig. 6.5. MSE distortion against increasing σ for various values of N using matched-moments linear MSE decoding, with $\sigma_x = 1$. The optimal value of σ is nonzero, which means that SR occurs, as expected. As N increases, the MSE distortion at the optimal σ gets closer to zero, and the optimal value of σ increases, other than for the uniform case, to a value larger than unity.

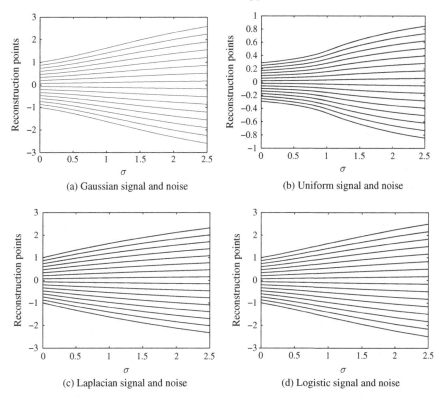

Fig. 6.6. Reconstruction points, $\hat{y}(n)$, against increasing σ, for $N = 15$ and $\sigma_x = 1$, using matched-moments linear MSE decoding. The qualitative behaviour is the same for each distribution; the difference between the maximum and minimum reconstruction points increases with increasing σ. Note that if $\sigma_x \neq 1$, the actual values of the reconstruction points scale in proportion. This is also the case for all similar plots in this chapter.

has support between $\pm\sigma_x/2$ is $\mathrm{var}[x] = \sigma_x^2/12$, whereas the variance for the other distributions is σ_x^2.

Figure 6.6 shows for each case and $N = 15$ the actual values of the decoded output, $\hat{y}(n)$, as a function of σ, with $\sigma_x = 1$. Again, the maximum and minimum values of $\hat{y}(n)$ are much smaller for the uniform case, for the same reasons as the MSE distortion. It can be seen that the reconstruction points are evenly spaced for any given value of σ.

Wiener optimal linear decoding

Section 6.3 derives an expression that minimizes the MSE distortion for a linear decoding. This decoding is a function of σ, and the value of c is derived in Eq. (6.37). As shown by Eq. (6.40), the MSE distortion is a function of the linear correlation coefficient. As with matched-moments decoding, given these expressions, and the expressions for $E[P_{1|x}^2]$ and $E[x\,P_{1|x}]$ derived earlier for matched Gaussian, uniform, and Laplacian signal and noise, the MSE distortion and the decoding, $\hat{y}(n)$, can be calculated exactly. For the logistic case, the MSE distortion and $\hat{y}(n)$ can be calculated numerically.

Figure 6.7 shows the optimal linear MSE distortion for each of the four matched signal and noise distributions considered in this chapter and $\sigma_x = 1$. Figure 6.8 shows for each case and $N = 15$ the actual values of the decoded output, $\hat{y}(n)$, as a function of σ. It can be seen that the reconstruction points are evenly spaced for any given value of σ.

As shown by Inequality (6.41), Wiener decoding always gives a smaller MSE distortion than matched-moments decoding. Furthermore, Wiener decoding was

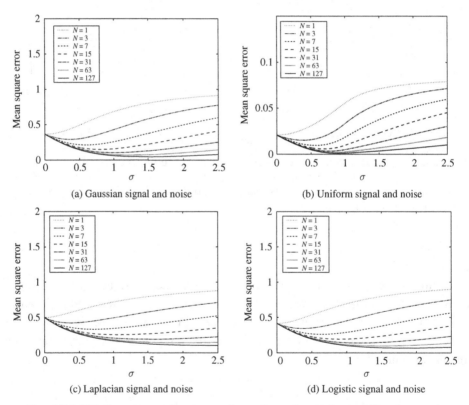

Fig. 6.7. MSE distortion against increasing σ for various values of N and $\sigma_x = 1$, using the optimal linear, or Wiener, MSE decoding. The optimal value of σ is nonzero, and thus SR occurs, as expected. As N increases, the MSE distortion at the optimal σ gets closer to zero, and the optimal value of σ increases, other than for the uniform case, to a value larger than unity.

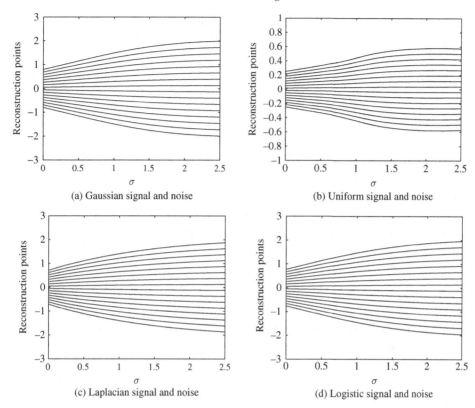

Fig. 6.8. Reconstruction points against increasing σ for $N = 15$ and $\sigma_x = 1$, using the optimal linear, or Wiener, MSE decoding. The qualitative behaviour is the same for each distribution; the difference between the maximum and minimum reconstruction points increases with increasing σ.

derived as the decoding that gives the minimum MSE distortion of all linear decoding schemes. However, as mentioned earlier, it is possible to obtain smaller MSE distortions by using nonlinear decoding schemes, that is, reconstruction points that are not a linear function of n, and are therefore not evenly spaced. This is the focus of the next section.

6.5 Nonlinear decoding schemes

Unlike the linear decoding schemes of Sections 6.3 and 6.4, the use of a nonlinear decoding scheme produces a reconstruction of the input signal with reconstruction points that are not equally spaced. We shall see that such a nonlinear scheme can provide a smaller MSE distortion than any linear scheme.

This section provides the theory behind three commonly used methods of using nonlinear decoding to estimate a signal from a nondeterministic measurement of it.

The applicability of these methods depends on whether knowledge of various probability distributions is available. One of the probability distributions required is the *backward conditional probability distribution*, and we first define this distribution, before considering each method, and their application to the decoding of the SSR model.

Backward conditional probabilities

Recall that the key distribution for calculating the mutual information or MSE distortion is the transition probabilities, $P_{y|x}(n|x)$ where $n = 0, \ldots, N$. This section will also use the *backward conditional probability distribution* (BCPD), $P_{x|y}(x|n)$. The BCPD is also known as the *a posteriori* distribution. The following formula gives the relationship between this distribution and the input and output distributions

$$f_x(x) = \sum_{n=0}^{N} P_{x|y}(x|n) P_y(n). \qquad (6.74)$$

In general, the only means available for calculating the BCPD is to use Eqs (4.3) and (4.4) from Section 4.3 in Chapter 4, after first calculating the transition probabilities. That is

$$P_{x|y}(x|n) = \frac{f_{xy}(x, y)}{P_y(n)} = \frac{f_x(x) P_{y|x}(n|x)}{P_y(n)} \qquad (6.75)$$

$$= \frac{f_x(x) P_{y|x}(n|x)}{\int_{-\infty}^{\infty} P_{y|x}(n|x) f_x(x) dx}. \qquad (6.76)$$

MAP decoding

The *maximum a posteriori* (MAP) criterion (Yates and Goodman 2005) is often used in estimation applications. Although it does not necessarily provide the optimal estimation, it can be convenient to use and often provides MSE distortions close to optimal.

The MAP criterion states that an estimate x, given observation $y = n$, should be chosen so that the estimate is

$$\hat{y}_n = \arg\max_x P_{x|y}(x|n). \qquad (6.77)$$

Given that the *a posteriori* distribution, $P_{x|y}(x|n)$, is given by Eq. (6.76), this criterion can be simplified to

$$\hat{y}_n = \arg\max_x f_x(x) P_{y|x}(n|x), \qquad (6.78)$$

that is, the MAP estimator, given observation $y = n$, is the value of x that maximizes the joint probability density function of x and y for $y = n$.

For SSR, $P_{y|x}(n|x)$ is given by the binomial formula as in Eq. (4.9), repeated here as

$$P_{y|x}(n|x) = \binom{N}{n} P_{1|x}^n (1 - P_{1|x})^{N-n} \quad n = 0, \ldots, N. \tag{6.79}$$

Finding the value of x that maximizes the joint distribution can be achieved by differentiating $f_{xy}(x, y) = P_{y|x}(n|x) f_x(x)$ with respect to x and setting the result to zero. The derivative of $P_{y|x}(n|x)$ with respect to x is given by Eq. (A1.1) in Appendix 1. Applying this formula leads to the following criterion for a stationary point of the joint density function, $f_{xy}(x, y)$, with respect to x

$$\frac{d}{dx} \left(f_x(x) P_{y|x}(n|x) \right) = P'(x) P_{y|x}(n|x)$$

$$+ f_x(x) P_{y|x}(n|x) \left(\frac{n - N P_{1|x}}{P_{1|x}(1 - P_{1|x})} \right) f_\eta(x)$$

$$= 0, \quad n = 0, \ldots, N. \tag{6.80}$$

For $f_x(x)$ and $f_\eta(x)$ with infinite tails, $P_{y|x}(n|x)$ cannot be zero and Eq. (6.80) reduces to

$$P'(x) + f_x(x) \left(\frac{n - N P_{1|x}}{P_{1|x}(1 - P_{1|x})} \right) f_\eta(x) = 0, \quad n = 0, \ldots, N. \tag{6.81}$$

In general there is no closed form solution to Eq. (6.81), although it is possible to simplify it for a specified $f_x(x)$ and solve numerically.

However, in practice, the simplest way to find the MAP decoding for SSR with given signal and noise distributions is to numerically calculate $f_{xy}(x, y) = f_x(x) P_{y|x}(n|x)$, and find the value of x for which $f_{xy}(x, y)$ is a maximum for each n. This value of x is the MAP decoding for SSR for each n. The result of this procedure for varying σ and N is shown in Fig. 6.9, which shows the MSE distortion, and Fig. 6.10, which shows the reconstruction points for $N = 15$ for the cases of Gaussian, Laplacian, and logistic signal and noise with $\sigma_x = 1$. The uniform case is not considered, as there is not always a unique maximum of the joint distribution.

Note that for the Laplacian case and $N = 1$, in Fig. 6.9(b), the MSE distortion saturates at unity for $\sigma \geq 1$. This is due to a general property of the MSE distortion, that a decoding can always be chosen such that MSE $= E[x^2]$, which in this case is unity. The decoding that achieves this is to set $\hat{y}_n = E[x] \ \forall \ n$; when $\sigma \geq 1$, the MAP criterion sets the decoding of both output states to be zero, so that the output is *always* zero. Hence, the MSE distortion is identical to the mean square value of the input signal. This is the only case we shall see where this happens, but note that

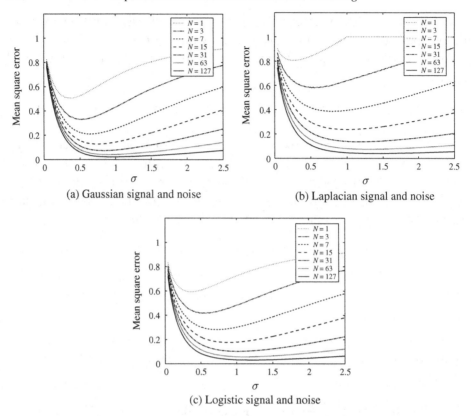

Fig. 6.9. MSE distortion against increasing noise intensity, σ, for various values of array size, N, and signal standard deviation set to $\sigma_x = 1$, using the maximum *a posteriori* decoding. Note that for the Laplacian case and $N = 1$, the MSE distortion saturates at unity for $\sigma \geq 1$. This is because the MAP criterion sets both output reproduction points to the signal mean of zero, so that the decoded output is always zero, meaning the MSE distortion is the mean square value of the input, that is, unity.

for σ sufficiently large, this situation will happen regardless of N and the signal and noise distributions.

Maximum likelihood decoding

In contrast with the MAP criterion, the maximum likelihood (ML) estimator (Yates and Goodman 2005) does not directly use the *a priori* distribution of the signal to be estimated, that is $f_x(x)$. The ML estimator for x given observation $y = n$ is given by

$$\hat{y}_n = \arg \max_x P_{y|x}(n|x), \tag{6.82}$$

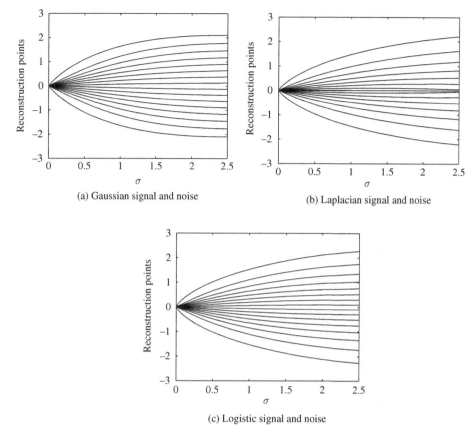

(a) Gaussian signal and noise

(b) Laplacian signal and noise

(c) Logistic signal and noise

Fig. 6.10. Reconstruction points against increasing noise intensity, σ, for $N = 15$, and the signal standard deviation set to $\sigma_x = 1$, using the maximum *a posteriori* decoding.

that is, the ML estimator, given observation $y = n$, is the value of x that maximizes the probability mass function of y given x.

As with the MAP estimator, the ML estimator cannot be shown to minimize the MSE distortion, but in practice often provides a fairly good approximation to it.

The values of the ML reconstruction points for SSR can be obtained by differentiating the transition probabilities, $P_{y|x}(n|x)$, with respect to x and setting the result to zero. The result, as given in Appendix 1 by Eq. (A1.1), is

$$\frac{n - N P_{1|x}}{P_{1|x}(1 - P_{1|x})} = 0, \quad n = 0, \ldots, N. \tag{6.83}$$

This reduces to $x = F_\eta^{-1}(n/N)$, and therefore the ML reconstruction points are

$$\hat{y}_n = F_\eta^{-1}\left(\frac{n}{N}\right), \quad n = 0, \ldots, N. \tag{6.84}$$

For $n = N$ and $n = 0$, the transition probabilities are maximized at $\pm\infty$ so that $\hat{y}_N = \infty$ and $\hat{y}_0 = -\infty$. Clearly this is not satisfactory, since for small σ, the largest output probabilities are $P_y(0)$ and $P_y(N)$. There is no entirely satisfactory way around this problem, although one possibility is to define a decoding which approximates the ML decoding for large N

$$\hat{y}_n = F_\eta^{-1}\left(\frac{n+1}{N+2}\right), \quad n = 0, \ldots, N. \tag{6.85}$$

Although this method does not give the ML reconstruction points, as N gets larger, the reconstruction points get closer and closer to the ML points.

A numerical approach could also provide an approximation to the ML solution. For example, if the reproduction points of Eq. (6.84) for $1 \leq N \leq N - 1$ are used, finite points for $n = 0$ and $n = N$ can be found by assuming $\hat{y}_0 = -\hat{y}_N$ and then varying $\hat{y}_N$ from $\hat{y}_{N-1}$ towards infinity and calculating the MSE distortion for each value of $\hat{y}_N$. The value of $\hat{y}_N$ that provides the minimum distortion is then used. However, this will turn out to be the optimal reproduction point for $n = 0$ and $n = N$, as discussed below. Due to these difficulties, we do not present results for a ML decoding in this chapter. However, Chapter 7, which discusses decoding for large N will show that, in certain circumstances, the optimal nonlinear decoding approaches the ML decoding for large N.

Minimum mean square error distortion decoding

We label the signal that results from an optimal MSE decoding as $\hat{x}$. As shown in Section A3.5 of Appendix 3, the decoding that gives the minimum possible MSE distortion is the decoding that consist of values

$$\hat{x}_n = E_x[x|n] = \int_x x P_{x|y}(x|n)dx = \frac{1}{P_y(n)}\int_x x P_{y|x}(n|x) f_x(x)dx, \quad n = 0, \ldots, N. \tag{6.86}$$

Like the MAP and ML decoding schemes, and unlike the linear decoding schemes considered earlier, this decoding varies for each value of the encoding, y. Since this is the theoretical best MSE distortion decoding, we call the resultant distortion the minimum mean square error (MMSE). Section A3.5 of Appendix 3 also shows that with this decoding, $E[x\hat{x}] = E[\hat{x}^2]$ and that, therefore, by inspecting Eq. (6.7), the MMSE is given by

$$\text{MMSE} = E[x^2] - E[\hat{x}_n^2]$$

$$= E[x^2] - \sum_{n=0}^{N} E[x|n]^2 P_y(n)$$

$$= E[x^2] - \sum_{n=0}^{N} \frac{1}{P_y(n)} \left(\int_x x P_{y|x}(n|x) f_x(x) dx \right)^2. \qquad (6.87)$$

Good references that discuss such a decoding include Gershenfeld (1999) and Yates and Goodman (2005). Given the initial motivation of SSR as a model for a population of neurons, note also that MMSE decoding has also previously been used in computational neuroscience research, for example Bethge *et al.* (2002). Just as we pointed out for the case of the optimal linear decoding, the encoded output, y, is uncorrelated with the error, $\epsilon = x - \hat{x}$ for MMSE decoding. This is proven in Section A3.5 of Appendix 3.

Section A3.5 of Appendix 3 also shows that $E[\hat{x}] = 0$. Hence, the correlation coefficient between the decoded output signal, $\hat{x}$, and the input signal, x, is

$$\rho_{x,\hat{x}} = \frac{E[x\hat{x}] - E[x]E[\hat{x}]}{\sqrt{\text{var}[x]\text{var}[\hat{x}]}}$$

$$= \frac{E[x\hat{x}]}{\sqrt{E[x^2]E[\hat{x}^2]}}$$

$$= \sqrt{\frac{E[\hat{x}^2]}{E[x^2]}}. \qquad (6.88)$$

Rearranging Eq. (6.88) shows that the MMSE can be written in terms of the correlation coefficient as

$$\text{MMSE} = E[x^2] \left(1 - \rho_{x,\hat{x}}^2 \right). \qquad (6.89)$$

Therefore, as with the optimal linear decoding scheme, given by Wiener decoding, the MMSE distortion can be written in terms of the mean square value of the input signal, and the correlation coefficient between the input signal and the decoded output signal. The difference between Eqs (6.89) and (6.40) is that Eq. (6.40) gives the MSE in terms of the correlation coefficient for *linear* decoding, and Eq. (6.89) gives the MSE in terms of the correlation coefficient for optimal *nonlinear* decoding.

Furthermore, as with optimal linear decoding, the SQNR for MMSE decoding can be written in terms of the correlation coefficient as

$$\text{SQNR} = \frac{1}{1 - \rho_{x\hat{x}}^2}. \qquad (6.90)$$

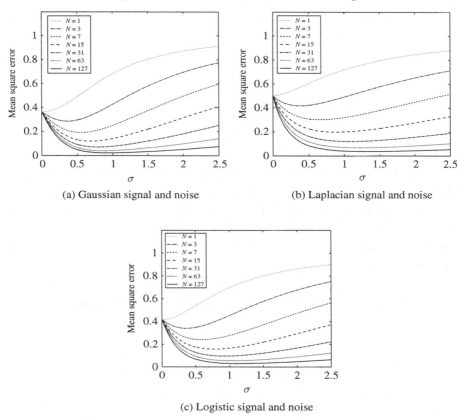

(a) Gaussian signal and noise

(b) Laplacian signal and noise

(c) Logistic signal and noise

Fig. 6.11. MSE distortion against increasing σ for various values of N using MMSE decoding and $\sigma_x = 1$. The optimal value of σ is nonzero, which means that SR occurs, as expected. As N increases, the MSE distortion at the optimal σ gets closer to zero, and the optimal value of σ increases, to a value larger than unity. For increasing $\sigma \geq 1$, the rate of decreases of the MMSE is quite slow, and gets slower for increasing N.

Figure 6.11 shows the MMSE distortion for various N and increasing σ, and Fig. 6.12 shows the reconstruction points against increasing σ for $N=15$, for the cases of Gaussian, Laplacian, and logistic signal and noise, and $\sigma_x = 1$. The case of uniform signal and noise will be considered shortly. As N increases, the MMSE distortion at the optimal σ gets closer to zero, and the optimal value of σ increases, to a value larger than unity. However, for increasing $\sigma \geq 1$, the rate of decrease of the MMSE is quite slow, and gets slower for increasing N.

If the reconstruction points shown in Fig. 6.12 are compared with those of the MAP decoding shown in Fig. 6.10, it can be seen that the points are almost identical, apart from the largest and smallest ones for any given σ, which correspond to $n = 0$ and $n = N$.

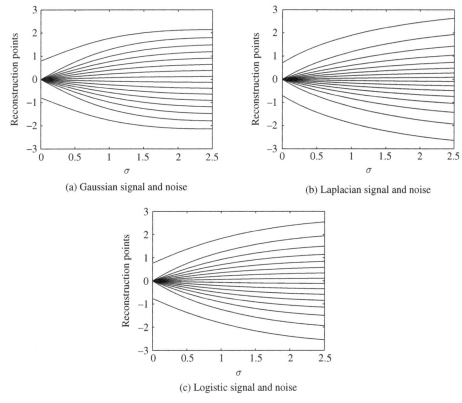

Fig. 6.12. Reconstruction points against increasing σ for various values of N using MMSE decoding and $\sigma_x = 1$. When compared with the reconstruction points obtained with the MAP decoding shown in Fig. 6.10, it can be seen that the points are almost identical, apart from the largest and smallest ones for any given σ, which correspond to $n = 0$ and $n = N$.

Fig. 6.13 shows the optimal correlation coefficient given by Eq. (6.88) for various N, against increasing σ, for Gaussian, Laplacian, and logistic signal and noise. As N increases, the correlation coefficient gets quite close to unity, especially for $\sigma \geq 1$. As with the MMSE distortion, the rate of decrease of the correlation coefficient as σ increases past unity is quite slow, and gets slower with increasing N. A comparison with the correlation coefficient for a linear decoding is presented shortly. First, however, we derive an exact result for MMSE decoding reconstruction points and MMSE distortion for uniform signal and noise, and $\sigma \leq 1$.

Exact MMSE decoding for uniform signal and noise

We now derive exact expressions for the MMSE distortion and reconstruction points for uniform signal and noise, and $\sigma \leq 1$. As discussed in Chapter 4, the output probability mass function, in this case, for $N > 1$, is

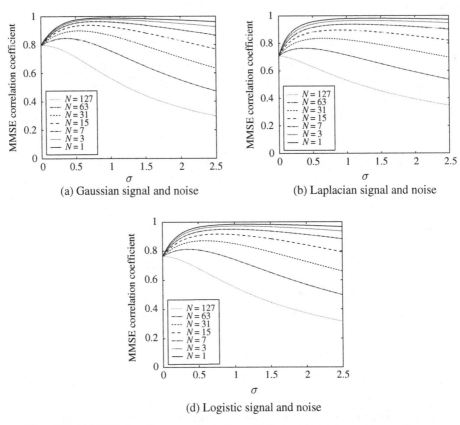

(a) Gaussian signal and noise   (b) Laplacian signal and noise

(d) Logistic signal and noise

Fig. 6.13. MMSE decoding correlation coefficient against increasing σ for various values of N. As N increases, the correlation coefficient gets quite close to unity, especially for $\sigma \geq 1$. As with the MMSE, the rate of decrease of the correlation coefficient as σ increases past unity is quite slow, and gets slower with increasing N.

$$P_y(n) = \frac{\sigma}{N+1}, \quad n = 1, \ldots, N-1 \tag{6.91}$$

and

$$P_y(0) = P_y(N) = \sigma\left(\frac{1}{N+1} - \frac{1}{2}\right) + \frac{1}{2} = \frac{1}{\sigma_x}F_\eta^{-1}\left(\frac{1}{N+1}\right) + \frac{1}{2}. \tag{6.92}$$

Note that if $N = 1$, we have $P_y(0) = P_y(N) = 0.5$.

The optimal reproduction points, $\hat{x}_n$, are given by Eq. (6.86). We define for convenience $A_n = \int_x x P_{y|x}(n|x) f_x(x) dx$, $n = 0, \ldots, N$. Using Eq. (4.9) from Chapter 4, and Eq. (A1.16) from Appendix 1, integration of the rhs of Eq. (6.86) gives

$$A_n = \frac{\sigma_x}{2}\sigma^2\left(\frac{2n-N}{N^2+3N+2}\right), \quad n = 1, \ldots, N-1, \quad N > 1 \qquad (6.93)$$

$$A_N = \int_x x\, P_{y|x}(N|x)\, f_x(x)\, dx = \frac{\sigma_x}{8}\left(1-\sigma^2\left(\frac{N^2-N+2}{N^2+3N+2}\right)\right), \qquad (6.94)$$

and

$$A_0 = \int_x x\, P_{y|x}(0|x)\, f_x(x)\, dx = -\frac{\sigma_x}{8}\left(1-\sigma^2\left(\frac{N^2-N+2}{N^2+3N+2}\right)\right). \qquad (6.95)$$

This gives

$$\hat{x}_n = \frac{\sigma_x}{2}\sigma\left(\frac{2n-N}{N+2}\right) = \frac{\sigma_\eta}{2}\left(\frac{2n-N}{N+2}\right), \quad n = 1, \ldots, N-1, \quad N > 1 \quad (6.96)$$

and

$$\hat{x}_N = -\hat{x}_0 = \frac{\sigma_x}{4}\left(\frac{1-\sigma^2\left(\frac{N^2-N+2}{N^2+3N+2}\right)}{1-\sigma\left(\frac{N-1}{N+1}\right)}\right). \qquad (6.97)$$

So for all n except $n = 0$ and $n = N$, the optimal decoding can be stated independently of either the signal variance or the noise variance, but not both. However, the ratio $\hat{x}/\sigma_x$ is a function of only σ, so the optimal reconstruction points scale proportionally to σ_x.

Rearranging Eq. (6.96) gives

$$\hat{x}_n = \sigma_\eta\left(\frac{n+1}{N+2}-\frac{1}{2}\right) = F_\eta^{-1}\left(\frac{n+1}{N+2}\right), \quad n = 1, \ldots, N-1, \quad N > 1. \qquad (6.98)$$

Thus, except for the cases of $y = 0$ and $y = N$, the optimal decoding is a linear decoding, as given by Eq. (6.18) with $c = \frac{N\sigma_\eta}{2(N+2)}$.

Note that when $\sigma = 1$, Eq. (6.97) reduces to

$$\hat{x}_N = -\hat{x}_0 = \frac{\sigma_x}{2}\left(\frac{N}{N+2}\right) = F_\eta^{-1}\left(\frac{N+1}{N+2}\right). \qquad (6.99)$$

Therefore, when $\sigma = 1$, $\hat{x}_n = F_\eta^{-1}\left(\frac{n+1}{N+2}\right)$ for all n, and MMSE decoding is a linear decoding, as given by Eq. (6.18) with $c = \frac{N\sigma_\eta}{2(N+2)}$. Thus, Wiener linear decoding must give the same MSE performance as MMSE decoding at $\sigma = 1$.

From Eq. (6.87), for $N > 1$ the MMSE distortion is

$$\text{MMSE} = \frac{\sigma_x^2}{12} - E[\hat{x}^2]$$

$$= \frac{\sigma_x^2}{12} - 2P_y(N)\hat{x}_N^2 - \sum_{n=1}^{N-1} P_y(n)\hat{x}_n^2$$

$$= \frac{\sigma_x^2}{12} - 2A_N\hat{x}_N - \sum_{n=1}^{N-1} A_n\hat{x}_n$$

$$= \frac{\sigma_x^2}{12} - 2A_N\hat{x}_N - \sum_{n=1}^{N-1} \frac{\sigma_x}{2}\sigma^2\left(\frac{2n-N}{N^2+3N+2}\right)\frac{\sigma_x}{2}\sigma\left(\frac{2n-N}{N+2}\right)$$

$$= \frac{\sigma_x^2}{12} - 2A_N\hat{x}_N - \frac{\sigma_x^2\sigma^3}{4}\frac{1}{(N+1)(N+2)^2}\sum_{n=1}^{N-1}(2n-N)^2$$

$$= \frac{\sigma_x^2}{12} - 2A_N\hat{x}_N - \frac{\sigma_x^2\sigma^3}{12}\frac{N(N-1)(N-2)}{(N+1)(N+2)^2}$$

$$= \frac{\sigma_x^2}{12}\left(1 - \frac{\frac{3}{4}\left(1 - \sigma^2\left(\frac{N^2-N+2}{N^2+3N+2}\right)\right)^2}{1 - \sigma\left(\frac{N-1}{N+1}\right)} - \sigma^3\frac{N(N-1)(N-2)}{(N+1)(N+2)^2}\right). \quad (6.100)$$

Figure 6.14 shows for uniform signal and noise the MMSE distortion given by Eq. (6.100), the corresponding MMSE correlation coefficient, and the values of the optimal MMSE reconstruction points given by Eqs (6.96) and (6.97), with $E[x^2] = 1/12$. Fig. 6.14 also shows these quantities calculated by numerical integration, to verify the exact results for $\sigma \leq 1$, and to show the continuation to $\sigma > 1$.

6.6 Decoding analysis

The purpose of this section is to compare the performance of all the different decoding schemes considered in this chapter. First, we shall compare the MSE distortion for matched-moments and Wiener linear decoding schemes, and the MAP and MMSE nonlinear decoding schemes, for the case of $N = 127$ and Gaussian, Laplacian, and logistic signal and noise. For the uniform case, we shall compare the MSE distortion of matched-moments and Wiener linear decoding, and MMSE nonlinear decoding for $N = 127$. In all cases we plot for $\sigma_x = 1$. We shall also

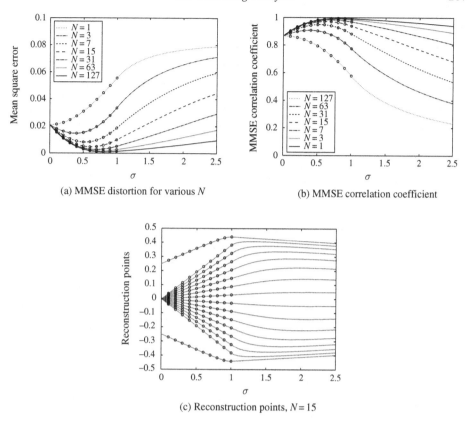

(a) MMSE distortion for various N

(b) MMSE correlation coefficient

(c) Reconstruction points, $N = 15$

Fig. 6.14. MMSE distortion, correlation coefficient, and reconstruction points for $N = 15$, for uniform signal and noise with $\sigma_x = 1$. The circles show the values calculated from the exact expressions given for $\sigma \leq 1$ in this section, and the lines show numerical values. The optimal value of σ is nonzero, and gets closer to unity as N increases.

compare the linear correlation coefficient with the nonlinear correlation coefficient obtained with MMSE decoding.

Secondly, we briefly look at the *average transfer function* of the decoded SSR model, for $N = 7$ and Gaussian signal and noise.

Distortion comparison

Fig. 6.15 shows the MSE distortion plotted against increasing σ, for the four decoding schemes for the cases of Gaussian, Laplacian, and logistic signal and noise. First, as expected, Wiener linear decoding gives a smaller MSE distortion for all σ than matched-moments decoding. The difference between the MSE distortion for these two decoding schemes is expressed by Eq. (6.41). Secondly, the optimal MSE distortion, obtained by MMSE decoding, is clearly verified as giving a smaller MSE distortion than the other decoding schemes.

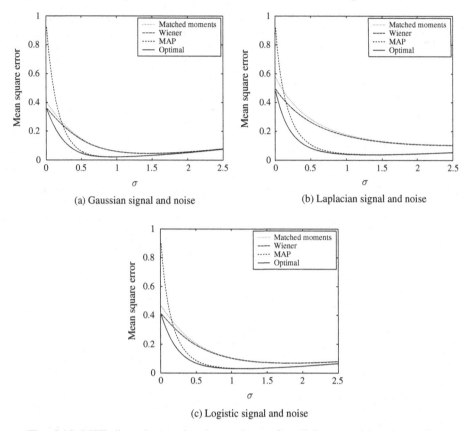

(a) Gaussian signal and noise

(b) Laplacian signal and noise

(c) Logistic signal and noise

Fig. 6.15. MSE distortion against increasing σ for all four considered decoding schemes, for $N = 127$ with $\sigma_x = 1$. This figure verifies that Wiener decoding is superior to matched-moments decoding, and that MMSE decoding is superior to all decoding schemes. It is also apparent that the MAP decoding is quite poor for small σ, but for larger σ has a performance very close to that of the optimal MMSE decoding. There are no remarkable differences between each of the signal and noise pairs.

Thirdly, MAP decoding can be seen to have quite poor distortion performance for small σ. However, for σ greater than about $0.1 - 0.2$, the MAP decoding MSE distortion is smaller than the MSE distortion for the linear decoding schemes and, although not visible in Fig. 6.15, actually gets closer and closer to the MMSE distortion as σ increases. This verifies the known empirical but as yet unproven fact that MAP decoding often provides a MSE distortion quite close to optimal.

Fig. 6.16 shows the linear and MMSE nonlinear correlation coefficients plotted against increasing σ. Clearly, the nonlinear correlation coefficient is always greater than the linear correlation coefficient. Interestingly, the difference between the two

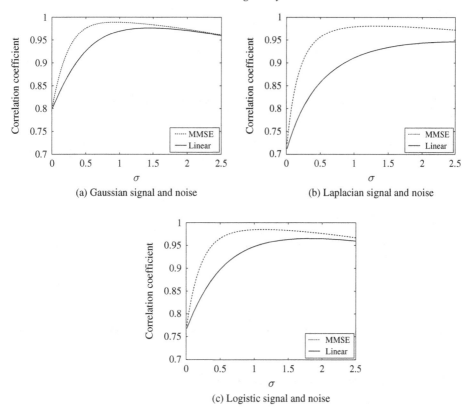

(a) Gaussian signal and noise

(b) Laplacian signal and noise

(c) Logistic signal and noise

Fig. 6.16. Comparison of the linear correlation coefficient with the MMSE decoding correlation coefficient, against increasing σ, for $N = 127$. This figure verifies that the correlation coefficient is larger for the nonlinear MMSE decoding than it is for any linear decoding, for all σ.

coefficients gets quite small for $\sigma > 2$ for the Gaussian case. The gap between each is far larger for the Laplacian and logistic cases.

Figure 6.17 compares the MSE distortion and correlation coefficient for uniform signal and noise and $N = 127$. Again, it is clear that MMSE decoding is optimal, and Wiener decoding is better than matched-moments decoding. However, as predicted in Section 6.5, for $\sigma = 1$, the performance of Wiener decoding and MMSE decoding is identical, since, at this point, MMSE decoding is a linear function of n.

Recall the definition of SQNR given by Eq. (6.9). Figures 6.18(a)–6.18(c) show the SQNR plotted in dB against increasing σ for $N = 127$. As we would expect, the SQNR is maximized for the same value of σ that minimizes the MSE distortion.

It is also possible to define an input SNR measure for the SSR model. If the input signal variance is σ_x^2 and the noise signal variance is σ_η^2, the input SNR in decibels is

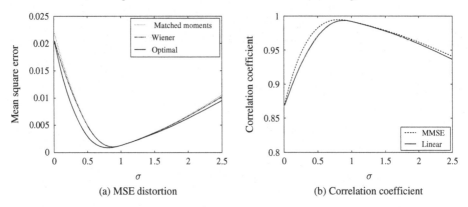

(a) MSE distortion   (b) Correlation coefficient

Fig. 6.17. MSE distortion and correlation coefficient against increasing σ for uniform signal and noise and $N = 127$. This figure verifies that Wiener decoding is superior to matched-moments decoding, and that MMSE decoding is superior to Wiener decoding for all σ. However, it can be seen that for $\sigma = 1$, the MMSE and Wiener decoding schemes have the same performance.

$$\text{SNR}_i = 10 \log_{10} \left(\frac{\sigma_x^2}{\sigma_\eta^2} \right) = -20 \log_{10} (\sigma) \quad \text{dB}. \tag{6.101}$$

Figures 6.18(d)–6.18(f) show the output SQNR in dB plotted against the input SNR in dB. Since the MSE distortion is minimized for σ near unity, this corresponds to an input SNR around zero dB. This is clearly illustrated in these subfigures.

Transfer functions

Recall that in Section 4.5 of Chapter 4 we plot the *average* transfer function of the SSR model's encoding, y, which is the expected value of y given x. We now plot the average transfer function after decoding of the SSR model.

Figure 6.19(a) shows the *average* transfer function for Wiener linear decoding, with $N = 7$, for various values of noise standard deviation, σ_η, and Gaussian noise. Figure 6.19(b) shows the average transfer function for MMSE nonlinear decoding. Also shown for comparison, with thick black lines, is the optimal deterministic transfer function for a three-bit quantization of a Gaussian source. This plot was obtained using the Lloyd Method I algorithm – see Section 8.5 in Chapter 8 for more details. Notice that the average transfer function of each decoding scheme can be seen to be almost identical; however, the maximum and minimum values of the average output are smaller in magnitude than for the optimal noiseless quantizer.

Figure 6.19(c) shows the variance corresponding to each value of σ_η as a function of x, for Wiener decoding, and Fig. 6.19(d) shows the variance for MMSE

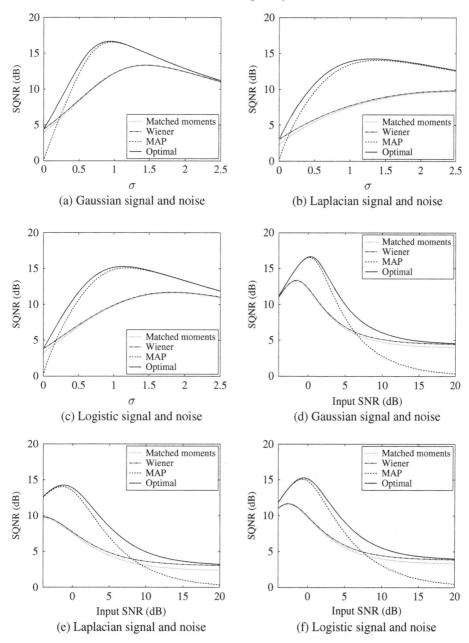

Fig. 6.18. SQNR in dB against increasing σ and the SQNR against increasing input SNR in dB for all four considered decoding schemes, for $N = 127$. As is obviously the case from its definition, the SQNR is maximized for the same value of σ that minimizes the MSE distortion, which when expressed as the input SNR is near zero dB.

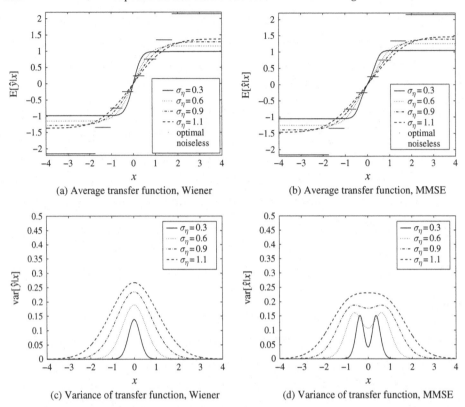

(a) Average transfer function, Wiener

(b) Average transfer function, MMSE

(c) Variance of transfer function, Wiener

(d) Variance of transfer function, MMSE

Fig. 6.19. Decoding transfer functions and variance. Figure 6.19(a) shows the *average* transfer function for Wiener linear decoding, with $N = 7$, for various values of noise standard deviation, σ_η, and Gaussian noise. Figure 6.19(b) shows the average transfer function for MMSE nonlinear decoding. Also shown for comparison with thick black lines is the optimal deterministic transfer function for a three-bit quantization of a Gaussian source. Figure 6.19(c) shows the variance corresponding to each value of σ_η as a function of x, for Wiener decoding; and Fig. 6.19(d) shows the variance for MMSE decoding. The average transfer function of each decoding scheme can be seen to be almost identical; however, the variance can be seen to be smaller for more values of x, particularly for values of x close to the signal mean.

decoding. This variance can be seen to be smaller for more values of x for MMSE decoding, particularly for values of x close to the signal mean.

This section and the previous two sections examined and compared various decoding schemes and the corresponding MSE distortions and correlation coefficients for SSR. The optimal decoding scheme was discussed, and verified to provide a smaller MSE distortion than any other decoding scheme. The next section investigates relationships between this optimal decoding scheme, and various ideas from estimation theory, including Fisher information, and the concept of bias.

6.7 An estimation perspective

This section examines the encoding and decoding of an input signal in the SSR model from the point of view of estimation theory. Before proceeding to some results, we first introduce the concepts of *estimation*, *bias*, *Fisher information*, the *Cramer–Rao bound*, and the *information bound*, and then show how they apply to the SSR model. The main result obtained is a method for calculating a lower bound on the MSE distortion for any given decoding of the SSR model.

Definitions

Estimators

Suppose we can make a measurement, z, of some scalar parameter, θ. If there is some random or systematic error in the measurement, then z is a point *estimate* of the parameter, θ (Lehmann and Casella 1998). The error of any given measurement is $\epsilon = z - \theta$.

Bias of an estimator

The *bias*, $b_z(\theta)$, of an estimator, z, for a parameter, θ, is the expected value of the error, ϵ, of the estimator. Suppose z is discretely valued, and has a probability mass function for a given θ of $P_z(z|\theta)$. Then the bias is

$$b_z(\theta) = \mathrm{E}[\epsilon|\theta]$$
$$= \mathrm{E}[z|\theta] - \theta$$
$$= \sum_z z P_z(z|\theta) - \theta. \tag{6.102}$$

An *unbiased estimator* is an estimator for which the bias is zero; that is, $\mathrm{E}[z|\theta] = \theta$ (Cover and Thomas 1991).

If the parameter, θ, is also a continuous random variable, with a PDF given by $P_\theta(\theta)$, the mean square value of the bias is

$$\mathrm{E}[b_z^2(\theta)] = \int_\theta b_z^2(\theta) P_\theta(\theta) d\theta. \tag{6.103}$$

Fisher information

Consider a random variable Z that estimates a parameter θ. Let $P_{Z|\theta}(Z = z|\theta)$ be the conditional probability density function of the random variable Z given measurement θ. The *Fisher information* is defined as the variance of the gradient of the log-likelihood function with respect to z (Cover and Thomas 1991). The gradient of the log-likelihood function is often called the *score function*

$$U(Z, \theta) = \frac{d \ln P_{Z|\theta}(z|\theta)}{d\theta} \tag{6.104}$$

and so the Fisher information is

$$J(\theta) = \text{var}[U(Z, \theta)] = \text{var} \left[\frac{d \ln P_{Z|\theta}(z|\theta)}{d\theta} \right]. \tag{6.105}$$

It can be seen that the Fisher information is a function of the parameter, θ. From this definition, it is possible to derive an expression for the Fisher information in terms of a second derivative with respect to θ as

$$J(\theta) = \text{E} \left[-\frac{d^2 \ln P_{Z|\theta}(z|\theta)}{d^2\theta} \right]. \tag{6.106}$$

Cramer–Rao bound

Fisher information can be related to the variance of an estimator by the *Cramer–Rao bound* (Cover and Thomas 1991). This bound holds for an unbiased estimator but is easily generalized to biased estimators, as discussed below. The Cramer–Rao bound states that the reciprocal of the Fisher information is a lower bound on the variance of the error of an unbiased estimator and thus gives the smallest possible variance on an unbiased estimator for θ. Note, however, that this smallest possible bound may not be achievable. The Cramer–Rao bound for an unbiased estimator is given by

$$\text{var}[\epsilon|\theta] \geq \frac{1}{J(\theta)}. \tag{6.107}$$

Section A3.7 of Appendix 3 gives a proof of the Cramer–Rao bound.

Efficient estimators

An unbiased estimator is said to be *efficient* if it meets the Cramer–Rao bound with equality.

Information bound for a biased estimator

A result analogous to the Cramer–Rao bound holds for a biased estimator (Cover and Thomas 1991, Lehmann and Casella 1998). A lower bound on the variance of the error of a biased estimator, z, for the parameter, θ, can be expressed in terms of the expected value of the estimate given θ and the Fisher information of θ as

$$\text{var}[\epsilon|\theta] \geq \frac{\left(\frac{d}{d\theta}\text{E}[z|\theta]\right)^2}{J(\theta)}. \tag{6.108}$$

This generalization of the Cramer–Rao bound is sometimes known as the *information bound* or *information inequality* (Lehmann and Casella 1998). Section A3.7 of Appendix 3 gives a proof of the information bound.

Noting that $\text{var}[z|\theta] = \text{var}[z|\theta] - \text{var}[\theta|\theta] = \text{var}[z - \theta|\theta] = \text{var}[\epsilon|\theta]$, inequality (6.108) is equivalent to stating

$$\text{var}[z|\theta] \geq \frac{\left(\frac{d}{d\theta}\text{E}[z|\theta]\right)^2}{J(\theta)}. \tag{6.109}$$

The information bound can be rewritten in terms of the MSE distortion for a given θ as

$$\text{E}[\epsilon^2|\theta] \geq \frac{\left(\frac{d}{d\theta}\text{E}[z|\theta]\right)^2}{J(\theta)} + \text{E}[\epsilon|\theta]^2$$
$$= \frac{(1 + \frac{db_z(\theta)}{d\theta})^2}{J(\theta)} + b_z^2(\theta). \tag{6.110}$$

This version of the information bound is a function of both the bias and the Fisher information. As with Eq. (6.4), the information bound states how a lower bound for the MSE distortion is a tradeoff between the mean square bias, and – via the Fisher information term – a term related to the conditional error variance, regardless of the actual decoding. Note that setting $b_z(\theta) = 0$ reduces inequality (6.110) to the Cramer–Rao bound, and thus the Cramer–Rao bound is a specific case of the more general information bound.

Application to the SSR model

We now apply the above theory to describing signal encoding and decoding in the SSR model.

Bias for the SSR model

Let the output of the SSR model, y, be an estimate for the input. Recalling that $\text{E}[y|x] = N P_{1|x}$, the bias is

$$b_y(x) = \text{E}[\epsilon(x)|x] = \text{E}[y|x] - x = \sum_{n=0}^{N} n P_{y|x}(n|x) - x = N P_{1|x} - x. \tag{6.111}$$

Thus, the SSR encoding, y, can be described as a biased estimator for the input signal, x.

The expected value of the bias, taken over all possible values of the parameter, x, is $\text{E}[b_y(x)] = \text{E}[y] - \text{E}[x] = N/2$. The mean square value of the bias is

$$\text{E}[b_y^2(x)] = \int_x f_x(x)b_y^2(x)dx = \int_x (\text{E}[y|x] - x)^2 f_x(x)dx. \tag{6.112}$$

Fisher information for the SSR model

Again, we consider the output of the SSR model, $y \in \{0, \ldots, N\}$, to be an estimate for the input, x. The score is the random variable

$$U(x, y) = \frac{d}{dx} \ln \left(P_{y|x}(n|x) \right), \quad n = 0, \ldots, N, \tag{6.113}$$

which is a function of both x and y, and is simply the gradient with respect to x of the log-likelihood function for each value of y. It is straightforward to show that the mean over all y of the score for a particular value of x is zero. Therefore the variance of the score for a particular value of x is the mean square value with respect to n. This defines the Fisher information for the SSR encoding

$$J(x) = E[U(x, y)^2]$$

$$= \sum_{n=0}^{N} P_{y|x}(n|x) U(x, n)^2$$

$$= \sum_{n=0}^{N} P_{y|x}(n|x) \left(\frac{d}{dx} \ln \left(P_{y|x}(n|x) \right) \right)^2. \tag{6.114}$$

It is also straightforward to show that the Fisher information can be written as

$$J(x) = E \left[-\frac{d^2 \ln \left(P_{y|x}(n|x) \right)}{dx^2} \right]. \tag{6.115}$$

Simplification of Eq. (6.114) gives

$$J(x) = \sum_{n=0}^{N} \frac{\left(\frac{d P_{y|x}(n|x)}{dx} \right)^2}{P_{y|x}(n|x)}. \tag{6.116}$$

Section A3.6 of Appendix 3 gives two derivations for the Fisher information of the SSR model in terms of $P_{1|x}$, the result being

$$J(x) = \left(\frac{d P_{1|x}}{dx} \right)^2 \frac{N}{P_{1|x}(1 - P_{1|x})}. \tag{6.117}$$

This is in agreement with the formula derived for the Fisher information in Hoch *et al.* (2003a) and Hoch *et al.* (2003b), as discussed in Section 5.4 of Chapter 5.

Eq. (6.117) can be simplified by recalling that for $\theta = 0$, and the noise PDF even about a mean of zero, $P_{1|x} = F_\eta(x)$. Hence

$$\left(\frac{d P_{1|x}}{dx} \right)^2 = f_\eta(x)^2, \tag{6.118}$$

and the Fisher information is

$$J(x) = \frac{N f_\eta(x)^2}{P_{1|x}(1 - P_{1|x})}.$$

(6.119)

It can be seen from Eq. (6.119) that the Fisher information increases linearly with N. The increase in information with N makes sense since, from the estimation point of view, N corresponds to the number of times a noisy measurement is taken of the same parameter x. With more measurements, more information about the parameter is obtained.

SSR encoding

Since the output of the SSR model, y, is a biased estimator for the input, x, the relevant inequality giving a bound on the variance of the error of y as an estimator is the information bound rather than the Cramer–Rao bound. Recalling that $E[y|x] = N P_{1|x}$, we have from inequality (6.108)

$$\text{var}[\epsilon|x] = \text{var}[y|x] \geq N^2 \frac{\left(\frac{dP_{1|x}}{dx}\right)^2}{J(x)} = N^2 \frac{f_\eta(x)^2}{J(x)}.$$

(6.120)

Substituting for the Fisher information given by Eq. (6.119) we have

$$\text{var}[y|x] \geq N^2 f_\eta(x)^2 \frac{P_{1|x}(1 - P_{1|x})}{N f_\eta(x)^2}$$

$$= N P_{1|x}(1 - P_{1|x})$$

$$= \text{var}[y|x],$$

(6.121)

where the last step follows by recalling the fact that $P_{y|x}(n|x)$ is given by Eq. (6.79) and the variance of such a binomial distribution is $N P_{1|x}(1 - P_{1|x})$. Thus, the SSR model provides an estimate for x with a conditional MSE distortion equal to its minimum possible MSE distortion, *given its bias*. Furthermore, the SSR model meets the information bound with equality, and although the definition of *efficient* applies only to unbiased estimators, the SSR encoding, y, can be said to be an *efficient biased estimator*. Discussion of such terminology, and relevant theoretical background, can be found in Lehmann and Casella (1998).

Despite the above result, we know that decoding the SSR model, by assigning reproduction points to its output, results in a smaller conditional MSE distortion for a given x than the raw conditional variance of y. This fact is accounted for once it is noted that, for a decoded output, the expected value of y given x in inequality (6.120) needs to be replaced by the expected value of the decoding, $\hat{y}$, given x. Assuming the decoded output is still a biased estimator, the bound on the conditional MSE distortion is

$$E[\epsilon^2|x] \geq \frac{\left(\frac{d}{dx}E[\hat{y}|x]\right)^2}{J(x)} + E[\epsilon|x]^2$$

$$= \frac{\left(\frac{d}{dx}E[\hat{y}|x]\right)^2}{J(x)} + \left(E[\hat{y}|x] - x\right)^2. \tag{6.122}$$

Note that the Fisher information is independent of the decoding. This can be seen by inspecting its definition, from which it is clear that it is only dependent on the transition probabilities, $P_{y|x}(n|x)$. In addition, from inequality (6.122), the bound on the conditional MSE distortion is entirely dependent on $E[\hat{y}|x]$. We next consider the information bound for both linear and nonlinear decoding of the SSR model.

SSR linear decoding

Bias of linear decoding

Recall that a zero mean linear decoding of the SSR model output can be written as in Eq. (6.18). The bias of this estimate is $b_{\hat{y}}(x) = E[\hat{y}|x] - x = \frac{2c}{N}E[y|x] - c - x$. Recalling that $E[y|x] = N P_{1|x}$, the bias can be written as

$$b_{\hat{y}}(x) = 2c P_{1|x} - c - x. \tag{6.123}$$

This is clearly nonzero in general, which means that $\hat{y}$ is a biased estimator, unless $P_{1|x} = x/2c + 0.5$; that is, $P_{1|x}$ is linear with x. This can occur for the case of uniform signal and noise, since for $\theta = 0$

$$P_{1|x} = \begin{cases} 0 & \text{for } x < \sigma_\eta/2, \\ x/\sigma_\eta + 1/2 & \text{for } -\sigma_\eta/2 \leq x \leq \sigma_\eta/2, \\ 1 & \text{for } x > \sigma_\eta/2. \end{cases} \tag{6.124}$$

In this situation, $\hat{y}$ is an unbiased estimator for all x when $c = \sigma_\eta/2$ and x is in the range $[-\sigma_\eta/2, \sigma_\eta/2]$, so that $\sigma_x \leq \sigma_\eta$. For other noise distributions, the bias may be close to zero for many values of x. For example, for Gaussian noise $P_{1|x}$ is given by the complementary error function, which is known to be approximately linear for values of x near its mean. In general, however, we assume that $\hat{y}$ will be biased, since the information bound holds whether an estimator is biased or unbiased.

Information bound for linear decoding

We now show that if a linearly decoded output is taken as an estimate for x, the estimate is an efficient biased estimator; that is, the estimate gives a MSE distortion that meets the information bound with equality.

Differentiating the bias given by Eq. (6.123) with respect to x and adding one gives

$$1 + b'_{\hat{y}}(x) = 2c\frac{d}{dx}P_{1|x} = 2cf_\eta(x). \tag{6.125}$$

Substituting from Eq. (6.125) into inequality (6.110) gives

$$E[\epsilon^2|x] \geq \frac{4c^2 f_\eta(x)^2}{J(x)} + 4c^2 P_{1|x}^2 - 4cP_{1|x}(c+x) + (c+x)^2. \tag{6.126}$$

Substituting from Eq. (6.119) into inequality (6.126) gives

$$E[\epsilon^2|x] \geq \frac{4c^2 P_{1|x}(1 - P_{1|x})}{N} + 4c^2 P_{1|x}^2 - 4cP_{1|x}(c+x) + (c+x)^2$$

$$= 4c^2 P_{1|x}(1 - P_{1|x})\left(\frac{1-N}{N}\right) - 4cP_{1|x}x + (c+x)^2. \tag{6.127}$$

The final expression is exactly that previously derived for the conditional MSE distortion in Eq. (6.44). Hence, a linearly decoded output of the SSR model, $\hat{y}$, gives an average conditional MSE distortion that is equal to the theoretical limit given by the information bound.

SSR nonlinear decoding

For nonlinear decoding no such result demonstrating biased efficiency, as in the linear case, could be found. However, the bound given by inequality (6.122) can be used to verify numerical calculations of the MSE distortion as follows. Notice that for any given decoding, $\hat{y}$, inequality (6.122) becomes

$$E[\epsilon^2|x] \geq \frac{\left(\frac{d}{dx}E[\hat{y}|x]\right)^2}{J(x)} + \left(E[\hat{y}|x] - x\right)^2. \tag{6.128}$$

Now

$$\frac{d}{dx}\left(E[\hat{y}|x]\right) = \sum_{n=0}^{N}\hat{y}_n\frac{d}{dx}\left(P_{y|x}(n|x)\right)$$

$$= \frac{f_\eta(x)}{P_{1|x}(1 - P_{1|x})}\sum_{n=0}^{N}\hat{y}_n P_{y|x}(n|x)(n - NP_{1|x}). \tag{6.129}$$

Therefore

$$
\frac{\left(\frac{d}{dx}\left(E[\hat{y}|x]\right)\right)^2}{J(x)} = \frac{\left(\sum_{n=0}^{N}\hat{y}_n P_{y|x}(n|x)(n - N P_{1|x})\right)^2}{N P_{1|x}(1 - P_{1|x})}
$$

$$
= \frac{\left(\sum_{n=0}^{N}\hat{y}_n P_{y|x}(n|x)(n - E[y|x])\right)^2}{N P_{1|x}(1 - P_{1|x})}
$$

$$
= \frac{\left(\mathrm{cov}[\hat{y}y|x]\right)^2}{\mathrm{var}[y|x]}
$$

$$
= \rho_{\hat{y}y|x}\mathrm{var}[\hat{y}|x]. \tag{6.130}
$$

For example, if $\hat{y} = y$ the rhs of Eq. (6.130) becomes simply the variance of y given x, and we have the case of the information bound being met with equality, as in Eq. (6.121). Furthermore, for a linear decoding, the correlation coefficient between y and $\hat{y}$ will be unity, and we have the same result as previously, that the information bound is met with equality. Note that Eq. (6.130) implies that

$$
\mathrm{var}[\epsilon|x] \geq \rho_{\hat{y}y|x}\mathrm{var}[\hat{y}|x]. \tag{6.131}
$$

Since $\mathrm{var}[\hat{y}|x] = \mathrm{var}[\epsilon|x]$, this inequality now states $\rho_{\hat{y}y|x} \leq 1$, which is a demonstration of the validity of our arguments, since the correlation coefficient is always smaller than unity. Furthermore, the bound is met with equality only if $\rho_{\hat{y}y|x} = 1$.

Average information bound

The main point of this section is to obtain an expression suitable for numerical calculation of a lower bound on the MSE distortion, and comparing this to the known MMSE distortion.

Taking the expected value of inequality (6.128) with respect to the input PDF, $f_x(x)$, gives a bound on the MSE distortion, in terms of two terms, the first term being the expected value of Eq. (6.130) and the second being the mean square value of the bias. Since we know that the best possible decoding is MMSE decoding, with reconstruction points given by $\hat{x}_n = E[x|n]$, we shall use $\hat{x}$ as the decoding in the rhs of inequality (6.128). Carrying this out gives

$$\text{MSE} \geq \text{E}\left[\frac{\left(\sum_{n=0}^{N} \hat{x}_n P_{y|x}(n|x)(n - N P_{1|x})\right)^2}{N P_{1|x}(1 - P_{1|x})}\right] + \text{E}[b_{\hat{x}}(x)^2]$$

$$= \int_x f_x(x)\left[\frac{\left(\sum_{n=0}^{N} \hat{x}_n P_{y|x}(n|x)(n - N P_{1|x})\right)^2}{N P_{1|x}(1 - P_{1|x})}\right] dx$$

$$+ \int_x f_x(x) b_{\hat{x}}(x)^2 dx. \tag{6.132}$$

The expression on the rhs of inequality (6.132) is a lower bound on the MSE distortion. It is a quantity smaller than, or equal to, the MMSE distortion that results from the optimal nonlinear decoding. Note that it has two components, one which we shall refer to as the *average error variance* component, and the other being the mean square bias. We shall call this expression the *average information bound* (AIB) and denote it as

$$\text{AIB} = \text{E}\left[\frac{\left(\sum_{n=0}^{N} \hat{x}_n P_{y|x}(n|x)(n - N P_{1|x})\right)^2}{N P_{1|x}(1 - P_{1|x})}\right] + \text{E}[b_{\hat{x}}(x)^2]. \tag{6.133}$$

The AIB is plotted from numerical calculations in Fig. 6.20 for the cases of matched Gaussian, Laplacian, and logistic signal and noise. When compared with the MMSE distortion as shown in Fig. 6.11, the MMSE distortion looks to be very close to the AIB. To verify this, Fig. 6.21 shows the percentage difference between the MMSE distortion and the AIB. Although the percentage difference is quite small – always less than 8 percent – it is clear that the AIB is always smaller than the MMSE distortion, as expected. Notice that for $N = 1$, the difference between the MMSE distortion and the AIB is constant for all σ and appears to be very nearly zero.

Fig. 6.21 also shows that as N increases, the peak percentage difference increases with N; however, the rate of increase appears to slow for larger N. For larger σ, the percentage difference is increasing with N for small N, but is decreasing with N for large N. These observations indicate that for sufficiently large N, the MMSE distortion might converge towards the AIB for all σ.

Note that the difference between the MMSE distortion and the AIB is solely due to the difference between the average error variance term – that is, the Fisher information term in the information bound – and the actual average error variance, since the mean square bias term is identical for both the MMSE distortion and the AIB.

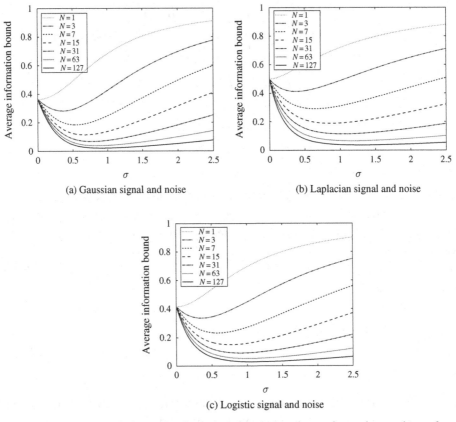

Fig. 6.20. Average information bound against increasing σ for various values of N, and the cases of matched Gaussian, Laplacian, and logistic signal and noise. When compared with the MMSE distortion as shown in Fig. 6.11, the MMSE distortion is very close to the AIB.

Tradeoff between error variance and bias

To illustrate the effects of the two different components of Eq. (6.133), the mean square bias is plotted in Fig. 6.22, and the average error variance in Fig. 6.23. The mean square bias can be seen to decrease with increasing N for all σ. It also decreases with increasing σ for small σ, but reaches a minimum value before increasing again as σ increases. For large N, the increase in the mean square bias with increasing σ greater than the minimizing σ is very slow. By contrast, apart from very small values of N, the average error variance decreases with increasing N but increases with increasing σ.

This behaviour is somewhat analogous to the mutual information for the SSR model, as discussed in Chapter 4. There, we saw that the mutual information has a maximum for nonzero noise, but consists of two components – the output entropy,

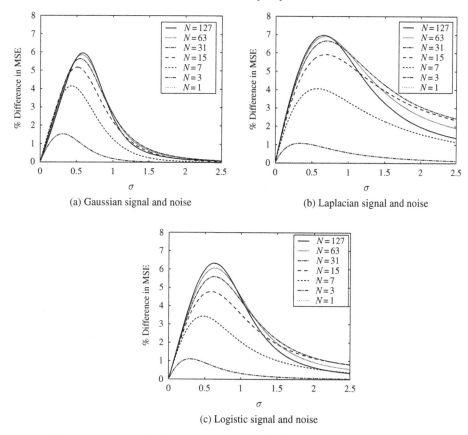

Fig. 6.21. Percentage difference between the MMSE distortion and the average information bound against increasing σ, for various values of N. As N increases, the peak percentage difference increases with N; however, the rate of increase appears to slow as N gets larger. For larger σ, the percentage difference is increasing with N for small N, but is decreasing with N for large N. These observations indicate that, for sufficiently large N, the MMSE might converge towards the AIB for all σ.

and the average conditional output entropy. We saw that the average conditional output entropy always increases with increasing σ, but the output entropy increases for small σ, reaches a maximum value at $\sigma = 1$, and then decreases again with increasing σ. This means that the optimal value of σ for the mutual information occurs when the slope of the output entropy with respect to σ equals the slope of the average conditional output entropy, indicating a tradeoff between output entropy and conditional output entropy. Here we have the minimizing value of σ for the AIB being given when the slope of the average error variance with respect to σ is equal to the negative of the slope of the mean square bias. This, again,

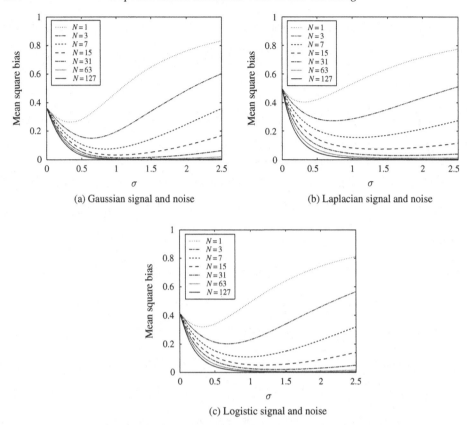

Fig. 6.22. Mean square bias against increasing σ for various values of N. The mean square bias can be seen to decrease with increasing N, for all σ, and have a minimum value at some value of σ near unity. As N increases, the rate of increase of the mean square bias for σ larger than the minimizing σ is very slow.

illustrates that the optimal MSE distortion is obtained for σ that gives the best tradeoff between a small error variance and a small bias.

Also of interest is the fact that the mean square bias seems to get very small for large N and σ larger than unity. This might indicate that for sufficiently large N, the mean square bias is negligible with respect to the average error variance, and the AIB becomes a function of only the average error variance. We shall return such large N behaviour of the MSE distortion in Chapter 7.

We now consider in detail previously proposed output SNR measures for the SSR model.

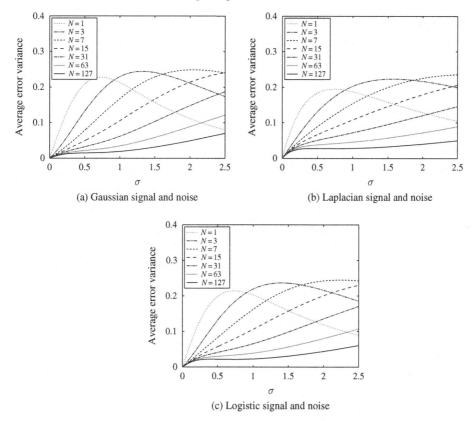

Fig. 6.23. The average error variance term against increasing σ for various values of N. Apart from very small values of N, the average error variance decreases with increasing N, but increases with σ.

6.8 Output signal-to-noise ratio

Previously in this chapter, we have measured the performance of the SSR stochastic quantizer by the SQNR measure. Although SQNR is commonly used in quantization theory, we saw in Chapter 2 that other SNR measures are often used in the SR literature, since often the output is not a quantization of the input, nor is the input signal a sequence of samples drawn from a stationary probability distribution.

However, a constant problem in using SNR to measure a nonlinear system's output performance is the question of how to separate the output signal from output noise. In linear systems, where the signal and noise are uncorrelated, noise power can be easily separated from signal power due to both signals being amplified or attenuated by the same ratio, so that the output signal and noise power sum to the total power. Furthermore, the output noise power can be measured as the power of the output when no signal is present at the input.

In nonlinear systems, the separation of noise from signal at the output of a nonlinear system is not so straightforward, as the input signal can interact with the input noise to yield extra output noise that is not present in the absence of a signal. In other words, if the signal and noise are correlated, their powers do not necessarily sum.

One solution to this problem is to suitably define the noise component of the output power, and then let the signal component be the total output power less the noise power. Alternatively, the signal component of the output power might be defined, and the noise power is the total power less the signal power. If this approach is taken, then the signal and noise powers sum, as carried out in McDonnell *et al.* (2004a). However, if the input signal and output noise are correlated, then does such a definition of an output signal really relate well to the actual input signal?

An approach that avoids this problem is that taken with the SQNR measure, which defines the output signal power as the *input* signal power. The output noise power is considered to be the power of the error signal, that is the variance of the error. Note that with these definitions, the output signal power and output noise power do not add up to the total output signal power, since the signal, x, and the error, ϵ, are correlated for a given value of x.

We are interested here in discussing whether other measures of SNR than SQNR are relevant for information-theoretic or estimation problems. The remainder of this section discusses this question of how to separate signal and noise in a nonlinear system, and argues that previous approaches taken for the SSR model – or equivalent systems analyzed using information theory – are inferior to the SQNR measure, other than in detection scenarios. This is because these measures do not take into account the correlation between output noise and input signal, and the associated fact that a small variance in the output for a given input can also result in a large mean error – that is, bias – for that input, and, conversely, a small mean error may have a large variance.

SNR in the DIMUS model

Recall that the Digital MUltibeam Steering (DIMUS) sonar array is very similar to the SSR model. One approach in the literature to measuring the SNR at the output of a DIMUS array relies on a method for separating signal and noise where the noise is considered to be the output response in the absence of a signal.

Recall that Eq. (6.53) is identical to the expression obtained in Remley (1966) for the output power, P_T, of the DIMUS sonar array for a Gaussian signal and Gaussian noise. The output of the DIMUS array is equivalent to linearly decoding the SSR model with $c = N$. For the output noise power, Remley (1966) uses the output variance when no signal is present. For the SSR model, this is equivalent to letting $\sigma = \infty$, and therefore $\rho_i = 0$ in Eq. (6.53), and the output noise power, $P_N = N$. The output signal power is taken to be $P_S = P_T - P_N$

With these definitions, the output SNR is

$$\text{SNR} = \frac{P_T - P_N}{P_N} = \frac{2(N-1)}{\pi} \arcsin\left(\frac{1}{\sigma^2 + 1}\right), \tag{6.134}$$

which – by realizing that Remley's (1966) input SNR variable, R_{in}, is the input SNR, $1/\sigma^2$ – is identical to Eq. (19) for the SNR in Remley (1966).

Note that Eq. (6.134) is strictly decreasing with increasing σ, and therefore SR does not occur in this SNR measure. Furthermore, when $\sigma = 0$ – that is, in the absence of input noise – the SNR attains its maximum value of $N - 1$. From an information-theoretic perspective, this does not make a lot of sense, as just as with the SSR model, at $\sigma = 0$ the output can only have two states. For DIMUS these are $\pm N$, with equal probability, and hence the total output power is large. This SNR measure does not take into account the fact that the noise power at $\sigma = \infty$ is significantly different from that of small values of σ, due to the need to consider the output noise to be correlated with the input signal, and this correlation varying as a function of σ. We note, however, that this SNR measure is intended to be one that is useful in a detection scenario, rather than an information theoretic one. In a detection scenario it does make sense that the SNR is maximum at $\sigma = 0$, since the optimal way of testing whether a signal is present in the absence of noise is simply to check whether the output of the array is constant or not.

Remley (1966), however, also points out that the case of most interest for DIMUS is when N is large, and R_{in} is small, that is σ is large. For sufficiently large σ, we have

$$\text{SNR} \simeq \frac{2(N-1)}{\pi(\sigma^2 + 1)}. \tag{6.135}$$

As noted by Remley (1966), if N is also large, and $\sigma^2 \gg 1$, $\text{SNR}_0 \simeq \frac{2N}{\pi\sigma^2}$.

Stocks' SNR measure for SSR

The final section of Stocks (2001a) considers the SSR model from an SNR point of view for the first time. The method of output signal and noise separation relies on defining the signal component of the overall output for every possible value of the input and can be described in the following manner. Since the aim is to consider detection performance, rather than how much information about a present signal can be transmitted, Stocks (2001a) does not define a decoding for the output, but simply considers the SNR of the encoded output, y.

For a given value of the input signal, x, the corresponding output is the random variable, $w = y|x$. Stocks (2001a) notes that the ensemble average of w is $E[w] = E[y|x]$, which can be considered to be the signal component of the output for the given input value, x. This conditional ensemble average has an uncertainty, or noise, that can be described by the variance of w, that is $\text{var}[w] = \text{var}[y|x]$.

For the SSR model, the conditional variance of the output is $\text{var}[y|x] = NP_{1|x}(1 - P_{1|x})$. Stocks (2001a) takes this to be the noise power for a given x, and hence the *average* noise power can be defined as the expected value of the *conditional* noise power, that is $\text{E}[\text{var}[y|x]] = \frac{N}{2} - N\text{E}[P_{1|x}^2]$. Note that since $\text{var}[y|x] = \text{var}[\epsilon|x]$, we have already come across this term in this chapter, and labelled it as the average error variance. Recall from Eq. (6.6) that the average error variance is only one component of the MSE distortion, the other component being the mean square bias.

Thus, Stocks (2001a) defines the average error variance as being the output noise power, N_p, and the total output power as the variance of the output, y, which is $\text{var}[y] = N(N - 1)\text{E}[P_{1|x}^2] - \frac{N(N-2)}{4}$. The difference between this total output power and the noise power is considered to be the output *signal* power, S_p.

With these definitions, the average output noise power for the undecoded SSR model is

$$N_p = \text{E}[\text{var}[y|x]] = \frac{N}{2} - N\text{E}[P_{1|x}^2], \tag{6.136}$$

and the average output signal power is

$$S_p = \text{var}[y] - \text{E}[\text{var}[y|x]] = N^2 \left(\text{E}[P_{1|x}^2] - \frac{1}{4} \right) = \frac{N^2 \rho_o}{4}, \tag{6.137}$$

where ρ_o is the correlation coefficient between the output of any two threshold devices, as given by Eq. (A3.20) in Appendix 3.

Hence, the output SNR is

$$\text{SNR}_o = \frac{S_p}{N_p} = \frac{N^2 \left(\text{E}[P_{1|x}^2] - \frac{1}{4} \right)}{\frac{N}{2} - N\text{E}[P_{1|x}^2]} = \frac{N(4\text{E}[P_{1|x}^2] - 1)}{2 - 4\text{E}[P_{1|x}^2]} = \frac{0.5 N \rho_o}{1 - 2\text{E}[P_{1|x}^2]}. \tag{6.138}$$

As shown in Stocks (2001a), unlike for the mutual information – or, as seen in this chapter, the SQNR – the SNR measure of Eq. (6.138) monotonically decreases with σ, which means that no SR effect is seen in this measure. The signal power, S_p, also decreases monotonically with increasing σ if $\theta = 0$, since, as we saw earlier, ρ_o decreases with increasing σ. Stocks (2001a) notes that the reason for this is that, in the noiseless case, the output is deterministic, and hence has zero noise power with these definitions. This implies an infinite output SNR for zero noise. For nonzero noise that is still small, the output is still nearly always in states zero and N, and therefore has little variance for a given x, meaning very little noise power, and a very high output SNR. As the input noise increases, the average output noise power also increases and therefore the output SNR decreases.

It is clear that while this definition of SNR is ideal for a detection scenario, it is deficient for information-theoretic or estimation analyses. Although the ideal situation is for the output to have no variance given the input, the ideal situation for an estimator is also for the bias to be zero, that is $E[w] = E[y|x] = x$. This is clearly not the case when the average conditional variance is minimized, which occurs at $\sigma = 0$. Defining the signal power in this manner ignores the fact that the ideal estimator of the input is one that finds the best tradeoff between bias and variance as illustrated by Eqs. (6.4) and (6.110).

Another problem with this approach is the assumption that the average output noise power and the average output signal power should sum to the total output power. The reason that the output signal and noise power do not sum to the total power is that the input signal and the output noise are correlated.

These problems with using a detection-based SNR for information-theoretic or estimation tasks have also been recognized by Martorell *et al.* (2005), who attempt to overcome them by modifying the noise power definition of Stocks (2001a) to take into account the error between the expected value of the output and some 'desired function,' as well as the variance. An example given by Martorell *et al.* (2005) for a 'desired function,' $q(x)$, is $q(x) = x$, so that the error considered is $E[y|x] - x$, which we recognize as being the *bias* for a given x. If the output is decoded so that it has a mean of zero, the modified noise power formula given by Martorell *et al.* (2005) is exactly the MSE distortion discussed in this chapter. However, the signal power of Martorell *et al.* (2005) is not modified from that used in Stocks (2001a), and the resultant SNR measure is neither that of Stocks (2001a) nor the SQNR. The SNR measure of Martorell *et al.* (2005) could perhaps be modified to allow the output signal power to be the same as the input signal power, and therefore be equivalent to the SQNR in the case where the desired function is x. Therefore, if the desired function is not x – for example, if a nonzero bias is considered ideal – the SNR measure of Martorell *et al.* (2005) would then be a generalization of the SQNR.

One final note is that for Gaussian signal and noise we have $E[P_{1|x}^2]$ given by Eq. (6.50) and therefore the SNR is

$$\text{SNR}_o = \frac{N \arcsin(\rho_i)}{\left(\frac{\pi}{2} - \arcsin(\rho_i)\right)} = \frac{N \arcsin(\rho_i)}{\arccos(\rho_i)}. \tag{6.139}$$

For sufficiently large σ, the SNR measure can be approximated by

$$\text{SNR}_o \simeq \frac{2N}{\pi(\sigma^2 + 1) - 2}. \tag{6.140}$$

Eq. (6.140) is very similar to the DIMUS SNR formula for large σ, given by Eq. (6.135). In fact, for large N and $\sigma \gg 1$, they are identical. This fact indicates that the detection-scenario SNR formulas from DIMUS and Stocks (2001a) may be valid for estimation problems for sufficiently large σ and N. We conjecture that, under these conditions, the output noise becomes almost uncorrelated with the input signal, and therefore the total output power is the sum of the average output noise power and the average output signal power.

Furthermore, recall from Eq. (6.16) that if a continuously valued noisy signal is averaged over N *iid* realizations, where the noise has variance σ_η^2, then the SNR is N/σ_η^2. This means that the scaling of this SNR with N and noise intensity is the same as the scaling in the DIMUS SNR of Eq. (6.135) and Stock's SNR of Eq. (6.140). This indicates that for large N and σ, the SSR model's output behaves like a continuously valued signal.

Gammaitoni's dithering formula

As briefly mentioned in Chapter 2, a similar situation occurs in Gammaitoni (1995a), which proposes a measure of comparing the average output of a dithered quantizer with a linear response. In Gammaitoni (1995a), the ideal output response is considered to be a linear transformation of the input signal. Hence, a measure of the quantizer's performance is given by

$$D_G = \sqrt{\int_0^1 (E[y|x] - x)^2 \, dx,} \qquad (6.141)$$

where a smaller D_G means a better performance.

Eq. (6.141) assumes all values of the input signal should be uniformly weighted. Generalizing Eq. (6.141) to the case considered throughout this chapter of signals being a random sequence of samples drawn from the distribution with PDF, $f_x(x)$, gives

$$D_G = \sqrt{\int_x (E[y|x] - x)^2 \, f_x(x) dx,} \qquad (6.142)$$

which, from Eq. (6.112), means that D_G^2 is the mean square bias, $E[b_y(x)^2]$.

Thus, the formula given in Gammaitoni (1995a) measures only the mean square bias in a quantizer, and ignores the average conditional error variance. This is the opposite to the SNR measure proposed in Stocks (2001a), which – if applied to an estimation task – ignores the bias, and measures only the average conditional error variance. The MSE distortion – or equivalently, the SQNR measure – used to measure a quantizer's performance, which takes the variance *and* bias into account, gives a more complete picture of a quantizer's performance than Eq. (6.141).

6.9 Chapter summary

The initial section of this chapter, Section 6.1, introduces the concept of decoding the SSR model, first introduced in Chapter 4. It defines an error between the input signal and the decoded output, as well as introducing appropriate measures of the average performance of that error, including the mean square error distortion, signal to quantization noise ratio, and correlation coefficient.

Section 6.2 shows how the signal-to-noise ratio of a signal can be reduced by averaging N independently noisy realizations of that signal.

Sections 6.3 through 6.6 then examine linear and nonlinear decoding schemes. Linear decoding schemes are ones for which the spacing between reconstruction points is constant for all points. Nonlinear ones are not necessarily constantly spaced. Standard theory is introduced to show that both optimal – in the sense of minimum mean square error distortion – linear and optimal nonlinear decoding schemes exist, and can be calculated for the SSR model. Such calculations always show that the mean square error distortion is minimized for some nonzero value of σ for the SSR model. Comparisons are made between each decoding, which verify that the optimal decoding is the nonlinear MMSE decoding.

Section 6.7 considers the SSR decoding from the point of view of estimation theory, and, using the biased-estimate generalization of the Cramer–Rao bound, finds an expression for a lower bound on the achievable mean square error distortion.

Finally, Section 6.8 discusses an SNR measure defined in Stocks (2001a), as well as a measure of a quantizer's performance defined in Gammaitoni (1995a), and argues that for information-theoretic or estimation problems, neither of these measures takes into account *both* the error variance and the bias, which are both necessary components of the standard mean square error distortion measure.

Chapter 6 in a nutshell

This chapter included the following highlights:

- The application of the concept of decoding to the SSR model, in order to provide a signal that approximates, or reconstructs, the input signal. Such a reconstruction can be measured with the mean square error distortion – as conventionally is the case for quantizers – as well as the correlation coefficient, and signal to quantization noise ratio.
- Analytical derivations of expressions for the mean square error distortion, signal to quantization noise ratio, and correlation coefficient for linear decoding schemes of the SSR model, in terms of the function $P_{1|x}$.
- Analytical derivations of expressions for the correlation coefficient for a linear decoding of the SSR model in terms of N and σ for the specific cases of Gaussian signal and noise, uniform signal and noise, and Laplacian signal and noise.
- Plots for the MSE distortion in the case of matched-moments and Wiener linear decoding schemes, and MAP and MMSE nonlinear decoding schemes, and the associated reconstruction points, and verification that MMSE decoding is the optimal decoding.

- Analytical derivation of an expression for the MMSE distortion and optimal recon-struction points for the SSR model for the case of uniform signal and noise and noise intensity, $\sigma \leq 1$.
- The application of the Cramer–Rao and information bounds to find a new lower bound, named here as the *average information bound*, on the MSE distortion for decoding of the SSR model. It is shown that a linear decoding for the SSR model is a biased efficient estimator, and that MMSE decoding is biased but does not meet the average information bound with equality. Analysis of the average information bound confirms that the value of noise intensity, σ, which minimizes the MSE distortion means finding the best tradeoff between average error variance and mean square bias. This tradeoff is analogous to the tradeoff between output entropy and average conditional output entropy required to maximize the mutual information.
- Arguments that the SNR measure developed for the DIMUS sonar array in Remley (1966) is not valid for small σ, as it ignores the fact that the output noise will be correlated with the input signal and varies with σ.
- Arguments are given that (i) the SNR measure used in Stocks (2001a) decreases monotonically with σ, because the definition of noise power used does not take bias into account, and therefore is incomplete, other than for detection scenarios; (ii) the measure used to describe quantization given in Gammaitoni (1995a) is also incomplete, because it only takes into account mean square bias, and not error variance. The MSE distortion and SQNR, as used by engineers in quantization theory, do take both of these factors into account.

Open questions

Possible future work and open questions arising from this chapter might include:

- Application of mean square error distortion decoding to one-sided signal and noise distributions, such as the Rayleigh or exponential distributions.
- Application of mean square error distortion decoding to mixed signal and noise distributions, for example a uniform signal subject to Gaussian noise.
- Consideration of distortion measures other than the mean square error distortion, such as the absolute error distortion.

This concludes Chapter 6, which studies the array of threshold devices in which SSR occurs as a quantization model, and discusses various output decod-ing schemes. Chapter 7 now examines decoding schemes and the MSE distortion for the SSR model under the assumption of a large number of threshold devices.

7

Suprathreshold stochastic resonance: large N decoding

The aim of this chapter is to find asymptotic large N approximations to the mean square error distortion for the suprathreshold stochastic resonance model. In particular, we are interested in how the distortion varies with noise intensity and how it scales with the number of threshold devices. That is, does the distortion become asymptotically small for large N?

7.1 Introduction

Chapter 6 developed the idea of treating the SSR model as a lossy source coding or quantization model. We saw that such a treatment requires specification of reproduction points corresponding to each of the $N + 1$ discrete output states. Once specified, an approximate reconstruction of the input signal can be made from a decoding, and the average error between this approximation and the original signal subsequently measured by the mean square error (MSE) distortion. We saw also in Chapter 5 that asymptotic approximations to the output probability mass function, $P_y(n)$, output entropy, average conditional output entropy, and mutual information can be found if the number of threshold devices, N, in the SSR model is allowed to become very large. The aim of this chapter is to again allow N to become very large, and develop asymptotic approximations to the MSE distortion for the cases of optimal linear and optimal nonlinear decodings of the SSR model.

Chapter structure

This chapter has three main sections. We begin in Section 7.2 by letting N become large in the formulas derived in Chapter 6 and analyzing the result. Next, Section 7.3 takes the same approach from the estimation theory perspective. Finally, we give a brief discussion in Section 7.4 on the concept of *stochastic resonance without tuning*, based on the results obtained in this chapter. Here, *tuning* refers to 'tuning to the optimal noise intensity'; that is, finding the optimal noise level. The term is used in an analogy with frequency resonance, where tuning to the right frequency gives the resonant response.

7.2 Mean square error distortion for large N

Recall that Chapter 6 considers two different classes of decoding, and their application to the SSR model, namely linear decoding and nonlinear decoding. Linear decoding refers to the situation where all reproduction points are linearly spaced, so that the decoding can be written in the form $\hat{y} = ay + b$, where a and b are constants for given signal and noise distributions, but may vary with the ratio of noise standard deviation to signal standard deviation, σ. Nonlinear decoding refers to the situation where the decoded output of the SSR model varies nonlinearly for each possible value of y. We saw that the optimal linear decoding is known as the Wiener decoding, and the optimal nonlinear decoding, which we refer to as minimum mean square error (MMSE) distortion decoding, has reproduction points given by $\hat{x}_n = E[x|y = n]$, $n = 0$, ..., N. This section examines the large N asymptotic behaviour of both Wiener and MMSE distortion decoding for the SSR model. As in previous chapters, we assume the signal PDF, $f_x(x)$, and the noise PDF, $f_\eta(\eta)$, are even functions with zero means and that the threshold values are all $\theta = 0$.

Wiener linear decoding

For zero-mean input signals and a linear decoding of the form $\hat{y} = ay + b$, we impose the condition that $E[\hat{y}] = 0$. Then $b = -aE[y]$, and, as discussed in Chapter 6, the optimal value of a is given by $a = \frac{E[xy]}{var[y]}$. Since for SSR we have $E[y] = N/2$, this means that the optimal linear reconstruction points are

$$\hat{y}(n) = \frac{E[xy]}{var[y]}\left(y - \frac{N}{2}\right),\tag{7.1}$$

and the Wiener MSE distortion can be expressed in terms of the linear correlation coefficient, ρ_{xy}, as

$$MSE = E[x^2](1 - \rho_{xy}^2).\tag{7.2}$$

Recall also that the correlation coefficient is identical for all possible linear decodings. In addition, for SSR, when the signal and noise PDFs are both even-valued functions about a mean of zero, the correlation coefficient can be expressed in terms of the probability that any given threshold device is 'on', $P_{1|x}$, as

$$\rho_{x\hat{y}} = \rho_{xy} = \frac{\sqrt{N}E[x P_{1|x}]}{\sqrt{E[x^2]\left((N-1)E[P_{1|x}^2] - \frac{(N-2)}{4}\right)}}. \tag{7.3}$$

Letting N approach infinity in Eq. (7.3) gives the large N correlation coefficient as

$$\lim_{N\to\infty} \rho_{xy} = \frac{E[x P_{1|x}]}{\sqrt{E[x^2]\left(E[P_{1|x}^2] - \frac{1}{4}\right)}}$$

$$= \frac{2E[x P_{1|x}]}{\sqrt{E[x^2]}\rho_o}, \tag{7.4}$$

where ρ_o is the correlation coefficient between the outputs of any two threshold devices in the SSR model, as given by Eq. (A3.20) in Appendix 3. By substitution of Eq. (7.4) into Eq. (7.2), the large N Wiener linear decoding MSE distortion is

$$\lim_{N\to\infty} MSE = E[x^2] - \frac{4E[x P_{1|x}]^2}{\rho_o}. \tag{7.5}$$

Note that the large N limits given by Eqs (7.4) and (7.5) are independent of N. Hence, as N increases, the correlation coefficient and MSE distortion get closer and closer to these expressions. This means that the MSE distortion does not necessarily approach zero for large N.

We now present exact results for the large N correlation coefficient and MSE distortion for specific matched signal and noise distributions. These expressions are found by letting N approach infinity in the corresponding equations stated in Chapter 6.

Gaussian signal and noise

For Gaussian signal and noise and large N, the linear decoding correlation coefficient is, from Eq. (6.59)

$$\lim_{N\to\infty} \rho_{xy} = \frac{1}{\sqrt{(\sigma^2 + 1) \arcsin\left(\frac{1}{\sigma^2+1}\right)}}$$

$$= \sqrt{\frac{\rho_i}{\arcsin(\rho_i)}}, \tag{7.6}$$

where $\sigma = \sigma_\eta/\sigma_x$ is the ratio of noise standard deviation to signal standard deviation, and ρ_i is the correlation coefficient between the inputs to any two devices in the SSR model, as defined in Eq. (A3.19) in Appendix 6. Using Eqs (7.2) and (7.6) we also have the Wiener decoding MSE distortion as

$$\lim_{N \to \infty} \text{MSE} = E[x^2] \left(1 - \frac{\rho_i}{\arcsin(\rho_i)} \right). \tag{7.7}$$

For $\sigma \to \infty$ we have $\rho_i/\sin(\rho_i) \to 1$, and therefore the correlation coefficient is strictly increasing with σ for large N, and asymptotically approaches unity for large σ, while the MSE distortion is strictly decreasing with σ, and asymptotically approaches zero for large σ.

Uniform signal and noise

For uniform signal and noise and large N, the linear decoding correlation coefficient is, from Eq. (6.64)

$$\lim_{N \to \infty} \rho_{x,y} = \begin{cases} \frac{3-\sigma^2}{2\sqrt{3-2\sigma}} & (\sigma \le 1), \\ 1 & (\sigma \ge 1), \end{cases} \tag{7.8}$$

and the Wiener decoding MSE distortion is therefore

$$\lim_{N \to \infty} \text{MSE} = \begin{cases} E[x^2] \left(\frac{(\sigma-1)^3(\sigma+3)}{8\sigma-12} \right) & (\sigma \le 1), \\ 0 & (\sigma \ge 1). \end{cases} \tag{7.9}$$

Thus, for uniform signal and noise, and infinite N, the correlation coefficient is exactly unity and the MSE distortion is exactly zero for $\sigma \ge 1$.

Laplacian signal and noise

For Laplacian signal and noise and large N, the linear decoding correlation coefficient is, from Eq. (6.68)

$$\lim_{N \to \infty} \rho_{xy} = \frac{(2\sigma + 1)\sqrt{(\sigma + 1)(\sigma + 2)}}{2(\sigma + 1)^2}, \tag{7.10}$$

and the Wiener decoding MSE distortion is therefore

$$\lim_{N \to \infty} \text{MSE} = E[x^2] \left(\frac{3\sigma + 2}{4(\sigma + 1)^3} \right). \tag{7.11}$$

For large σ, the linear decoding correlation coefficient asymptotically approaches unity, and the Wiener MSE distortion asymptotically approaches zero.

Logistic signal and noise

As discussed in Chapter 6, we do not have an exact expression for the correlation coefficient or Wiener MSE distortion for logistic signal and noise. However, since the quantities $E[x P_{1|x}]$ and ρ_o can be calculated numerically, the large N

correlation coefficient and Wiener MSE distortion can be found numerically using Eqs. (7.4) and (7.5).

We now graphically illustrate the behaviour of the above formulas and briefly discuss their main features.

Results

Figure 7.1 shows the large N Wiener decoding MSE distortion and linear correlation coefficient for the four different matched signal and noise situations considered. It is clear that in all cases the MSE distortion is strictly decreasing with increasing σ and the correlation coefficient is strictly increasing towards unity with increasing σ. This behaviour is discussed further in Section 7.4.

We now consider the case of optimal MMSE distortion decoding.

MMSE optimal decoding

As discussed in Chapter 6, and as shown in Section A3.5 of Appendix 3, the reproduction points that give the minimum possible MSE distortion are

$$\hat{x}_n = \mathrm{E}[x|n] = \int_x x P_{x|y}(x|n)dx = \frac{1}{P_y(n)}\int_x x P_{y|x}(n|x) f_x(x)dx, \quad n = 0, \ldots, N.$$

$$(7.12)$$

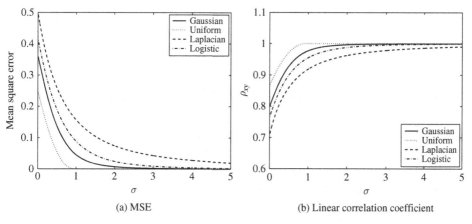

(a) MSE   (b) Linear correlation coefficient

Fig. 7.1. MSE distortion and correlation coefficient for Wiener decoding and large N. Figure 7.1(a) shows the large N Wiener decoding MSE distortion and Fig. 7.1(b) shows the large N linear correlation coefficient, for the cases of matched Gaussian, uniform, Laplacian, and logistic signal and noise. For small noise intensity, σ, the MSE distortion is clearly nonzero. However, as σ increases, the MSE distortion decreases asymptotically towards zero. Likewise, the correlation coefficient approaches unity for increasing σ.

The MSE distortion that results from this decoding is

$$\text{MMSE} = E[x^2] - E[\hat{x}_n^2]$$

$$= E[x^2] - \sum_{n=0}^{N} E[x|n]^2 P_y(n)$$

$$= E[x^2] - \sum_{n=0}^{N} \frac{1}{P_y(n)} \left(\int_x x P_{y|x}(n|x) f_x(x) dx \right)^2. \qquad (7.13)$$

Convergence to maximum likelihood decoding

Recall from Chapter 5 that for matched signal and noise distributions, and PDFs that are even functions about a mean of zero, the following large N approximation holds

$$P_y(n) \simeq \frac{f_Q(\frac{n}{N})}{N}, \quad n = 0, \ldots, N. \qquad (7.14)$$

As derived in Chapter 4, $f_Q(\tau)$ is a PDF defined with support $\tau \in [0, 1]$, and for $\theta = 0$ and $f_x(x)$ and $f_\eta(\eta)$ even, and the support of P contained in the support of R, is given by

$$f_Q(\tau) = \frac{f_x(x)}{f_\eta(x)} \Big|_{x=F_\eta^{-1}(\tau)}, \qquad (7.15)$$

where $F_\eta^{-1}(\cdot)$ is the inverse cumulative distribution function (ICDF) of the noise. For finite N, Eq. (7.14) becomes an approximation, and is most accurate near $\sigma = 1$, and exact at $\sigma = 1$. We saw also in Chapter 5 that

$$f_{\bar{y}}(\bar{y}) = \frac{f_Q(\frac{\bar{y}}{N})}{N}, \quad \bar{y} \in [0, 1], \qquad (7.16)$$

is the exact PDF of the continuously valued random variable, $\bar{y}(x)$, describing the expected value of the SSR encoding, y, given the input, x.

We also used in Chapter 5 a large N approximation to the transition probabilities, which from Eqs. (5.55) and (5.56) can be written as

$$P_{y|x}(n|x) \simeq \frac{\delta \left(P_{1|x} - \left(\frac{n}{N} \right) \right)}{N}. \qquad (7.17)$$

Substituting from Eqs. (7.17) and (7.14) into Eq. (7.12) gives

$$\hat{x}_n \simeq \frac{1}{f_Q(\frac{n}{N})} \int_x x \delta \left(P_{1|x} - \left(\frac{n}{N} \right) \right) f_x(x) dx, \quad n = 0, \ldots, N. \qquad (7.18)$$

Making a change of variable to $\tau = P_{1|x} = F_{\eta}(x)$ in Eq. (7.18) gives a large N approximation of $\hat{x}_n$ as

$$\hat{x}_n \simeq \frac{1}{f_Q\left(\frac{n}{N}\right)} \int_{\tau=0}^{\tau=1} F_{\eta}^{-1}(\tau)\delta\left(\tau - \left(\frac{n}{N}\right)\right) f_Q(\tau)d\tau$$

$$= \frac{1}{f_Q\left(\frac{n}{N}\right)} F_{\eta}^{-1}\left(\frac{n}{N}\right) f_Q\left(\frac{n}{N}\right)$$

$$= F_{\eta}^{-1}\left(\frac{n}{N}\right), \quad n = 0, \ldots, N. \tag{7.19}$$

Recall from Chapter 6 that this is exactly the maximum likelihood decoding, and that therefore the MMSE decoding converges with large N to the maximum likelihood decoding. This result is well known in estimation theory; the maximum likelihood estimate achieves minimum variance as the number of samples approaches infinity (Lehmann and Casella 1998). Thus, the main result here is that the MMSE distortion reconstruction points converge to the maximum likelihood points, for large N.

However, unlike the optimal linear Wiener reconstruction points, which depend on σ_x, the maximum likelihood decoding reconstruction points are completely independent of the signal PDF. This is an indication that the result above may not always be accurate. However, provided that $f_{\eta}(\eta)$ has infinite tails, we might expect that it is only the reconstruction points corresponding to n near zero and N that are significantly different from the maximum likelihood points, and that should depend on the variance of the signal distribution. We now find evidence for this assertion.

Large N MMSE

Substituting the maximum likelihood decoding reconstruction points into Eq. (7.13) gives the MMSE distortion for large N as

$$\text{MMSE} \simeq \text{E}[x^2] - \sum_{n=0}^{N} P_y(n) F_{\eta}^{-1}\left(\frac{n}{N}\right)^2. \tag{7.20}$$

Substituting from Eq. (7.14) into Eq. (7.20) gives

$$\text{MMSE} \simeq \text{E}[x^2] - \frac{1}{N} \sum_{n=0}^{N} f_Q\left(\frac{n}{N}\right) F_{\eta}^{-1}\left(\frac{n}{N}\right)^2. \tag{7.21}$$

Suppose we replace the summation in Eq. (7.21) by an integral as

$$\text{MMSE} \simeq \text{E}[x^2] - \frac{1}{N} \int_{n=0}^{N} f_Q\left(\frac{n}{N}\right) F_{\eta}^{-1}\left(\frac{n}{N}\right)^2 dn. \tag{7.22}$$

With a change of variable from n to $s = \frac{n}{N}$, Eq. (7.22) becomes a limiting expression

$$\lim_{N \to \infty} \text{MMSE} \simeq E[x^2] - \int_{s=0}^{s=1} f_Q(s) F_\eta^{-1}(s)^2 ds. \qquad (7.23)$$

With a further change of variable from s to $x = F_\eta^{-1}(s)$, we have

$$\lim_{N \to \infty} \text{MMSE} \simeq E[x^2] - \int_{x=-\infty}^{x=\infty} f_x(x) x^2 dx = 0. \qquad (7.24)$$

The result is that, provided the summation in Eq. (7.21) can be approximated by the integral, the MMSE approaches zero for large N. Note that we would expect the integral to more closely approximate the summation as N increases.

As stated in Section A2.2 of Appendix 2, a more accurate approximation of a summation by an integral is given by the Euler–Maclaurin summation formula (Spiegel and Liu 1999). This formula has remainder terms, which are significant if the 0th and Nth terms are significant. For SSR, since $P_y(0)$ and $P_y(N)$ are significant for $\sigma \le 0$, it is difficult to justify the approximation given in Eq. (7.22). Hence, we do not expect Eq. (7.24) to hold for $\sigma \le 1$. This confirms our previous discussion regarding the maximum likelihood reconstruction points not being dependent on the signal distribution.

In order to further investigate this, we now examine letting N become large in our known exact results for MMSE decoding for uniform signal and noise.

Exact result for uniform signal and noise

Recall that Chapter 6 presents exact analytical expressions for the MMSE distortion decoding reproduction points, and the MMSE distortion for uniform signal and noise and $\sigma \le 1$. From Eq. (6.96), the reproduction points for large N are

$$\hat{x}(n) \simeq \frac{\sigma_\eta}{2} \left(2 \left(\frac{n}{N} \right) - 1 \right), \quad n = 1, \ldots, N, \qquad (7.25)$$

and from Eq. (6.97), $\lim_{N \to \infty} \hat{x}(N) = -\hat{x}(0) = \frac{\sigma_x}{4}(1 + \sigma)$. From Eq. (6.100), the MMSE distortion reduces to

$$\lim_{N \to \infty} \text{MMSE} = \frac{\sigma_x^2}{48}(1 - \sigma)^3. \qquad (7.26)$$

It is clear that for large N, the MMSE distortion is strictly decreasing for $\sigma \le 1$, and is zero at $\sigma = 1$. We saw for a linear decoding that the MSE distortion for large N is exactly zero for $\sigma \ge 1$ for infinite N, since it scales with $1/N$. The result of Eq. (7.26) therefore means that no decoding scheme can make the MSE distortion asymptotically approach zero for $\sigma < 1$ for uniform signal and noise.

From Table 4.1 in Chapter 4, the ICDF of the noise is given by $F_\eta^{-1}(w) = \frac{\sigma_\eta}{2}(2w - 1)$. Therefore, for $n = 1, \ldots, N - 1$ the MMSE reproduction points, for large N, given by Eq. (7.25) are exactly the maximum likelihood reproduction points given by $F_\eta^{-1}\left(\frac{n}{N}\right)$. However, for $n = 0$ and $n = N$, the MMSE and maximum likelihood reproduction points do not coincide, apart from at $\sigma = 1$. This fact illustrates why the MMSE decoding is said to asymptotically converge to the maximum likelihood decoding, since it only converges in the event of an infinite number of observations. In such a situation, the 0th and Nth reproduction values become insignificant, and can be ignored in calculations of the MSE distortion, regardless of the probability of the 0th and Nth state occurring, provided $\sigma \neq 0$.

We now consider the large N distortion performance of the SSR model from the point of view of estimation theory.

7.3 Large N estimation perspective

Recall that Chapter 6 discusses decoding of the SSR model from the point of view of estimation theory. This discussion includes the introduction of the definitions of *bias* and *Fisher information*, and their application to the concept of the *Cramer–Rao bound*, which gives a lower bound on the MSE distortion for an unbiased estimator, and the *information bound*, which gives a lower bound on the MSE distortion of biased estimators. Furthermore, Chapter 6 uses such estimation theory to show that the MSE distortion performance of the SSR model depends strongly on the tradeoff between bias and error variance. This section extends this discussion to the case of a large number of threshold devices in the SSR model, which is equivalent to the case of a large number of independently noisy observations of a parameter in estimation theory. We begin by introducing some estimation terminology that applies in such large N situations, before illustrating the asymptotic behaviour of the SSR model from this perspective.

Asymptotic large N theory

Suppose we can make a measurement, z, of some parameter, θ. If there is some random or systematic error in the measurement, then the error of any given measurement is $\epsilon(z) = z - \theta$.

Asymptotically unbiased estimators

An estimator is called *asymptotically unbiased* if in the limit of an infinite number of observations, the estimate becomes unbiased, that is

$$\lim_{N \to \infty} \mathrm{E}[\epsilon(z)|\theta] = 0. \tag{7.27}$$

Consistent estimators

An estimator is called *consistent* if its MSE distortion approaches zero in the limit of an infinite number of observations, that is

$$\lim_{N\to\infty} \mathrm{E}[\epsilon^2(z)|\theta] = D(\theta) = 0, \tag{7.28}$$

where $D(\cdot)$ is our notation for the conditional MSE distortion.

Mean asymptotic square error

Suppose for large N an estimator is asymptotically unbiased. Then, since the bias is zero, the mean square bias is also zero, and the derivative of $\mathrm{E}[z|\theta]$ is unity. Thus, from Eq. (6.110) in Chapter 6, the MSE distortion for a given θ is lower bounded by the Cramer–Rao bound, rather than the information bound, so that

$$D(\theta) = \mathrm{var}[\epsilon(z)|\theta] \geq \frac{1}{J(\theta)}. \tag{7.29}$$

If θ has PDF $f_\theta(\cdot)$, when averaged over all possible values of θ, the result is a lower bound on the MSE distortion as

$$\mathrm{MSE} \geq \mathrm{E}\left[\frac{1}{J(\theta)}\right] = \int_\tau f_\theta(\tau)\frac{1}{J(\tau)}d\tau. \tag{7.30}$$

Asymptotic efficiency

If the Cramer–Rao bound is met with equality, then the estimator, z, is said to be *efficient*. If efficiency is met asymptotically with increasing N, then the estimator is said to be *asymptotically efficient*. When this occurs for all θ, the lower bound on the MSE distortion given by Inequality (7.30) becomes an equality, which we call the mean asymptotic square error (MASE), that is

$$\mathrm{MASE} = \mathrm{E}\left[\frac{1}{J(\theta)}\right] = \int_\tau f_\theta(\tau)\frac{1}{J(\tau)}d\tau, \tag{7.31}$$

which is simply the expected value of the reciprocal of the Fisher information. Note that such a quantity is discussed by Bethge *et al.* (2002) and Bethge (2003), in the context of neural population coding.

For situations where the asymptotic behaviour of an estimator's bias or efficiency is unknown, Eq. (7.31) can still be useful, as it provides a lower bound on the MSE distortion when only the Fisher information is known, without the need for specifying a decoding.

Application to SSR

When using the SSR output encoding, y, as an estimate for the input, x, it does not make sense to consider asymptotic unbiasedness or consistency, as $\mathrm{E}[y|x]$

ranges between zero and N, whereas we assume that the signal mean is zero. However, once the output is decoded, it is possible that asymptotic unbiasedness and consistency can occur. The purpose of this section is to examine this possibility.

The Fisher information for SSR is given in Chapter 5 by Eq. (5.72). Assuming that for large N there exists an asymptotically unbiased and efficient estimator, the MASE is

$$\text{MASE} = \frac{1}{N} \int_x f_x(x) \frac{P_{1|x}(1 - P_{1|x})}{R^2(x)} dx. \tag{7.32}$$

If the integral in Eq. (7.32) does not diverge, then the MASE exists and decreases with $\frac{1}{N}$, and we have a consistent estimator for all x. However, it is possible that the integral diverges. For example, in the case of logistic signal and noise, we have $f_\eta(x) = P_{1|x}(1 - P_{1|x})/b_\eta$ and therefore the MASE can be written as

$$\text{MASE} = \frac{b_\eta}{N} \int_x \frac{f_x(x)}{f_\eta(x)} dx, \tag{7.33}$$

which diverges for $\sigma \leq 1$.

In the event that the MASE diverges, then it is likely that an asymptotically unbiased and efficient estimator does not exist. The remainder of this section examines first the question of whether an asymptotically unbiased estimator exists for the SSR model, and second the question of efficiency.

Linear decoding

We have already seen that for a linear decoding, as N approaches infinity and σ approaches infinity, the MSE distortion approaches zero. Therefore, under these conditions, the Wiener linear decoding is a consistent estimator.

The large N Wiener decoding reconstruction points are given by

$$\hat{y}(n) \simeq c \left(2 \left(\frac{n}{N} \right) - 1 \right), \tag{7.34}$$

where $c = 2\text{E}[x P_{1|x}]/\rho_o$.

We have also, by taking the conditional expectation of Eq. (7.1), that

$$\text{E}[\hat{y}|x] = c \left(2P_{1|x} - 1 \right). \tag{7.35}$$

Therefore, the bias is asymptotically zero for a linear decoding, if and only if

$$c \left(2P_{1|x} - 1 \right) - x = 0. \tag{7.36}$$

This condition means that $P_{1|x} = \frac{x}{2c} + \frac{1}{2}$, which holds for the case of uniform noise when $\sigma_\eta = 2c$. However, in general, the bias for a linear decoding is not asymptotically equal to zero for all x in the SSR model.

Nonlinear decoding

Rearranging the large N reconstruction points of Eq. (7.19) gives $y(n) = N P_{1|x=\hat{x}_n}$. Since the average transfer function of the SSR model is $\bar{y}(x) = \mathrm{E}[y|x] = N P_{1|x}$, then this implies that for large N the bias of the MMSE decoding asymptotically approaches zero for all x. This is in agreement with the known result that the maximum likelihood decoding is asymptotically unbiased (Lehmann and Casella 1998).

Consequently

$$\lim_{N\to\infty} \mathrm{E}[\hat{x}|x] = \sum_{n=0}^{N} F_\eta^{-1}\left(\frac{n}{N}\right) P_{y|x}(n|x) = x. \qquad (7.37)$$

Since these arguments indicate that the large N bias is asymptotically zero for the maximum likelihood decoding, then how is it possible for the MASE to diverge, when we know from Chapter 6 that the MMSE distortion is smaller than the Wiener MSE distortion, for which large N expressions were presented in Section 7.2? The answer to this problem is related to the fact that for $\sigma < 1$, the most probable output states of the SSR model are states zero and N. Recall also from Chapter 5 that the average transfer function PDF, $f_Q(\tau)$, is infinite at zero and unity.

The maximum likelihood reconstruction points that correspond to these states are $F_\eta^{-1}(0)$ and $F_\eta^{-1}(1)$, which for noise PDFs with infinite tails are at $\pm\infty$. With such large values, if $P_y(0)$ and $P_y(N)$ do not approach zero, then the MSE distortion will be very large. This appears to be the case when the MASE diverges.

Also note that for the MMSE reconstruction points at $n = 0$ and $n = N$ to be equal to $\pm\infty$, N truly needs to be infinite. If it is not, then there is some large error between the MMSE points and the maximum likelihood points, and if $P_y(0)$ and $Py(N)$ are not small, a significant difference between the MSE distortion obtained with the MMSE reconstruction points and maximum likelihood reconstruction points will occur.

These issues mean that no meaningful analytical or numerical results can be obtained for the MASE for small σ, say $\sigma < 1$, and that no efficient unbiased estimator exists under these conditions. The best we can do is numerically calculate the MMSE reconstruction points and MMSE distortion for specific signal and noise distributions for as large an N as is practicable. Such results show the same qualitative behaviour as Fig. 7.1.

7.4 Discussion on stochastic resonance without tuning

We have seen that for large N the correlation coefficient increases with increasing σ and approaches unity as σ approaches infinity. Likewise, the MSE distortion decreases with increasing σ towards zero. Thus, there is a broadening of the system response, meaning near optimal performance occurs for a broad range of noise intensities. Furthermore, other than the case of uniform signal and noise, for infinite N the optimal noise intensity is at infinity, a strongly counterintuitive notion, even given an understanding of stochastic resonance. However, such a result has previously been published under the title of *stochastic resonance without tuning* (Collins *et al.* 1995b).

The model examined in Collins *et al.* (1995b) is quite similar to the SSR model, in that the outputs of N independently noisy individual neuron models are summed to give an overall system output in response to aperiodic stimuli. By contrast to SSR, however, Collins' stimuli are always subthreshold. Collins *et al.* (1995b) use the correlation coefficient to show that SR occurs in this system, and emphasize the fact that as the number of elements, N, increases, the peak of the correlation coefficient with noise intensity broadens, in the same manner as the results presented here. Note that the interpretations of Collins *et al.* (1995b) have been criticized or explained in terms of well-known statistical results in several papers (Noest 1995, Petracchi 2000, Greenwood *et al.* 2003). The general theme is that the reported result should come as no surprise to those familiar with the averaging of independently noisy data. As is shown in Chapter 6, if a signal is averaged without thresholding, the MSE distortion is reduced by a factor of $\frac{1}{N}$. This means that regardless of the original SNR, the SNR of the averaged signal can be made arbitrarily small by increasing N. This is essentially the same result as that given in Collins *et al.* (1995b) and here for large N and σ.

The difference between SSR and noise reduction by averaging is that quantization of the noisy input signal is performed in the SSR model, and this adds a component to the error between an input signal and the resultant output approximation. For sufficiently large noise, the quantization loss can be overcome by averaging. However for small noise intensities, there is no decoding scheme that can overcome the mismatch of SSR quantization. To explain this further, we saw in Chapter 5 that for small noise, the most probable output states are states close to zero and N. By contrast, for signals with PDFs with infinite tails, the most likely values are close to the mean. The results presented here indicate that even infinite N is not sufficient to overcome this mismatch between the shape of the output probability mass function and the input PDF. By contrast, as the noise intensity increases, this mismatch lessens, and, at the optimum value of noise, the MSE distortion can be asymptotically reduced to zero by increasing N. However, for large σ, using large values of N is also very inefficient, because nearly all values of the SSR model's output, y, become highly improbable, other than those close to $y = N/2$. This has some analogies with *rate-distortion* theory (Berger and Gibson 1998). Information theorists understand that decreasing the average distortion in a quantization scheme can be achieved by increasing the *rate* – here, *rate* can be interpreted to mean the number of output states, N. We shall further discuss rate-distortion theory in Chapter 9.

7.5 Chapter summary

Section 7.1 briefly introduces the aims of this Chapter, which are to examine the large N performance of the SSR model from the point of view of signal quantization and reconstruction. Hence, we re-examine the results presented in Chapter 6, with the goal of finding large N limiting expressions.

Section 7.2 then presents large N asymptotic expressions for the Wiener decoding – that is, optimal linear decoding – reconstruction points, MSE distortion, and correlation coefficient. Exact analytical formulas for these expressions are found in terms of σ for the cases of matched Gaussian, uniform, and Laplacian signal and noise. Section 7.2 then goes on to consider MMSE nonlinear decoding, and finds an approximation that states that the large N reconstruction points are the maximum likelihood reconstruction points, and points out that with these points, an expression for the large N MMSE distortion is available. Verification of this result is given for the case of uniform signal and noise, and $\sigma \leq 1$, for which we have an exact expression for the large N reconstruction points. It is, however, also pointed out that such an approximation breaks down for SSR and small σ, due to the approximation being more inaccurate for large and small n, since large and small n are also the most probable states.

Section 7.3 introduces large N asymptotic estimation theory, including the ideas of asymptotic unbiasedness and consistent estimators, before presenting an expression giving a lower bound in terms of Fisher information for the MSE distortion of an asymptotically unbiased estimator. The Fisher information for the SSR model is then shown to be such that the integral component of this lower bound does not always converge, meaning that asymptotically small MSE distortions are not achievable for small σ in the SSR model, even for infinite N.

Finally, Section 7.4 discusses the results presented in Section 7.2 which show that the optimal noise intensity appears to be infinite for infinite N, and why this result is somewhat misleading.

Chapter 7 in a nutshell

This chapter included the following highlights:

- Derivation of large N expressions for the optimal linear mean square error distortion, reconstruction points, and linear correlation coefficient.
- Discussion of the MMSE distortion for large N, and the mean asymptotic square error, and how these expressions cannot be made asymptotically small for small noise intensities in the SSR model, even for infinite N.
- Discussion of the fact that, for large N, the correlation coefficient and MSE distortion appear to be optimized for infinite noise intensity.

Open questions

Possible future work and open questions arising from this chapter might include:

- More rigorous analysis and justification of results showing that the MSE distortion cannot be made arbitrarily small for small noise intensities.

This concludes Chapter 7, which studies the decoding of the SSR model for large N. Chapter 8 introduces a new variation to the SSR model, by allowing the threshold values in each device in the model to vary independently.

8

Optimal stochastic quantization

As described and illustrated in Chapters 4–7, a form of stochastic resonance called suprathreshold stochastic resonance can occur in a model system where more than one identical threshold device receives the same signal, but is subject to independent additive noise. In this chapter, we relax the constraint in this model that each threshold must have the same value, and aim to find the set of threshold values that either maximizes the mutual information, or minimizes the mean square error distortion, for a range of noise intensities. Such a task is a stochastic optimal quantization problem. For sufficiently large noise, we find that the optimal quantization is achieved when all thresholds have the same value. In other words, the suprathreshold stochastic resonance model provides an optimal quantization for small input signal-to-noise ratios.

8.1 Introduction

The previous four chapters consider a form of stochastic resonance, known as suprathreshold stochastic resonance (SSR), which occurs in an array of identical noisy threshold devices. The noise at the input to each threshold device is independent and additive, and this causes a randomization of effective threshold values, so that all thresholds have unique, but random, effective values. Chapter 4 discusses and extends Stocks' result (Stocks 2000c) that the mutual information between the SSR model's input and output signals is maximized for some nonzero value of noise intensity. Chapter 6 considers how to reconstruct an approximation of the input signal by decoding the SSR model's output signal. It is shown that the mean square error (MSE) distortion between the original signal and the reconstruction is minimized for some nonzero values of noise. Chapters 5 and 7 examine the large N behaviour of the mutual information and MSE distortion, and show that stochastic resonant behaviour persists for large N, and can be described by asymptotic formulas.

248

In this chapter, the constraint in the SSR model, that all thresholds are identical, is discarded, and we examine how to set the threshold values to optimize the performance of the resultant array of threshold devices for a range of noise intensities. Other than this modification, all other conditions are identical to the SSR model. The motivation for this modification is that – in the absence of noise – it is straightforward to show that setting all thresholds in such a model to identical values, as is the case for the SSR model, is not an optimal situation. To illustrate this, as discussed in Section 8.4, the maximum mutual information – obtained with *distributed* fixed threshold values – in the absence of noise is $\log_2 (N + 1)$ bits per sample. In contrast, for SSR, we saw in Chapter 5 that the maximum mutual information is of the order of $0.5 \log_2 (N)$.

Given that in the SSR case the mutual information is only one bit per sample in the absence of noise, it is clear that for small noise SSR is very far from optimal. The unresolved question is whether the SSR situation of all identical thresholds is optimal for any range of noise intensities. The obvious way to address this question is to attempt to find the optimal thresholds as the noise intensity increases from zero. This is the focus of this chapter, and can be described as an optimal stochastic quantization problem.

The mathematical treatment in this chapter follows along the lines of McDonnell *et al.* (2002c), McDonnell and Abbott (2004a), McDonnell *et al.* (2005c), McDonnell *et al.* (2005d), McDonnell *et al.* (2005b), and McDonnell *et al.* (2006b).

Optimal quantization literature review

Deterministic optimal quantization

Theoretical results on optimal quantization tend to focus on the conventional situation where quantization is performed by fixed deterministic thresholds, and dithering is not considered. A comprehensive reference for such results is the textbook by Gersho and Gray (1992). Such research has usually been published in the electronic engineering literature, either in the field of information theory or in that of communications theory.

Optimal stochastic quantization

There have been very few previous studies on the problem of optimal quantization in the presence of *independent* threshold noise. There has been a large amount of prior research into dithering,[1] however without considering how to optimize

---

[1] Note that unlike the SSR effect, or what we refer to as 'stochastic quantization', conventional dithering – as well as many studies of SR in single-threshold systems – involves adding random or pseudo-random noise signals to an input signal prior to quantization – whereas for SSR, the noise is *independent* for each threshold.

the thresholds. Other studies into noisy quantization either do not consider input signal-to-noise ratios that are as large as those required for SSR to occur, do not consider optimal quantization in the presence of *independent* threshold noise, or do not consider optimal quantization as the input signal-to-noise ratio decreases from very large to very small values. For completeness, listed below are some references that, while at first glance might consider stochastic quantization in the same manner as we do here, are actually about different topics:

- The existing work perhaps most closely related to this chapter considers the topics of 'random reference' correlation (Castanie *et al.* 1974) and 'random reference quantizing' (Castanie 1979, Castanie 1984). In particular, in Castanie (1979) and Castanie (1984), the 'transition points' – that is, threshold values – in a quantization operation are independent random variables. However, unlike here, it is also assumed that the set of threshold values remains unique and ordered. In our notation, this means that the random variables corresponding to the ith and $(i + 1)$th threshold values are such that $\theta_i + \eta_i < \theta_{i+1} + \eta_{i+1} \, \forall \, i$. We do not impose such restrictions here, and therefore there is no ordering of threshold values. Furthermore, no result presented in Castanie *et al.* (1974), Castanie (1979), and Castanie (1984) considers a quantizer's performance as a function of the threshold noise intensity.
- The 'stochastic quantization' and 'randomized quantization' referred to in Berndt and Jentschel (2001) and Berndt and Jentschel (2002) do not refer to the case of independently randomized thresholds.
- The term 'random quantization' used in Bucklew (1981) and Zador (1982) refers to random reconstruction points, rather than random thresholds, while the term 'randomized quantizer' in Zamir and Feder (1992) refers to conventional dithering.
- There is a large body of research that considers noisy source coding, where a noisy signal needs to be compressed or coded, for example Ayanoğlu (1990). However, generally, the noise is considered to be added to the signal prior to arriving at a quantizer. This has the effect of making the set of threshold values a single random variable, rather than being independent random variables, just as with dithering.
- The term 'randomized quantizers' is also used in the context of statistical signal detection theory, but refers to a random choice between a number of deterministic quantizers (Tsitsiklis 1993, Blum 1995, Warren and Willett 1999).

Stochastic resonance, dithering and quantization

In the SR literature, quantization by a multithreshold system has only been considered in the context of dithering (Gammaitoni 1995b, Wannamaker *et al.* 2000a), and not from the point of view of optimizing the threshold values. Furthermore, the work on multithreshold systems contained in Gammaitoni (1995b) restricts the

---

Furthermore, as discussed in Section 3.4 of Chapter 2, dither signals are usually assumed to have a small dynamic range when compared with the input signal. All thresholds in a quantizer are subjected to the same dither signal, rather than independent noise.

input signal to be entirely subthreshold, which in the language of quantization theory means that the input signal is always smaller than the quantizer bin-size. The noise signal is also not independent for each threshold. We point out that if an attempt had been made to optimize the threshold values in Gammaitoni (1995b), then the signal would no longer be subthreshold, and SR would not occur, due to the lack of noise independence.

By contrast, quantization by a single threshold has been discussed in many papers, as pointed out elsewhere in this book, for example Bulsara and Zador (1996).

However, the extension of the SSR model to an optimal quantization problem is briefly explored, for the first time, in Stocks (2000a). This paper first discusses optimal quantization in the absence of noise, and points out that an exact solution to the optimal noiseless thresholds exists for maximum mutual information. It also comments on the fact that it seems there have been no previous studies on optimal quantization when all thresholds are subject to internal noise. Stocks (2000a) does not attempt to solve this problem by finding the optimal thresholds in the presence of noise, but does plot the mutual information for a range of σ – that is, the ratio of noise standard deviation to signal standard deviation, and various N, when the thresholds are set to maximize the mutual information in the absence of noise. These results show that the mutual information decreases monotonically with increasing σ, so that SR does not occur. However, they also show that, for sufficiently large σ, the mutual information obtained with the SSR situation of all thresholds equal to the signal mean is greater than the mutual information obtained with the optimal noiseless thresholds. For small σ, SSR is very far from optimal.

Similar approaches examining independently noisy and distributed thresholds, as an extension of the SSR model, were subsequently undertaken in McDonnell *et al.* (2002a) and also Rousseau and Chapeau-Blondeau (2005), without attempting to find the optimal thresholds values.

A very interesting observation made in Stocks (2000a) is that although better performance than SSR in the absence of noise – or with very small noise – is obtained by a quantizer with optimal thresholds, the SSR situation may be more robust to nonstationary signal distributions, since, if the signal distribution changes, the optimal noiseless thresholds will change.

Later, Stocks (2001b) analyzes the simple case of $N = 3$ and Gaussian signal and noise, by finding the optimal thresholds for a range of values of σ. Stocks (2001b) states the assumption that by symmetry we would expect that one of three threshold values would be zero, say $\theta_2 = 0$, and that $\theta_1 = -\theta_3$. With these assumptions, maximizing the mutual information depends only on a single variable, that is θ_1. The mutual information can be calculated for various values of θ_1 to find the optimal value. Stocks (2001b) does not plot the optimal value of θ_1 against σ, but does plot the mutual information corresponding to the optimal θ_1, which decreases monotonically with σ.

Neither Stocks (2000a) nor Stocks (2001b) discusses how the mutual information is calculated for the case of nonidentical threshold values; however, as is discussed in Section 8.2, numerical calculations are reasonably straightforward, especially for small N such as $N = 3$.

The sparse nature of previous work on optimal quantization in the presence of independent threshold noise demonstrates that the area is wide open for study, and this chapter now comprehensively investigates this problem.

Chapter structure

Section 8.2 describes the stochastic quantization model, and the differences between this model and the SSR model. It also mathematically describes the optimal stochastic quantization problem as an unconstrained optimization problem. Of particular importance is the outline of a method for recursively generating the transition probabilities for arbitrary thresholds and signal and noise distributions. Section 8.3 then describes relevant solution methods for optimization problems such as discussed in Section 8.2. Next, Sections 8.4 and 8.5 present the results of finding the optimal threshold values for various numbers of threshold devices, signal and noise distributions, and the two different objectives of maximizing the mutual information, and minimizing the MSE distortion. Section 8.6 then discusses the results of Sections 8.4 and 8.5 and explains some of the key features of these results. Section 8.7 briefly examines how the optimality of SSR depends on N. The chapter is then summarized in the concluding section.

8.2 Optimal quantization model

A schematic model of the system we examine in this chapter is given by Fig. 4.1 in Chapter 4. Note that when we use this model in the SSR configuration, we set all thresholds to the same value, θ. Figure 4.1 can be seen to be more general than SSR, since it explicitly shows that each threshold device may have different threshold values, labelled $\theta_1, \ldots, \theta_N$.

As with the SSR model, the system consists of N threshold devices, which all receive the same signal, x. This signal is a sequence of *iid* samples drawn from a continuously valued distribution with probability density function (PDF), $f_x(x)$. The ith threshold device is subject to continuously valued *iid* additive noise, η_i, $(i = 1, \ldots, N)$ with PDF $f_\eta(\eta)$. Each noise signal is also independent of the input signal. The output from each comparator, y_i, is unity if the input signal plus the noise, η_i, are greater than the threshold, θ_i, of that device and zero otherwise. The output from threshold i, y_i, is summed to give the overall output signal, y. Hence, y is a discrete signal, which can have integer values between zero and N.

Unlike in Chapter 4, we need to label the output of threshold device i in terms of the ith threshold value, for a given x, as

$$y_i(x) = \begin{cases} 1 & \text{if} \quad x + \eta_i \geq \theta_i, \\ 0 & \text{otherwise.} \end{cases} \tag{8.1}$$

The overall output of the array is still $y(x) = \sum_{i=1}^{N} y_i(x)$, which can be expressed in terms of the signum (sign) function as

$$y(x) = \frac{1}{2} \sum_{i=1}^{N} \text{sign}[x + \eta_i - \theta_i] + \frac{N}{2}. \tag{8.2}$$

As is the case in Chapter 4, the joint input–output PDF can be written as

$$f_{xy}(x, y) = P_{y|x}(y = n|x) f_x(x) \tag{8.3}$$
$$= f_{x|y}(x|y = n) P_y(n), \tag{8.4}$$

where the $P_{y|x}(y = n|x)$ are the *transition probabilities*, and $P_y(n)$ is the output probability mass function, which can be written as

$$P_y(n) = \int_{-\infty}^{\infty} P_{y|x}(y = n|x) f_x(x) dx, \quad n = 0, \ldots, N. \tag{8.5}$$

We shall always assume knowledge of $f_x(x)$, and from this point forward abbreviate our notation for the transition probabilities to $P_{y|x}(y = n|x) = P_{y|x}(n|x)$. Furthermore, in this chapter the PDFs studied, that is $f_x(x)$ and $f_\eta(x)$, are always even functions about a mean of zero.

Recall from Table 4.1 in Chapter 4 that for the distributions considered, the variance of the signal is a function of σ_x and the variance of the noise is a function of σ_η. It was shown in Chapter 4 that for such distributions, when $\theta = 0$, the mutual information is always a function of the ratio $\sigma = \sigma_\eta/\sigma_x$. In this chapter, we shall again parameterize the noise intensity with this same inverse signal-to-noise ratio, σ. However, we also briefly pointed out in Chapter 4 that, if $\theta \neq 0$, the mutual information will also be a function of θ/σ_x as well as σ and θ independently. We might therefore expect that, for arbitrary thresholds, the mutual information will also depend on these factors. This makes sense, as σ_x is generally a measure of the 'width' of the PDF $f_x(x)$. Therefore, if both σ_η and σ are fixed, then σ_x must change. If this happens, then we would intuitively expect the optimal thresholds to also vary with θ_i/σ_x. Given this, we arbitrarily set $\sigma_x = 1$ in the remainder of this chapter. Therefore, results plotted against σ are equivalent to results plotted against σ_η, and results given for optimal thresholds and optimal reconstruction points can be taken to be optimal values of θ_i/σ_x and $\hat{x}_n/\sigma_x$.

To allow for non-identical thresholds, we generalize the notation used in Stocks (2000c) by letting $P_{1|x,i}$ be the probability of device i being 'on' – that is, the probability that the sum of the input signal and noise exceeds the threshold value, θ_i – given the input signal x. Then

$$P_{1|x,i} = \int_{\theta_i - x}^{\infty} f_\eta(\eta) d\eta = 1 - F_\eta(\theta_i - x) \quad i = 1, \ldots, N, \tag{8.6}$$

where $F_\eta(\cdot)$ is the cumulative distribution function (CDF) of the noise. Since we assume $f_\eta(\eta)$ is an even function of η, then

$$P_{1|x,i} = F_\eta(x - \theta_i). \tag{8.7}$$

Given a noise density, $f_\eta(\eta)$, and threshold value, θ_i, $P_{1|x,i}$ can be calculated exactly for any value of x from Eq. (8.7).

Moment generating function

The *moment generating function* (MGF) of a random variable, X, is defined (Yates and Goodman 2005) as

$$\phi_X(s) = E[\exp(Xs)], \tag{8.8}$$

where s is real-valued. The main use of the MGF is to derive the moments of a random variable, since it can be shown that the pth moment of X is

$$E[X^p] = \frac{d^p \phi_X(s)}{ds^p} \bigg|_{s=0}. \tag{8.9}$$

Now consider the conditional output of the array of threshold devices

$$y(x) = \sum_{i=1}^{N} y_i(x). \tag{8.10}$$

The $y_i(x)$ in the summation are independent random variables. It can be shown in such a situation that the MGF of $y(x)$ is the product of the MGFs of each $y_i(x)$ (Yates and Goodman 2005) and we have

$$\phi_y(s) = \prod_{i=1}^{N} \phi_{y_i}(x). \tag{8.11}$$

However, we have also the probability mass function of y given x as $P_{y|x}(n|x)$. Therefore, from Eq. (8.8)

$$\phi_y(s) = E[\exp(ys)]$$

$$= \sum_{n=0}^{N} P_{y|x}(n|x) \exp(ns). \qquad (8.12)$$

We have also for each individual $y_i(x)$ that

$$\phi_{y_i}(s) = (1 - P_{1|x,i}) \exp(0s) + P_{1|x,i} \exp(1s)$$

$$= 1 - P_{1|x,i} + P_{1|x,i} \exp(s). \qquad (8.13)$$

Substituting from Eq. (8.13) into Eq. (8.11) and equating with Eq. (8.12) gives a relationship between the set of $\{P_{y|x}(n|x)\}$ and the set of $\{P_{1|x,i}\}$ as

$$\phi_y(s) = \prod_{i=1}^{N}(1 - P_{1|x,i} + P_{1|x,i}\exp(s)) = \sum_{n=0}^{N} P_{y|x}(n|x)\exp(ns), \qquad (8.14)$$

which holds for arbitrary threshold values.

Note that instead of defining the MGF in terms of $\exp(s)$, it is sometimes defined in terms of z^{-1}, where $z = \exp(-s)$. Using this approach, $P_{y|x}(n|x)$ can be described as the coefficient of z^{-n} in the power series expansion of $\prod_{i=1}^{N}[1 - P_{1|x,i} + z^{-1}P_{1|x,i}]$. That is

$$\prod_{i=1}^{N}[1 - P_{1|x,i} + z^{-1}P_{1|x,i}] = \sum_{n=0}^{N} P_{y|x}(n|x)z^{-n}. \qquad (8.15)$$

Conditional output moments

Since each $y_i(x)$ is independent, the expected value of y given x is the sum of the expected values of each y_i given x

$$E[y|x] = \sum_{i=1}^{N} P_{1|x,i}. \qquad (8.16)$$

For the special case of SSR, each $P_{1|x,i} = P_{1|x}$ and $E[y|x] = NP_{1|x}$, which agrees with the result for SSR given in Chapter 6.

Similarly, an expression for the variance of y given x can be derived as the sum of the N individual variances. This fact is due to the random variable y given x being the sum of the N individual threshold outputs, y_i, each of which is an independent random variable for a given x (Yates and Goodman 2005). Thus, the covariance between y_i and y_j, $i \neq j$, is zero and

$$\text{var}[y|x] = \sum_{i=1}^{N} P_{1|x,i}(1 - P_{1|x,i}). \tag{8.17}$$

We saw in Chapter 6 that the variance of y given x for a binomial distribution is $N P_{1|x}(1 - P_{1|x})$, which also precisely agrees with Eq. (8.17) for the SSR case.

These results for the conditional mean and variance can be easily verified from the expression for the MGF of Eq. (8.14) and the relation of Eq. (8.9).

Transition probabilities

Recall from Chapter 4 that the transition probabilities for the SSR model are given by the binomial distribution, as in Eq. (4.9). Although the previous section derives a formula for the MGF of the conditional output distribution, this formula is not useful for specifying the distribution itself for arbitrary thresholds. However, it is possible to find $P_{y|x}(n|x)$ numerically.

This requires making use of the probability that any given threshold device is 'on', given x. Assuming $P_{1|x,i}$ has been calculated for desired values of x, a convenient way of numerically calculating the probabilities $P_{y|x}(n|x)$ for a given number of threshold devices, N, is as follows.

Let $T_j^k(x)$ denote the probability that j of the thresholds 1, ..., k are 'on', given x. Then $T_0^1(x) = 1 - P_{1|x,1}$ and $T_1^1(x) = P_{1|x,1}$ and we have a set of recursive formulas

$$
\begin{aligned}
T_0^{k+1}(x) &= (1 - P_{1|x,k+1})T_0^k(x), \\
T_j^{k+1}(x) &= P_{1|x,k+1}T_{j-1}^k(x) + (1 - P_{1|x,k+1})T_j^k(x), \quad j = 1, \ldots, k, \\
T_{k+1}^{k+1}(x) &= P_{1|x,k+1}T_k^k(x).
\end{aligned} \tag{8.18}
$$

The recursion ends when $k + 1 = N$ and we have $P_{y|x}(n|x)$ given by $T_j^N(x) = T_n^N(x)$. We can rewrite Eq. (8.18) in matrix form as

$$
\begin{bmatrix}
T_0^{k+1}(x) \\
T_1^{k+1}(x) \\
T_2^{k+1}(x) \\
\vdots \\
T_k^{k+1}(x) \\
T_{k+1}^{k+1}(x)
\end{bmatrix}
=
\begin{bmatrix}
0 & T_0^k(x) \\
T_0^k(x) & T_1^k(x) \\
T_1^k(x) & T_2^k(x) \\
\vdots & \vdots \\
T_{k-1}^k(x) & T_k^k(x) \\
T_k^k(x) & 0
\end{bmatrix}
\begin{bmatrix}
P_{1|x,k+1} \\
1 - P_{1|x,k+1}
\end{bmatrix}. \tag{8.19}
$$

Thus, to derive the transition probabilities, $P_{y|x}(n|x)$, for a given x and any arbitrary threshold settings and noise PDF, it suffices to first calculate the set of $P_{1|x,i}$ from Eq. (8.6), and then to apply the recursive formula of Eq. (8.19).

Note that this approach also works well in the special case of SSR where all thresholds are zero. Although in this case $P_{y|x}(n|x)$ can be calculated from the binomial formula, for large N using the binomial formula directly requires calculations of very large numbers, since $\binom{N}{n}$ can be very large. Such large numbers cannot be accurately represented in a computer. The alternative approach of using the recursive formulas of Eq. (8.19) has the benefit that $P_{y|x}(n|x)$ can be found with excellent accuracy for very large values of N, since no such large numbers need to be calculated. In fact, since each T_j^k represents a probability, all numbers used in the calculation of each $P_{y|x}(n|x)$ are numbers between zero and unity.

The order of complexity of a software implementation of the recursive formula of Eq. (8.19) is $O(N^2)$, for any given x. If N is doubled, then the runtime of an implementation will be approximately quadrupled. More specifically, if $P_{1|x,i}$ and $1 - P_{1|x,i}$ are pre-computed, there will be $0.5N(N-1)$ additions and $(N+2)(N-1)$ multiplications to find the transition probabilities.

Appendix 1 gives expressions for a number of probability distributions, and for each of these gives expressions for $P_{1|x}$ obtained directly from Eq. (8.6) for threshold value, θ.

We are now ready to mathematically describe our optimal stochastic quantization problem.

Optimization problem formulation

We denote a vector of threshold values $(\theta_1, \theta_2, \ldots, \theta_N)$ as $\vec{\theta}$. Note that for the purpose of optimization, the order of the values of the vector components is not important. The aim is to find the vector of thresholds, $\vec{\theta}^*$, that maximizes some *objective function*, $f(\sigma, \theta_1, \theta_2, \ldots, \theta_N) = f(\sigma, \vec{\theta})$, as the noise parameter, σ, varies. The optimal value of the objective function for a given σ is then $f(\sigma, \vec{\theta}^*)$. Such a problem can be formulated for a given value of σ as the nonlinear – since the objective function will be nonlinear – optimization problem

$$\text{Find:} \quad \max_{\vec{\theta}} f(\sigma, \vec{\theta})$$

$$\text{subject to:} \quad \vec{\theta} \in \mathbb{R}^N. \tag{8.20}$$

Such a problem can be solved by standard optimization techniques, as will be described in Section 8.3.

8.3 Optimization solution algorithms

Unconstrained multidimensional optimization

The optimization problem of (8.20) simply describes an N-dimensional unconstrained function optimization. The function $f(\sigma, \vec{\theta})$ is, as we shall see, a nonlinear
function of $\vec{\theta}$. Solving such a problem is amenable to many standard techniques (Nocedal and Wright 1999), most of which rely on calculations of the
gradient of the objective function with respect to the free variables, which in this
case is the vector $\vec{\theta}$,

$$\nabla f(\sigma, \vec{\theta}) = \left(\frac{\partial f}{\partial \theta_1}, \frac{\partial f}{\partial \theta_1}, \ldots, \frac{\partial f}{\partial \theta_N} \right). \tag{8.21}$$

Critical points – maxima, minima, or saddle points – occur when the gradient
vector is equal to $\vec{0}$.

Standard optimization techniques usually start with some arbitrary initial solu-
tion, say $\vec{\theta}_0$, and then apply some method of finding a new solution, $\vec{\theta}_1$, from $\vec{\theta}_0$, that
increases the value of the objective function. This process is repeated in an iterative
fashion, until no further increase in the objective function can be found. Finding
a new vector that increases the value of the objective function is often achieved
by finding the gradient vector at a current solution, $\vec{\theta}_k$, and then performing a
one-dimensional search for the maximum in the direction of the gradient.

The method used to obtain the results presented in this chapter is a method called
the Broyden–Fletcher–Goldfarb–Shanno (BFGS) algorithm. This algorithm falls
into the sub-category of optimization algorithms known as *quasi-Newton methods* (Press *et al.* 1992). The BFGS method was found to be superior in speed
and convergence to other similar methods such as the Fletcher–Reeves–Polak–
Ribiere conjugate gradient method, and the Powell conjugate gradient method.
Further information on these algorithms can be found in Nocedal and Wright
(1999) or Press *et al.* (1992).

Note that although this is not relevant for an implementation of such an optimization procedure, changing the threshold vector will induce a new set of transi-
tion probabilities, $P_{y|x}(n|x)$, which will give a new value of the objective function.
This means it is equivalent to write the optimization problem of (8.20) in terms of
an optimization over the $N + 1$ correlated functions of x that describe the transition
probabilities. Such formulations are often solved using the calculus of variations.

For example, our optimization problem is similar to previous work on clustering
and neural coding problems solved using a combination of the calculus of variations and a method known as deterministic annealing (Rose *et al.* 1990, Rose *et al.*
1992, Rose 1994, Rose 1998, Tishby *et al.* 1999, Dimitrov and Miller 2001, Dimitrov *et al.* 2003). In particular, the formulation reached in Dimitrov and Miller

(2001) can be expressed in a fashion identical to Problem (8.20) with one exception. Here, we have *structural constraints* on how the transition probabilities, $P_{y|x}(n|x)$, are formed, since – as expressed by Eq. (8.19) – each $P_{y|x}(n|x)$ is a function of the set of probabilities, $P_{1|x,i}$. Due to this difference, the solution method used in Dimitrov and Miller (2001) to find the optimal conditional distribution, $\{P_{y|x}(n|x)\}$, cannot be used here, and instead we concentrate on optimizing the only free variable, the vector of threshold values, $\vec{\theta}$.

Dealing with local optima

One of the major problems with standard deterministic optimization methods, such as the BFGS method, is that they do not cope well when applied to objective functions that possess more than one local minimum or maximum. If the objective function is known to be convex, this is not an issue. However, if it is not convex, then these standard methods will converge towards a local optimum, but give no guarantees about whether or not this optimum is the global optimum. The local optimum found depends on the initial solution used. Hence, one possibility for trying to find a good local optimum is to run a standard optimization procedure many times, using many different initial solutions, and then pick the solution found with the largest value – if the problem is a maximization – of the objective function.

Numerical experiments show that for the objective functions considered in this chapter, there are many local maxima. In particular, it seems that the SSR situation (all thresholds zero) is always a local optimum. This means that the gradient of the mutual information with respect to $\vec{\theta}$ is always the zero vector, $\vec{0}$, at $\vec{\theta} = \vec{0}$.

However, we shall see that there is a special structure to the local optima, which makes it theoretically possible to find all local optima. In particular, numerical experiments indicate that there are at most two locally optimal solutions corresponding to each possible ordered clustering of N thresholds to unique values. For example, in the case of $N = 3$, there are four locally optimal situations: (1) the case of all thresholds being equal to the same value; (2) the case of all three threshold values being unique; (3) the case of two thresholds being equal to the same positive value, and the third threshold being negative; and (4) the case of one threshold being positive, and the other two thresholds being equal to the same negative value. Note that the third and fourth cases are effectively equivalent, as the optimal threshold values in the third case are the negative of the optimal threshold values in the fourth case, and give the same function maximum.

Thus, in order to find the globally optimal solution, all possible 'clusters' of N thresholds can be trialled. The dimension of the optimization problem in each case is the number of unique threshold values. The locally optimal thresholds only change very slightly for a small change in σ. Hence, for each possible local solution, once the optimal thresholds are found for a particular value of σ, these values can be used as the initial values for a new value of σ a very small increment larger.

This procedure works well for the results presented in this chapter where N is no greater than five; however, the number of local optima increases combinatorially with N, and it is not practical to consider all local optima for N any larger than five. As mentioned, one way of dealing with this is to simply try a number of random initial solutions, and pick the best. Another way is to employ random search optimization algorithms such as simulated annealing (Kirkpatrick *et al.* 1983) or genetic algorithms (Beasley *et al.* 1993). Unlike deterministic methods, such algorithms are known to be able to 'climb' out of local optima, and provide a greater chance of finding a global optimum. However, for the small N results presented in this chapter, it was found that the combination of the BFGS and initial solution method outlined above is far superior to random search methods in runtime and for ensuring all local optima are tracked.

Having now described the solution methods we use, the next two sections give results for the optimal stochastic quantization for maximum mutual information and minimum MSE distortion.

8.4 Optimal quantization for mutual information

Recall from Chapter 4 that the mutual information between the input and output signals of the array of threshold devices in Fig. 4.1 is given by

$$I(x, y) = H(y) - H(y|x)$$

$$= -\sum_{n=0}^{N} P_y(n) \log_2 P_y(n)$$

$$- \int_{-\infty}^{\infty} f_x(x) \sum_{n=0}^{N} P_{y|x}(n|x) \log_2 P_{y|x}(n|x) dx, \qquad (8.22)$$

where $H(y)$ is the entropy of the output signal, and $H(y|x)$ is the average conditional output entropy. Given the procedure described in Section 8.2 for numerically calculating the transition probabilities for any value of x, the mutual information can be calculated for arbitrary thresholds and $f_x(x)$ by numerical integration of Eq. (8.22). Note that this integration requires calculations of $P_{y|x}(n|x)$ from Eq. (8.19) for many values of x.

Given the aim of maximizing the mutual information, the optimization problem is

Find: $\max_{\vec{\theta}} I(x, y)$

subject to: $\vec{\theta} \in \mathbb{R}^N.$ (8.23)

Absence of noise

As pointed out in Stocks (2000a), an exact solution to Problem (8.23) can be found in the absence of noise, that is when $\sigma = 0$. In this case, the average conditional output entropy, $H(y|x)$, is zero, and therefore the mutual information is the output entropy, $H(y)$. Maximizing the mutual information reduces to maximizing the output entropy. It is well known that for a discrete random variable this occurs when all states are equally likely (Cover and Thomas 1991), which means $I(x, y) = \log_2 (N + 1)$. In the absence of noise, we therefore require that the output probability mass function is

$$P_y(n) = \frac{1}{N+1}, \quad n = 0, \ldots, N. \tag{8.24}$$

If we let $\theta_n > \theta_{n-1}$, then – since there is no threshold noise – the output, y, is in state n if and only if $x \in [\theta_{n-1}, \theta_n]$. This means that there is no ambiguity about what output state corresponds to a given value of x, and that therefore $P_{y|x}(n|x) = 1 \ \forall \ x \in [\theta_{n-1}, \theta_n]$. More precisely, since $P_y(n) = \int_x f_x(x) P_{y|x}(n|x) dx$, this requires that

$$\int_{-\infty}^{\theta_1} f_x(x) dx = \frac{1}{N+1},$$

$$\int_{\theta_i}^{\theta_{i+1}} f_x(x) dx = \frac{1}{N+1}, \quad i = 1, \ldots, n-1$$

$$\text{and} \quad \int_{\theta_N}^{\infty} f_x(x) dx = \frac{1}{N+1}. \tag{8.25}$$

This implies that

$$F_x(\theta_i) = \frac{i}{N+1}, \quad i = 1, \ldots, N, \tag{8.26}$$

where $F_x(\cdot)$ is the CDF of the input signal, and that therefore

$$\theta_i = F_x^{-1}\left(\frac{i}{N+1}\right), \quad i = 1, \ldots, N, \tag{8.27}$$

where $F_x^{-1}(\cdot)$ is the inverse cumulative distribution function (ICDF) of the input signal.

This analysis shows that a simple formula for the optimal thresholds for mutual information is available in the absence of noise and that all N optimal thresholds are uniquely valued.

Optimal decoding for maximized I (x, y) in the absence of noise

If the thresholds are set to maximize the noiseless mutual information, then from Eq. (8.40) the minimum mean square error (MMSE) reconstruction points are given by

$$\hat{x}_0 = (N + 1) \int_{-\infty}^{\theta_1} x f_x(x) dx,$$

$$\hat{x}_n = (N + 1) \int_{\theta_n}^{\theta_{n+1}} x f_x(x) dx, \quad n = 1, \ldots, N - 1$$

$$\hat{x}_N = (N + 1) \int_{\theta_N}^{\infty} x f_x(x) dx, \tag{8.28}$$

since all output states are equally probable. Substituting from Eq. (8.27) into Eq. (8.28) gives the exact optimal MMSE reconstruction points for the thresholds that maximize the noiseless information.

We can simplify this substitution by making a change of variable. Let $x = F_x^{-1}(\frac{\tau}{N+1})$. Then $\tau = (N + 1) F_x(x)$ and $d\tau = (N + 1) f_x(x) dx$. Carrying this out in Eq. (8.28), and using Eq. (8.26) gives

$$\hat{x}_n = \int_{\tau=n}^{\tau=n+1} F_x^{-1} \left(\frac{\tau}{N + 1} \right) d\tau, \quad n = 0, \ldots, N. \tag{8.29}$$

The integral in Eq. (8.29) can be solved exactly in some cases. For example, for a uniformly distributed signal we get

$$\hat{x}_n = \sigma_x \left(\frac{2n + 1}{2N + 2} - \frac{1}{2} \right), \quad n = 0, \ldots, N, \tag{8.30}$$

and for a logistically distributed signal we get

$$\hat{x}_n = \frac{\sqrt{3}\sigma_x}{\pi} ((n + 1) \ln(n + 1) - n \ln(n) + (N - n) \ln(N - n) -$$

$$(N + 1 - n) \ln(N + 1 - n)), \quad n = 0, \ldots, N. \tag{8.31}$$

Results in the presence of noise

Results for N = 2, ..., 5

We now present the results of solving Problem (8.23) for nonzero σ, and various matched signal and noise distributions. The optimal thresholds as found by the deterministic algorithm explained in Section 8.3, are plotted for $N = 2, 3, 4$, and 5, in Figs 8.1, 8.2, 8.3, and 8.4, and the optimal reconstruction points for the thresholds of the $N = 5$ case in Fig. 8.5. Discussion of these results is left for Section 8.6.

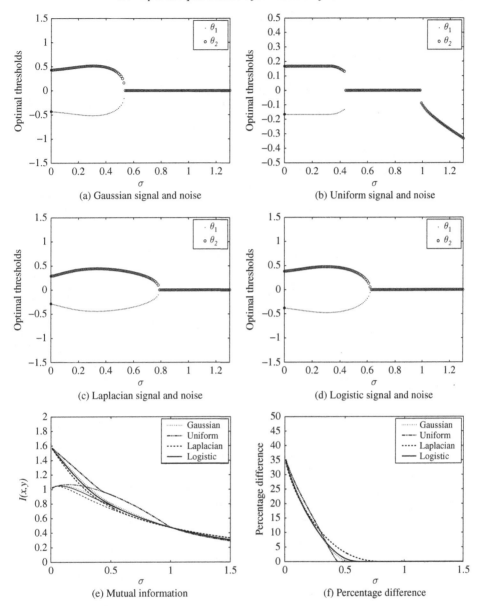

Fig. 8.1. Optimal thresholds for mutual information, $N = 2$. Figures 8.1(a), 8.1(b), 8.1(c), and 8.1(d) show plots of the optimal thresholds for $N = 2$ for four different matched signal and noise pairs against increasing noise intensity, σ. The optimal noiseless threshold values calculated from Eq. (8.27) are shown by large black dots. Figure 8.1(e) shows the mutual information obtained with these optimal threshold settings against increasing σ, as well as the mutual information for SSR shown with dotted lines. Figure 8.1(f) shows the percentage difference between the mutual information obtained by optimally setting the thresholds, and that for SSR, against increasing σ. For σ greater than some critical value, σ_c, the SSR situation can be seen to be optimal.

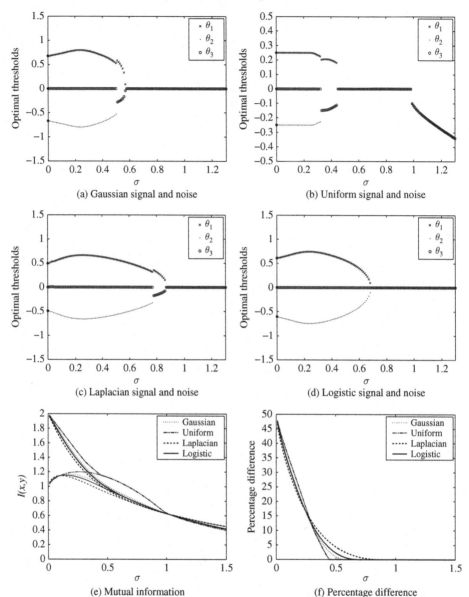

Fig. 8.2. Optimal thresholds for mutual information, $N = 3$. Figures 8.2(a), 8.2(b), 8.2(c), and 8.2(d) show plots of the optimal thresholds for $N = 3$ for four different matched signal and noise pairs against increasing noise intensity, σ. The optimal noiseless threshold values calculated from Eq. (8.27) are shown by large black dots. Figure 8.2(e) shows the mutual information obtained with these optimal threshold settings against increasing σ, as well as the mutual information for SSR shown with dotted lines. Figure 8.2(f) shows the percentage difference between the mutual information obtained by optimally setting the thresholds, and that for SSR, against increasing σ. For σ greater than some critical value, σ_c, the SSR situation can be seen to be optimal.

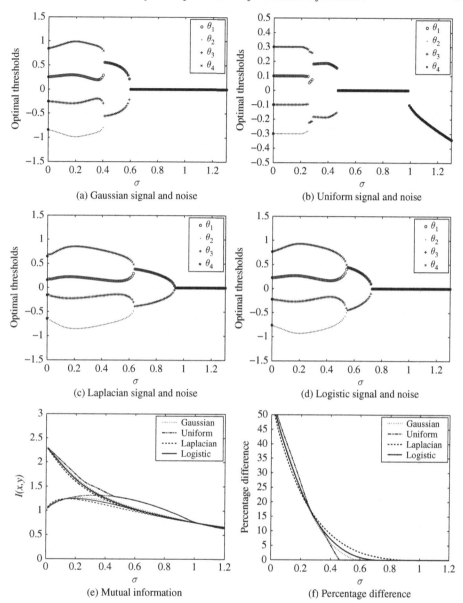

Fig. 8.3. Optimal thresholds for mutual information, $N = 4$. Figures 8.3(a), 8.3(b), 8.3(c), and 8.3(d) show plots of the optimal thresholds for $N = 4$ for four different matched signal and noise pairs against increasing noise intensity, σ. The optimal noiseless threshold values calculated from Eq. (8.27) are shown by large black dots. Figure 8.3(e) shows the mutual information obtained with these optimal threshold settings against increasing σ, as well as the mutual information for SSR shown with dotted lines. Figure 8.3(f) shows the percentage difference between the mutual information obtained by optimally setting the thresholds, and that for SSR, against increasing σ. For σ greater than some critical value, σ_c, the SSR situation can be seen to be optimal.

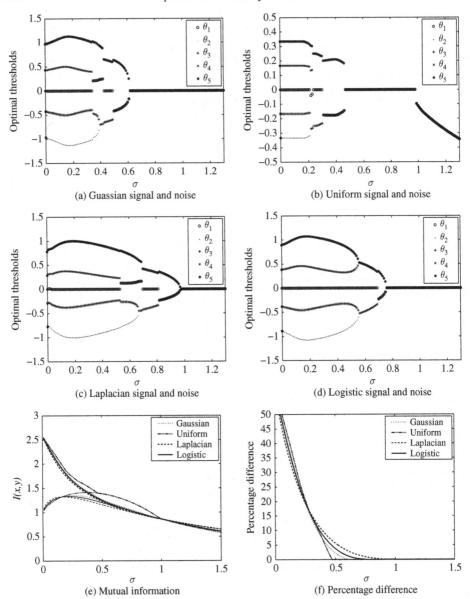

(a) Guassian signal and noise

(b) Uniform signal and noise

(c) Laplacian signal and noise

(d) Logistic signal and noise

(e) Mutual information

(f) Percentage difference

Fig. 8.4. Optimal thresholds for mutual information, $N = 5$. Figures 8.4(a), 8.4(b), 8.4(c), and 8.4(d) show plots of the optimal thresholds for $N = 5$ for four different matched signal and noise pairs against increasing noise intensity, σ. The optimal noiseless threshold values calculated from Eq. (8.27) are shown by large black dots. Figure 8.4(e) shows the mutual information obtained with these optimal threshold settings against increasing σ, as well as the mutual information for SSR shown with dotted lines. Figure 8.4(f) shows the percentage difference between the mutual information obtained by optimally setting the thresholds, and that for SSR, against increasing σ. For σ greater than some critical value, σ_c, the SSR situation can be seen to be optimal.

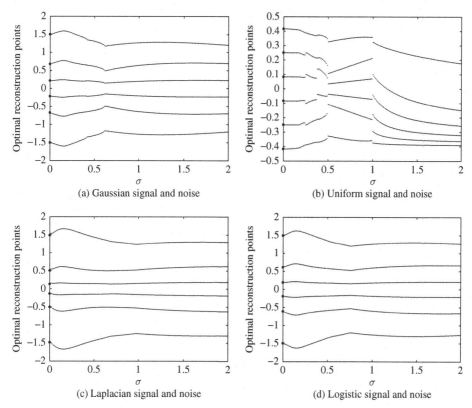

Fig. 8.5. Optimal reconstruction points for mutual information, $N = 5$. Figures 8.5(a), 8.5(b), 8.5(c), and 8.5(d) show plots of the optimal reconstruction points for $N=5$ for four different matched signal and noise pairs against increasing noise intensity, σ. The optimal noiseless reconstruction points calculated from Eq. (8.29) are shown by large black dots.

Results for $N = 15$

Figure 8.6 shows the optimal thresholds and mutual information for $N = 15$ and Gaussian signal and noise. The same qualitative behaviour as for each $N \leq 5$ case can be seen. However, we cannot be completely sure that the thresholds shown provide the globally optimal solution, as the number of local optima is now very large. However, provided only near optimal thresholds are required, we can find an approximation to the optimal thresholds for reasonably large N by applying the BFGS method for various initial conditions and can still be reasonably confident that the threshold values shown provide values of the mutual information very close to the optimal solution.

The next section looks at optimal quantization for the objective of minimizing the MSE distortion.

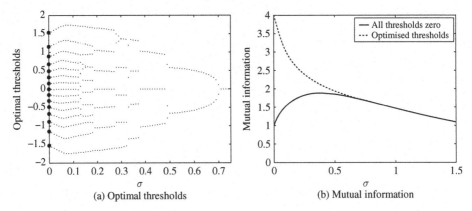

(a) Optimal thresholds   (b) Mutual information

Fig. 8.6. Optimal thresholds for mutual information for $N = 15$ and Gaussian signal and noise against increasing noise intensity, σ. The optimal noiseless threshold values calculated from Eq. (8.27) are shown by large black dots. Figure 8.6(a) shows the mutual information for the thresholds shown in Fig. 8.6(b), as well as the mutual information obtained in the SSR case. As with the $N = 2, \ldots, 5$ cases, for σ greater than some critical value, σ_c, the SSR situation can be seen to be optimal. Adapted with permission from McDonnell *et al.* (2006b). Copyright Elsevier 2006.

8.5 Optimal quantization for MSE distortion

Linear decoding

Recall from Chapter 6 that, for zero-mean input signals and a linear decoding of the form $\hat{y} = ay + b$, if the condition $E[\hat{y}] = 0$ is imposed, then $b = -aE[y]$. Note that, while for SSR the expected value of the output is $E[y] = N/2$, this will certainly not necessarily be the case for arbitrary thresholds.

As discussed in Chapter 6, with such a linear decoding, the optimal value of a is given by $a = \dfrac{E[xy]}{\text{var}[y]}$, so that the optimal linear reconstruction points are

$$\hat{y} = \frac{E[xy]}{\text{var}[y]}\,(y - E[y]). \tag{8.32}$$

Such a decoding is the optimal linear decoding, and is known as the Wiener decoding. The MSE distortion with such a decoding can be written in terms of the correlation coefficient of a linear decoding, ρ_{xy}, as

$$\text{MSE} = E[x^2](1 - \rho_{xy}^2), \tag{8.33}$$

where the correlation coefficient can be expressed as

$$\rho_{xy} = \frac{E[xy]}{\sqrt{E[x^2]\text{var}[y]}}. \tag{8.34}$$

Minimizing the linear decoding MSE is equivalent to maximizing the linear decoding correlation coefficient. Since we also assume knowledge of the input signal PDF, and therefore of its mean square value, $E[x^2]$, minimizing the linear decoding MSE distortion is equivalent to solving the optimization problem

$$\text{Find:} \quad \max_{\vec{\theta}} \frac{E[xy]}{\sqrt{\text{var}[y]}}$$
$$\text{subject to:} \quad \vec{\theta} \in \mathbb{R}^N. \tag{8.35}$$

Nonlinear decoding

Recall from Chapter 6 that the optimal MSE decoding is the nonlinear decoding given by $\hat{x}_n = E[x|n]$. This decoding results in the minimum possible MSE distortion for given transition probabilities, which is given by

$$\text{MMSE} = E[x^2] - E[\hat{x}^2]. \tag{8.36}$$

Since $E[x^2]$ is known, if the aim is to find the thresholds that minimize the MSE distortion resulting from using the MMSE decoding, the optimization problem is

$$\text{Find:} \quad \max_{\vec{\theta}} E[\hat{x}^2]$$
$$\text{subject to:} \quad \vec{\theta} \in \mathbb{R}^N. \tag{8.37}$$

Absence of noise

Unlike maximizing the mutual information, in general no simple formula exists for the threshold values that minimize the MSE distortion in the absence of noise for an arbitrary signal PDF. However, the optimal noiseless thresholds can be easily found numerically for a given $f_x(x)$, and the reconstruction points shown to be unique.

In the absence of noise, assume all thresholds are unique, and that therefore a given value of the output, $y = n$, can only be achieved by values of x that lie between consecutive thresholds, say θ_{i-1} and θ_i. The optimal MMSE reconstruction points are given by

$$\hat{x}_n = E_x[x|n] = \int_x x P_{x|y}(x|n)dx, \tag{8.38}$$

and since $P_{y|x}(n|x)$ is unity for $x \in [\theta_n, \theta_{n+1}]$

$$P_y(0) = \int_{-\infty}^{\theta_1} f_x(x)dx,$$

$$P_y(n) = \int_{\theta_n}^{\theta_{n+1}} f_x(x)dx, \quad n = 1, \ldots, N-1$$

$$P_y(N) = \int_{\theta_N}^{\infty} f_x(x)dx, \tag{8.39}$$

and

$$\hat{x}_0 = \frac{1}{P_y(0)} \int_{-\infty}^{\theta_1} x f_x(x)dx,$$

$$\hat{x}_n = \frac{1}{P_y(n)} \int_{\theta_n}^{\theta_{n+1}} x f_x(x)dx, \quad n = 1, \ldots, N-1$$

$$\hat{x}_N = \frac{1}{P_y(N)} \int_{\theta_N}^{\infty} x f_x(x)dx. \tag{8.40}$$

Therefore, each MMSE reconstruction point is the centroid of the corresponding partition of the input PDF. The MMSE distortion is

$$\text{MMSE} = E[x^2] - \sum_{n=0}^{N} \frac{1}{P_y(n)} \left(\int_{\theta_n}^{\theta_{n+1}} x f_x(x)dx \right)^2. \tag{8.41}$$

In the absence of noise, finding the optimal thresholds for the MMSE distortion can be achieved by a simple iterative procedure known as the *Lloyd Method I algorithm* (Lloyd 1982), which is a commonly used technique for finding the optimal quantization for a given source PDF. A second algorithm known as the Lloyd Method II algorithm is also called the Lloyd–Max algorithm, due to its rediscovery by Max (1960).[2] See also Gersho and Gray (1992) and Gray and Neuhoff (1998) for more details. The Lloyd Method I algorithm begins with an initial guess for the reconstruction points, and then finds the optimal thresholds for those points, which for MSE distortion, and any given decoding, can be shown simply to be the midpoints of the reconstruction points. Given these new thresholds, a new set of reconstruction points is found from Eq. (8.40), and the new MMSE distortion from Eq. (8.41). This new MMSE distortion can be shown to be smaller than the previous MMSE distortion. This iteration is repeated until the MMSE distortion no longer decreases, at which point the optimal noiseless thresholds and reconstruction points have been found.

---

[2] Note that both of Lloyd's algorithms were first described in an unpublished Bell Laboratories technical report in 1957 (Gersho and Gray 1992), and not published in the open literature until 1982 (Lloyd 1982).

The Lloyd Method I algorithm can also be used for the case of finding the optimal thresholds for a linear decoding. The only difference is that instead of calculating the new optimal reconstruction points at each iteration from Eq. (8.40), the optimal linear reconstruction points are calculated from Eq. (8.32), and the resulting MSE distortion from Eq. (8.33). The update of the optimal thresholds at each iteration remains as the midpoint between the current reconstruction points.

Results in the presence of noise

We now present the results of solving Problems (8.35) and (8.37) for nonzero σ and various matched signal and noise distributions. Due to the results having the same qualitative behaviour as those found in the case of maximizing the mutual information, we present optimal thresholds only for the case of $N = 5$ for minimized linear decoding MSE distortion (Fig. 8.7), and $N = 2$ and $N = 5$ for minimized MMSE distortion (Figs 8.8 and 8.9), and the optimal reconstruction points for $N = 5$ and MMSE distortion (Fig. 8.10). Since the uniform distribution has a different mean square value from the other distributions for the same value of σ, we plot the signal-to-quantization-noise ratio (SQNR) rather than the MSE distortion.

8.6 Discussion of results

Observations

First, we see from the figures above that the mutual information, or MSE distortion, obtained with the optimal thresholds is strictly monotonic with increasing σ. This means that no SR effect is seen for optimized thresholds. For small σ, optimizing the thresholds gives a substantial increase in performance. However, as σ increases, the difference between the SSR case and optimized thresholds decreases, until, for sufficiently large σ, SSR is optimal. While the fact that the performance decreases with increasing σ means that there is no advantage to be gained by increasing the noise level in such an optimized system, the fact that SSR is optimal for sufficiently large noise intensity is still highly significant. Unlike a single threshold device, where the optimal threshold value is always at the signal mean regardless of the noise level and therefore SR can never occur, this result shows that when the noise is large, changing the thresholds from a situation where SR can occur gains no advantage.

Second, inspection of all the figures showing optimal threshold values against σ in the previous two sections indicates several common features of the optimal threshold configuration. For very small noise, the optimal thresholds and reconstruction points are consistent with the optimal noiseless values. There does

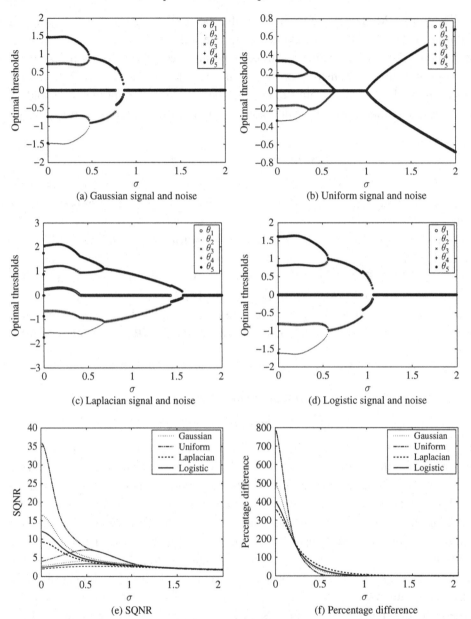

Fig. 8.7. Optimal thresholds for linear decoding MSE distortion, $N = 5$. Figures 8.7(a), 8.7(b), 8.7(c), and 8.7(d) show plots of the optimal thresholds for $N = 5$ for four different matched signal and noise pairs against increasing noise intensity, σ, and the objective of minimized linear decoding MSE distortion. The optimal noiseless thresholds, as calculated by the Lloyd Method I algorithm, are shown with large black dots. Figure 8.7(e) shows the SQNR obtained with these optimal threshold settings against increasing σ, as well as the linear decoding SQNR for SSR, which is shown with dotted lines. Figure 8.7(f) shows the percentage difference between the MSE distortion obtained by optimally setting the thresholds, and that for SSR, against increasing σ. For σ greater than some critical value, σ_c, the SSR situation can be seen to be optimal.

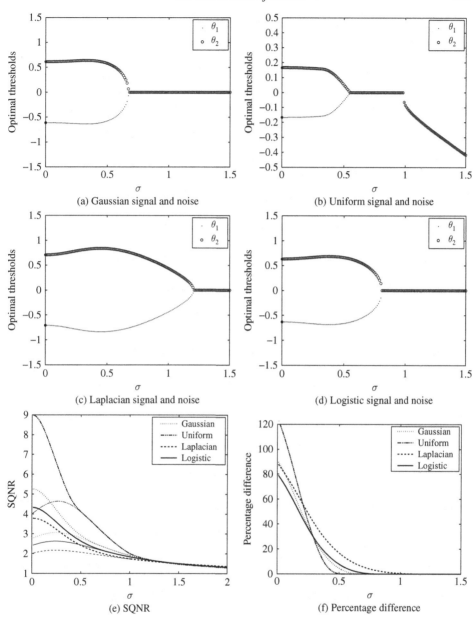

Fig. 8.8. Optimal thresholds for MMSE distortion, $N = 2$. Figures 8.8(a), 8.8(b), 8.8(c), and 8.8(d) show plots of the optimal thresholds for $N = 2$ for four different matched signal and noise pairs against increasing noise intensity, σ, and the objective of minimized MMSE distortion. The optimal noiseless thresholds, as calculated by the Lloyd Method I algorithm, are shown with large black dots. Figure 8.8(e) shows the SQNR obtained with these optimal threshold settings against increasing σ, as well as the SQNR for SSR, which is shown with dotted lines. Figure 8.8(f) shows the percentage difference between the MMSE distortion obtained by optimally setting the thresholds, and that for SSR, against increasing σ. For σ greater than some critical value, σ_c, the SSR situation can be seen to be optimal.

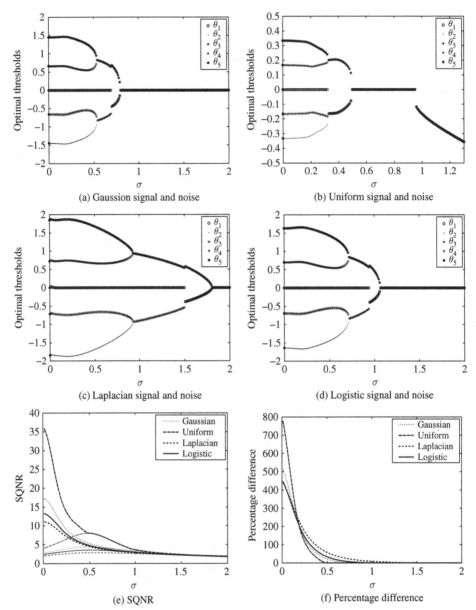

Fig. 8.9. Optimal thresholds for MMSE distortion, $N = 5$. Figures 8.9(a), 8.9(b), 8.9(c), and 8.9(d) show plots of the optimal thresholds for $N = 5$ for four different matched signal and noise pairs against increasing noise intensity, σ, and the objective of minimized MMSE distortion. The optimal noiseless thresholds, as calculated by the Lloyd Method I algorithm, are shown with large black dots. Figure 8.9(e) shows the SQNR obtained with these optimal threshold settings against increasing σ, as well as the SQNR for SSR, which is shown with dotted lines. Figure 8.9(f) shows the percentage difference between the MMSE distortion obtained by optimally setting the thresholds, and that for SSR, against increasing σ. For σ greater than some critical value, σ_c, the SSR situation can be seen to be optimal.

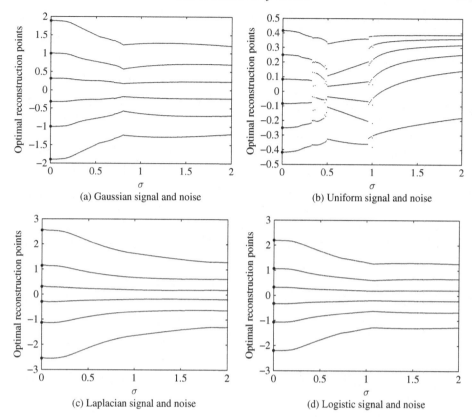

(a) Gaussian signal and noise

(b) Uniform signal and noise

(c) Laplacian signal and noise

(d) Logistic signal and noise

Fig. 8.10. Optimal reconstruction points for minimized MMSE, $N = 5$. Figures 8.10(a), 8.10(b), 8.10(c), and 8.10(d) show plots of the optimal reconstruction points for $N = 5$ for four different matched signal and noise pairs against increasing noise intensity, σ. The optimal noiseless reconstruction points, as calculated after applying the Lloyd Method I algorithm from Eq. (8.40) are shown with large black dots.

not appear to be a discontinuity in the optimal thresholds as the noise intensity increases from zero to some small nonzero value.

The most striking feature is the fact that *bifurcations* are present. Consider first the simplest case of maximized mutual information and $N = 2$. For each signal and noise pair, for σ between zero and some critical value greater than zero, σ_c, the optimal placements of the two thresholds are at $\pm A$, where $A > 0$. However, for $\sigma > \sigma_c$, the optimal thresholds both have the same value. Apart from the uniform case, this value is always the signal mean of zero, which is simply the SSR situation. For the uniform case, we see that for $\sigma \in [\sigma_c, 1]$, we have the SSR situation. However for $\sigma > 1$, both thresholds remain identical, but via a discontinuous bifurcation this identical value is no longer zero.

The behaviour for small σ and large σ seen in the $N = 2$ case persists for larger N; that is, for sufficiently small σ, the optimal threshold values are all unique, and for sufficiently large σ – with the exception of the case of uniform signal and noise, $N = 5$, and minimized linear decoding MSE distortion – the optimal threshold values are identical for each threshold.

Most importantly, apart from the special situation of the uniform case, it is evident in all cases that above a certain value of σ the SSR situation is optimal; that is, the optimal quantization for large noise is to set all thresholds to the signal mean.

For $N > 2$, we see that there are also regions of σ where some fraction of the optimal thresholds tend to cluster to particular identical values. We shall refer to the number of thresholds with the same value as the *size* of a cluster, and the actual threshold value of those thresholds as the *value* of a cluster.

This tendency of the optimal thresholds to form clusters at identical values leads to regions of asymmetry about the x-axis, since if, for example, $N = 5$ and there are two clusters of size three and two, then the value of the cluster of size two is larger in magnitude than the value of the cluster of size three. Note that in such regions of asymmetry, there are two globally optimal threshold vectors. The second global solution is simply the negative of the set of thresholds in the first global solution, that is $f(\vec{\theta}^*) = f(-\vec{\theta}^*)$. This result stems from the fact that both the signal and noise PDFs are even functions.

We can also see that it is quite common for bifurcations to occur, so that the number of clusters suddenly decreases with increasing σ. Sometimes a continuous bifurcation occurs as more than one cluster converges to the same value, as σ increases, to form a larger, merged, cluster. On other occasions a discontinuous bifurcation occurs, and two clusters with completely different values merge to form a larger cluster with a value somewhere between the two values of the two merging clusters. It does not appear possible for the number of clusters to increase with increasing σ, other than, again, for the case of uniform signal and noise, and minimized linear decoding MSE distortion.

However, further bifurcations can occur within a region of σ with k clusters, where the order of the size of the clusters changes with respect to the values of those clusters. For example, for Gaussian signal and noise and $N = 5$, Fig. 8.4 shows that there are three distinct clusters at $\sigma = 0.42$, and three at $\sigma = 0.45$. However, there is a bifurcation between $\sigma = 0.42$ and $\sigma = 0.43$, since for $\sigma = 0.42$ the optimal solution is to have clusters of size 2, 2, 1, in that order from smallest cluster values to largest, while for $\sigma = 0.43$ the optimal solution is to have clusters of size 2, 1, 2.

Figure 8.11 further illustrates the behaviour of the clusters for this case, for six different values of σ. Each sub-panel shows the size of each cluster on the y-axis, and the value of each cluster on the x-axis. For noise greater than the final bifurcation point, we have the SSR region occurring; that is, the optimal solution is for all thresholds to be equal to the signal mean of zero.

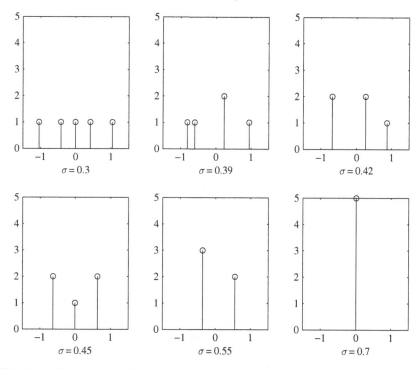

Fig. 8.11. The number of thresholds in each cluster for various values of σ in the case of Gaussian signal and noise and $N = 5$, and maximized mutual information.

The bifurcational structure is quite surprising, but appears to be fundamental to the problem type, since the same qualitative behaviour occurs whether we are maximizing the mutual information, or minimizing the MSE distortion. In Chapter 9, we shall see the same behaviour for constrained mutual information maximization. We can also see that the pattern is qualitatively the same for each signal and noise pair considered. Furthermore, numerical experiments find that very similar patterns appear for mixed signal and noise distributions, such as a uniform signal subject to Gaussian noise.

However, there are some anomalies that make it difficult to generalize. For example consider the case of logistic signal and noise, and $N = 3$ shown in Fig. 8.2(d). In this case, there is only one bifurcation, since at some value of σ between 0.68 and 0.69, the optimal threshold solution changes from all three thresholds being unique, to all three thresholds being zero, without an intermediate region with a cluster of size two, and a cluster of size one. Furthermore, while in the Gaussian case of $N = 5$ there are five bifurcations, one of which occurs within a region of three clusters, in the logistic case of $N = 5$ there are only three bifurcations, since there is no region where there are four clusters.

A clue to the reason for the bifurcations comes from inspection of the plots of the optimal reconstruction points corresponding to the optimal thresholds. Despite the bifurcational structure in the optimal thresholds, in all cases but the uniform case the optimal reconstruction points appear to change very smoothly with σ. This indicates that at values of σ where discontinuous bifurcations occur in the optimal thresholds more than one local optimum gives the same optimal reconstruction points.

Mathematical description of optimal thresholds

Quantizer point density function

We now describe mathematically the observations made above. For the purposes of optimization, the ordering of the optimal threshold vector, $\vec{\theta}^*$, is not important. However, to simplify the mathematical description, we now introduce an ordered sequence notation for the optimal thresholds. Specifically, we label the ith optimal threshold value as θ_i^*, so that the sequence $(\theta_i^*)_{i=1}^N$ is nondecreasing. As we saw in Eq. (8.27), in the absence of noise, it is straightforward to show for the goal of maximum mutual information that each optimal threshold is given by
$\theta_i^* = F_x^{-1}\left(\frac{i}{N+1}\right)$.

We now introduce a concept used in the theoretical analysis of high resolution quantizers in information theory – that of a quantizer *point density function*, $\lambda(x)$, defined over the same support variable as the source PDF, $f_x(x)$ (Gray and Neuhoff 1998). The point density function has the property that $\int_x \lambda(x)dx = 1$, and usually is used only in the context where the number of thresholds is very large. In this situation, the point density function gives the density of thresholds across the support of the signal PDF.

For any given N, and some nonzero values of σ, we observe from the plots of optimal thresholds that our empirically optimal threshold sequence, $(\theta_i^*)_{i=1}^N$, can have at most $k(\sigma)$ unique values, where $1 \leq k \leq N$. When bifurcations occur as σ increases, $k(\sigma)$ may either decrease or – in the situation where the ordering of clusters changes – remain constant.

We now denote $v(j, \sigma)$ as the fraction of the total thresholds in the jth cluster, at noise intensity σ, where $j \in \{1, \ldots, k(\sigma)\}$, so that $\sum_{j=1}^{k(\sigma)} v(j, \sigma)=1$. Thus, $v(j, \sigma)$ is the size of the jth cluster divided by N.

Denote the value of the jth cluster as Θ_j, so that the size of the cluster at $x = \Theta_j$ is $Nv(j, \sigma)$. As with the ordered optimal threshold sequence, we can define an ordered sequence of cluster values as $(\Theta_j)_{j=1}^{k(\sigma)}$. Unlike the optimal threshold sequence, this sequence is strictly increasing.

Table 8.1. *Parameters for the optimal threshold point density function for N = 5, Gaussian signal and noise, and maximized mutual information for several values of σ. The value of k for each σ and the corresponding cluster sizes, Nv(j, σ), and the approximate cluster values, $(\Theta_j)_{j=1}^{k(\sigma)}$, are shown.*

σ	$k(\sigma)$	$\{Nv(j, \sigma)\}$	$(\Theta_j)_{j=1}^{k(\sigma)}$
0.3	5	$\{1, 1, 1, 1, 1\}$	$(-1.0410, -0.4164, 0.0, 0.4164, 1.0410)$
0.39	4	$\{1, 1, 2, 1\}$	$(-0.8075, -0.5967, 0.2288, 0.9489)$
0.42	3	$\{2, 2, 1\}$	$(-0.6783, 0.2549, 0.8770)$
0.45	3	$\{1, 2, 1\}$	$(-0.6319, 0.0, 0.6319)$
0.55	2	$\{3, 2\}$	$(-0.3592, 0.5513)$
0.7	1	$\{5\}$	(0.0)

We are now able to write a point density function as a function of σ to describe our empirically optimal threshold configuration. This is

$$\lambda(x, \sigma) = \sum_{j=1}^{k(\sigma)} v(j, \sigma)\delta(x - \Theta_j), \tag{8.42}$$

where $\delta(\cdot)$ is the delta function. We note also that $\int_{x=-\infty}^{a} \lambda(x, \sigma)dx$ is the fraction of thresholds with values less than or equal to a, and that $\int_{x=-\infty}^{\infty} \lambda(x, \sigma)dx = 1$.

For the special case of $\sigma = 0$ and maximized mutual information, we can use Eq. (8.27) to write the analytically optimal point density function as

$$\lambda(x, 0) = \sum_{j=1}^{N} v(j, 0)\delta(x - \Theta_j) = \sum_{i=1}^{N} \frac{1}{N}\delta\left(x - F_x^{-1}\left(\frac{i}{N+1}\right)\right). \tag{8.43}$$

As an example for nonzero σ, consider the case of $N = 5$ and Gaussian signal and noise shown for maximized mutual information and various σ in Fig. 8.11. The value of k for each σ and the corresponding cluster sizes, $Nv(j, \sigma)$, and the approximate cluster values, $(\Theta_j)_{j=1}^{k(\sigma)}$, are shown in Table 8.1.

So far, we have point density functions consisting only of singularities. In high resolution quantization theory, point density functions are generally continuous functions, analogous to PDFs. Due to the small N considered here, our point density functions are analogous to discrete probability mass functions, rather than PDFs. Discussion of the behaviour of this description of the optimal thresholds for large N is left for future work.

Conditional output moments

Using the notation introduced above, we are able to rewrite our previous expressions for the conditional mean and variance of the output encoding, y, for arbitrary thresholds.

From Eq. (8.16), the expected value of y given x is

$$E[y|x] = \sum_{i=1}^{N} P_{1|x,i} = N \sum_{j=1}^{k} v_j P_{1|x,j} = N \sum_{j=1}^{k} v_j F_\eta(x - \Theta_j) \tag{8.44}$$

and from Eq. (8.17), the variance of y given x is

$$\mathrm{var}[y|x] = \sum_{i=1}^{N} P_{1|x,i}(1 - P_{1|x,i}) = N \sum_{j=1}^{k} v_j P_{1|x,j}(1 - P_{1|x,j})$$

$$= N \sum_{j=1}^{k} v_j F_\eta(x - \Theta_j)(1 - F_\eta(x - \Theta_j)). \tag{8.45}$$

We shall use these results in Section 8.6.

Local maxima

Partitions of integers

A *partition* of a positive integer is a way of writing that integer as a sum of smaller positive integers. The order of the integers in the sum is not considered, and conventionally a partition is written in the order of largest to smallest integers in the sum. For example, there are seven partitions of the integer 5. These are {5}, {4, 1}, {3, 2}, {3, 1, 1}, {2, 2, 1}, {2, 1, 1, 1}, and {1, 1, 1, 1, 1}. The theory of partitions of integers is a rich area of number theory, and was of interest to such well-known number theorists as Ramanujan, Hardy, and Littlewood (Andrews 1976). The number of partitions of the integer N increases very rapidly with N.

If ordering is taken into account, then extra partitions are possible, namely {1, 3, 1}, {2, 1, 2}, and {1, 2, 1, 1}. The reverse of all the ten partitions listed are also feasible. Hence, there are 19 possible ordered partitions of the integer 5.

Description of local maxima

Explaining the presence of discontinuous bifurcations in the optimal threshold figures at first seems very difficult. However, most of the discontinuous bifurcations are actually due to the presence of many locally optimal threshold configurations. In fact, numerical experiments find that for every value of σ, there is at least one locally optimal solution – that is, a set of threshold values giving a gradient vector of zero – corresponding to every possible *partition* of N. For each partition,[3] there are as many locally optimal solutions as there are unique

---

[3] Here, however, due to considering only PDFs that are even functions, there will be some symmetry such that we can ignore all partitions that are the reverse order, and in the example of the integer 5, we need only consider ten possible partitions.

orderings of that partition. For small σ, all of these local optima are unique. As σ increases, more and more of these local optima bifurcate continuously to be equal to other local optima. For example, a local optimum corresponding to $k = 3$ clusters, with $\{Nv(j, \sigma)\} = \{2, 2, 1\}$, might have Θ_2 and Θ_3 converge to the same value with increasing σ. At the point of this convergence, a bifurcation occurs, and the local optimum becomes one consisting of $k = 2$ clusters, with $\{Nv(j, \sigma)\} = \{2, 3\}$.

Again, the exception to this rule of thumb seems to be the uniform case, in which bifurcations can occur discontinuously in locally optimal solutions.

To illustrate this effect, in the simplest possible case of $N = 3$, Fig. 8.12 shows the optimal thresholds for all three locally optimal solutions for maximum mutual information, for all four matched signal and noise cases. Notice that apart from the uniform case when $\sigma > 1$, the SSR situation is always a local optimum. Fig. 8.13 shows the mutual information achieved by each of the three locally optimal solutions. Notice that for small σ, there is a significant difference between the mutual information in each case. However, as σ increases, the difference between each case decreases.

Notice in Fig. 8.12 that both the $\{1, 1, 1\}$ and $\{2, 1\}$ cases change smoothly with increasing σ until they both converge to the SSR case for sufficiently large noise. This is in contrast to the globally optimal thresholds for Gaussian signal and noise shown in Fig. 8.2(a), where there is a discontinuous bifurcation. This is due to the optimal solution switching from being the $\{1, 1, 1\}$ case to the $\{2, 1\}$ case. Fig. 8.14 shows the difference in mutual information between these two cases, in the region of σ where the bifurcation occurs. It is clear that the difference between cases is very small, and therefore the fact that the optimal thresholds correspond to the $\{2, 1\}$ case rather than the $\{1, 1, 1\}$ case is really only of academic interest. The more important fact is that, for sufficiently large σ, having all thresholds identical is optimal – since this implies a reduction in complexity for the specification of the optimal quantizer. This is because only one threshold value is required, rather than N, which is the case for no noise.

An estimation perspective

Fisher information

Recall from Chapter 6 that the Fisher information for the SSR model can be calculated and used in a formula giving a lower bound on the MSE distortion. This result can also be applied to the arbitrary threshold model considered in this chapter.

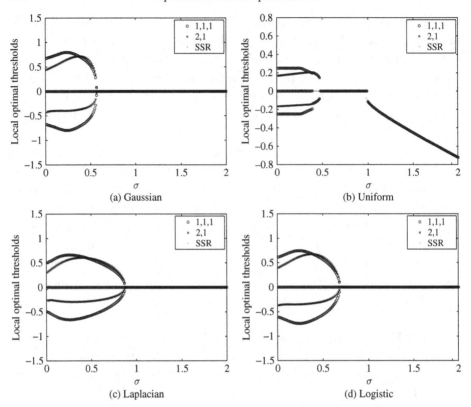

Fig. 8.12. The three locally optimal solutions for $N = 3$ and maximum mutual information. Apart from the uniform case when $\sigma > 1$, the SSR situation is always a local optimum. For the uniform case when $\sigma > 1$, although SSR is no longer a local optimum, the optimum is for all thresholds to have the same value.

The output of the array of threshold devices, y, provides a biased estimate of the input, x. Recall from Chapter 6 that the *information bound* states that

$$\text{var}[y|x] \geq \frac{\left(\frac{d}{dx}E[y|x]\right)^2}{J(x)}, \tag{8.46}$$

where $J(x)$ is the Fisher information, which for arbitrary thresholds is given by

$$J(x) = \sum_{n=0}^{N} \frac{\left(\frac{dP_{y|x}(n|x)}{dx}\right)^2}{P_{y|x}(n|x)}. \tag{8.47}$$

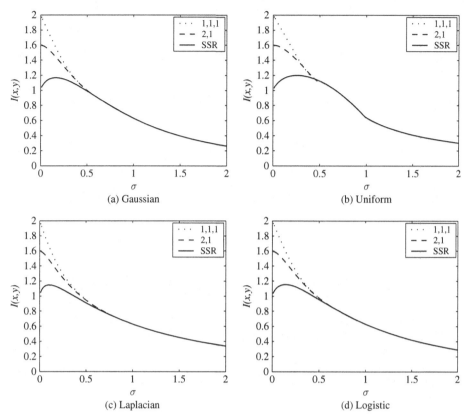

Fig. 8.13. Mutual information for the three locally optimal solutions for $N=3$. It is clear that for small σ, all thresholds unique gives much larger mutual information than the other cases. However, for sufficiently large σ, the differences in mutual information become smaller and smaller, until all three local solutions converge to the same solution corresponding to SSR.

From Eq. (8.16), we have for the arbitrary threshold model that

$$\frac{d}{dx}E[y|x] = \sum_{i=1}^{N} \frac{d P_{1|x,i}}{dx}$$

$$= \sum_{i=1}^{N} \frac{d F_\eta(x - \theta_i)}{dx}$$

$$= \sum_{i=1}^{N} f_\eta(x - \theta_i)$$

$$= N \sum_{j=1}^{k} v_j f_\eta(x - \Theta_j). \qquad (8.48)$$

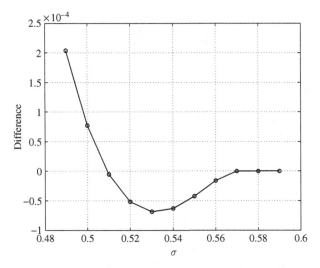

Fig. 8.14. Difference in mutual information between the locally optimal solutions with $v = \{1, 1, 1\}$ and $v = \{2, 1\}$ for Gaussian signal and noise and $N = 3$. Clearly, $v = \{2, 1\}$ is larger than $v = \{1, 1, 1\}$ only for $\sigma \in [0.51, 0.57]$, and is larger only by an amount of the order of 0.5×10^{-4} bits per sample, which is very small. However, this changeover from $v = \{1, 1, 1\}$ being optimal to $v = \{2, 1\}$ being optimal, is the reason that a *discontinuous* bifurcation appears in the *globally* optimal solution shown in Fig. 8.2(a). For $\sigma \geq 0.57$, the SSR situation is optimal, and both the $v = \{1, 1, 1\}$ and the $v = \{2, 1\}$ situation bifurcate at this point to the SSR case.

Substituting from Eqs (8.48) and (8.45) into Inequality (8.46) and rearranging gives an inequality for the Fisher information as

$$
\begin{aligned}
J(x) &\geq \frac{\left(\frac{d}{dx} E[y|x]\right)^2}{\text{var}[y|x]} \\
&= \frac{\left(\sum_{i=1}^{N} f_\eta(x - \theta_i)\right)^2}{\sum_{l=1}^{N} P_{1|x,i}(1 - P_{1|x,i})} \\
&= \frac{N\left(\sum_{j=1}^{k} v_j f_\eta(x - \Theta_j)\right)^2}{\sum_{j=1}^{k} v_j P_{1|x,j}(1 - P_{1|x,j})}.
\end{aligned}
\tag{8.49}
$$

For the special case of SSR, we have $k = 1$, $v_k = 1$ and $\Theta_j = 0$, which when substituted into Inequality (8.49) gives a rhs that is exactly the Fisher information for SSR expressed by Eq. (6.119) in Section 6.7 of Chapter 6. The Fisher information for SSR is derived from first principles in Section A3.6 of Appendix 3.

Thus, for SSR, Inequality (8.49) becomes an equality and, as discussed in Chapter 6, the SSR Fisher information meets the information bound with equality, in the case of no decoding. However, for arbitrary thresholds, a derivation of the Fisher information from first principles using Eq. (8.47) is not a trivial task – since we do not have an analytic expression for $P_{y|x}(n|x)$ – and therefore we do not know whether the information bound is also met with equality in the general case.

However, the Fisher information for arbitrary thresholds and signal and noise distributions can be calculated numerically from Eq. (8.47), and compared with numerical calculations of the rhs of Inequality (8.49). Experiments with such calculations indicate that the bound does not hold exactly, apart from the SSR situation, but that, even for small N, the bound has a maximum error when compared to the exact Fisher information, in the order of 1 percent. It is possible this error is attributable to numerical errors, and that the bound does hold with equality. In any case, we are able to state

$$J(x) \simeq \frac{\left(\sum_{i=1}^{N} f_\eta(x - \theta_i)\right)^2}{\sum_{1=1}^{N} P_{1|x,i}(1 - P_{1|x,i})} = \frac{N\left(\sum_{j=1}^{k} v_j f_\eta(x - \Theta_j)\right)^2}{\sum_{j=1}^{k} v_j P_{1|x,j}(1 - P_{1|x,j})}. \tag{8.50}$$

Future work may be able to justify making Inequality (8.49) a strict equality under certain conditions, using Brunel and Nadal (1998) as a starting reference.

Average information bound

From inspection of Eq. (8.47), the Fisher information for x is unchanged if the output y is decoded to $\hat{y}$, since $J(x)$ depends only on $P_{y|x}(n|x)$. However, the bias changes for a decoding, as does the conditional variance. The information bound for the decoding, $\hat{y}$, is then

$$\text{var}[\hat{y}|x] \geq \frac{\left(\frac{d}{dx}E[\hat{y}|x]\right)^2}{J(x)}. \tag{8.51}$$

Substituting from Eq. (8.50) into Inequality (8.51) gives

$$\text{var}[\hat{y}|x] \geq \frac{\sum_{1=1}^{N} P_{1|x,i}(1 - P_{1|x,i})\left(\frac{d}{dx}E[\hat{y}|x]\right)^2}{\left(\sum_{i=1}^{N} f_\eta(x - \theta_i)\right)^2}$$

$$= \frac{\left(\sum_{1=1}^{N} P_{1|x,i}(1 - P_{1|x,i})\right)\left(\sum_{n=0}^{N} \hat{y}_n \frac{d}{dx} P_{y|x}(n|x)\right)^2}{\left(\sum_{i=1}^{N} f_\eta(x - \theta_i)\right)^2}. \tag{8.52}$$

Thus, a lower bound on the conditional MSE distortion is

$$D(x) \geq \frac{\left(\sum_{1=1}^{N} P_{1|x,i}(1 - P_{1|x,i})\right)\left(\sum_{n=0}^{N} \hat{x}_n \frac{d}{dx} P_{y|x}(n|x)\right)^2}{\left(\sum_{i=1}^{N} f_\eta(x - \theta_i)\right)^2} + b_{\hat{x}}(x)^2, \tag{8.53}$$

where the decoding is the optimal decoding, $\hat{x}_n = E[x|n]$, and $b_{\hat{x}}(x) = E[\hat{x}|x] - x$ is the bias of the decoding.

Multiplying the rhs of Inequality (8.53) by $f_x(x)$ and then integrating over all x gives the average information bound (AIB) introduced in Chapter 6. The AIB is a lower bound on the MMSE distortion, and is therefore

$$
\text{AIB} = \int_x f_x(x) \left(\frac{\left(\sum_{1=1}^N P_{1|x,i}(1 - P_{1|x,i}) \right) \left(\sum_{n=0}^N \hat{x}_n \frac{d}{dx} P_{y|x}(n|x) \right)^2}{\left(\sum_{i=1}^N f_\eta(x - \theta_i) \right)^2} \right) dx
$$
$$
+ E[b_{\hat{x}}(x)^2], \tag{8.54}
$$

where $E[b_{\hat{x}}(x)^2]$ is the mean square bias.

As carried out in Chapter 6, it is instructive to numerically calculate the average information bound, and its two components, the mean square bias and the average error variance.

Figure 8.15 shows the AIB, and its components, the mean square bias and the average error variance, for the case of Gaussian signal and noise, and $N = 3$, for all three locally optimal solutions when the thresholds are optimized to minimize the MMSE distortion. Figure 8.15(b) shows that the AIB is less than 2 percent different from the actual optimal MMSE distortion in each case. As we saw earlier, it is clear that optimizing the thresholds results in a decrease in the MMSE distortion when compared to SSR. Figures 8.15(c) and 8.15(d) show that this decrease is a result of optimizing the thresholds to offset a small increase in the average conditional variance against a larger decrease in the mean square bias. Optimizing the thresholds means optimizing the tradeoff between bias and variance. The end result of this is a MMSE distortion and AIB that are strictly increasing with increasing σ.

8.7 Locating the final bifurcation

We have seen that for sufficiently large σ, the SSR situation of all thresholds equal to the signal mean becomes optimal. We have also seen that the largest value of σ, which we shall call σ_b, for which SSR is not optimal, increases with increasing N. It appears also that in the region of σ just smaller than σ_b, there are two clusters. For even N, these each have size $N/2$, and values $\Theta_1 = -\Theta_2$. If we assume that this will always be the case, it becomes a straightforward task to numerically find the value of σ_b as a function of increasing N. It requires only to set $\Theta_1 = -\Theta_2 = \epsilon$, where ϵ is small, say 0.001, and for a given N to find the value of σ at which the SSR mutual information or MMSE distortion changes from being smaller or larger than the mutual information with two threshold clusters at $\pm\epsilon$.

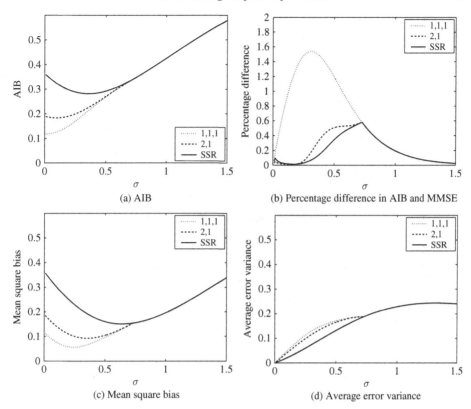

Fig. 8.15. Average information bound (AIB), Gaussian signal and noise, $N = 3$.
Figure 8.15(a) shows the AIB of Eq. (8.54) plotted against increasing σ for the
three different local optima. Figure 8.15(b) shows that the percentage difference
between the AIB and the MMSE distortion for each local optimum is very small,
indicating that the AIB gives a bound that is very close to the actual MMSE
distortion. Figures 8.15(c) and 8.15(d) show the two components of the AIB, the
mean square bias, and the average error variance. Notice how, for small σ, the
mean square bias for the optimal threshold situation of $\{1, 1, 1\}$ is far smaller
than for SSR, while the average error variance is slightly larger. This illustrates
how optimizing the thresholds for small σ optimizes the tradeoff between bias
and variance. For sufficiently large σ, the optimal tradeoff is provided by the SSR
situation.

The result of carrying this out is shown in Fig. 8.16 for mutual information, and
Fig. 8.17 for MMSE distortion. The value of N, which we shall refer to as N_b,
plotted for each value of σ, is the smallest even-valued N for which the mutual
information or MMSE distortion with $\Theta_1 = -\Theta_2 = 0.001$ gives better performance than the SSR situation. Thus, for each value of σ, if $N < N_b$, then SSR is
optimal.

It is clear that, as σ increases, N_b also increases, and increases very rapidly
near some critical value of σ. It appears likely that, for large N, N_b asymptotically

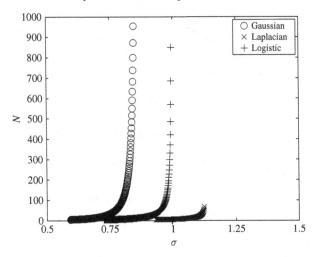

Fig. 8.16. Final bifurcation point, mutual information. The smallest even-valued N for which the mutual information with $\Theta_1 = -\Theta_2 = 0.001$ gives better performance than the SSR situation. Thus, the value of σ corresponding to each N is the approximate final bifurcation point. Note that the Laplacian situation is only plotted with N up to 84. This is due to numerical calculations using the Laplacian PDF being less accurate than the Gaussian and logistic cases, due to the Laplacian PDF having a non-differentiable point at $x = 0$.

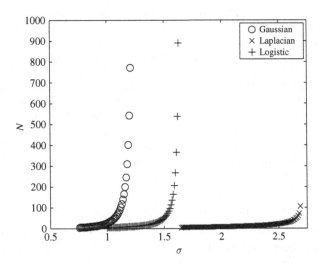

Fig. 8.17. Final bifurcation point, MMSE distortion. The smallest even-valued N for which the MMSE distortion with $\Theta_1 = -\Theta_2 = 0.001$ gives better performance than the SSR situation. Thus, the value of σ corresponding to each N is the approximate final bifurcation point.

converges towards some fixed value, σ_b^*. This means that, for sufficiently large σ, SSR is always optimal, regardless of the size of N. The value of σ_B^* does, however, depend on the measure used, and the signal and noise distribution. Further results on such asymptotic large N behaviour for arbitrary thresholds is left for future work.

8.8 Chapter summary

The introductory section of this chapter discusses the SSR case of all thresholds being identical, and sets the context for removing this restriction. When this restriction is removed, and all thresholds are free variables, it is of interest to optimally set the threshold values. In the presence of independent threshold noise, such a problem is an optimal stochastic quantization problem. Section 8.1 therefore also briefly reviews the literature on optimal quantization and points out that the problem addressed in this chapter does not appear to have been previously studied.

Section 8.2 mathematically describes the generalization of the SSR model to arbitrary thresholds, outlines a method for recursively calculating the transition probabilities, and mathematically formulates the optimal stochastic quantization problem we aim to solve. Section 8.3 then describes the solution method we use in solving these problems.

Results are presented in Sections 8.4 and 8.5, and discussed in Section 8.6. The key features of our results are that in general, for sufficiently large σ, the SSR situation is optimal, while, for smaller σ, the optimal thresholds tend to cluster to identical values, with the number of clusters decreasing with increasing σ. We introduce also notation to describe these optimal thresholds, and briefly consider the information bound, and an associated lower bound on the MSE distortion. As N increases, the value of σ at which SSR becomes optimal also increases.

Finally, Section 8.7 shows that, for sufficiently large σ, it appears that SSR is always optimal, regardless of the magnitude of N.

Chapter 8 in a nutshell

This chapter includes the following highlights:

- A statement of a relationship between the set of $\{P_{1|x,i}\}$ and the transition probabilities, via the moment generating function for the array of threshold devices.
- Derivation of a very general recursive formula, with $O(N^2)$ computational complexity, that makes it straightforward to numerically calculate the transition probabilities, $P_{y|x}(n|x)$, for any given threshold values, and noise distribution.
- Mathematical formulation of the optimal stochastic quantization problem, for the array of threshold devices, in terms of the vector of optimal thresholds, $\vec{\theta}$.

- Presentation of numerical solutions to the stochastic optimal quantization problems of maximizing the mutual information, and minimizing the MSE distortion. The optimal thresholds found are consistent with the optimal noiseless thresholds, in that for very small noise the optimal thresholds are very close to the optimal noiseless ones.
- Discovery and discussion of the unexpected bifurcation pattern in the optimal stochastic quantization results.
- Numerical validation that the SSR situation of all thresholds equal to the signal mean is in fact optimal for sufficiently large noise intensity, for a range of signal and noise distributions, and both the mutual information and MSE distortion measures.
- Derivation of an approximation to the Fisher information for arbitrary thresholds, and its application to a calculation of a lower bound on the MSE distortion. Optimally setting the thresholds is shown to be related to finding the optimal tradeoff between the two components of this lower bound, the mean square bias, and the average error variance.
- Numerical validation that SSR remains optimal for sufficiently large noise intensity, even if N becomes very large.

Open questions

Possible future work and open questions arising from this chapter might include:

- Discussion of other signal and noise distributions than those considered here, including mixed signal and noise distributions, deterministic signals, and distributions with one-sided PDFs such as the Rayleigh distribution.
- Mathematical proofs of the fact that SSR is optimal for large N, and further mathematical analysis of the bifurcational structure.
- Extension of the optimal quantization problems considered here – with only simple on–off threshold devices – to more realistic neural models, for example the FitzHugh–Nagumo neuron model.
- Rigorous justification of the fact that Inequality (8.49) can be approximated as an equality under certain conditions. A good starting point for this research question is the material contained in Brunel and Nadal (1998).

This concludes Chapter 8, which studies the extension of the SSR model to a model with arbitrary thresholds. Chapter 9 now examines a further extension of the SSR model to incorporate constraints on energy and information.

9

SSR, neural coding, and performance tradeoffs

Engineered systems usually require finding the right tradeoff between cost and per-formance. Communications systems are no exception, and much theoretical work has been undertaken to find the limits of achievable performance for the transmission of information. For example, Shannon's celebrated channel capacity formula and coding theorems say that there is an upper limit on the average amount of information that can be transmitted in a channel for error-free communication. This limit can be increased if the power of the signal is increased, or the bandwidth in the channel is increased. However, nothing comes for free, and increasing either power or bandwidth can be expensive; hence there is a tradeoff between cost and performance in such a communications system – performance (measured by bit rates) can be increased by increasing the cost (power or bandwidth). This chapter discusses several problems related to the tradeoff between cost and performance in the SSR model. We are interested in the SSR model as a channel model, from an energy efficient neural coding point of view, as well as the lossy source coding model, where there is a tradeoff between rate and distortion.

9.1 Introduction

Chapter 8 introduces an extension to the suprathreshold stochastic resonance (SSR) model by allowing all thresholds to vary independently, instead of all having the same value. This chapter further extends the SSR model by introducing an energy constraint into the optimal stochastic quantization problem. We also examine the tradeoff between rate and distortion, when the SSR model is considered as a stochastic quantizer.

The mathematical treatment in this chapter follows along the lines of McDonnell *et al.* (2004b), McDonnell *et al.* (2005a), McDonnell *et al.* (2005c), and McDonnell *et al.* (2005d).

Recall that the initial work on the SSR model (Stocks 2000c, Stocks and Mannella 2001) was partly motivated by its relevance to neural coding. The constraint we consider in this chapter is also motivated by this fact. If, as discussed in Chapter 4, the issue is that of information transmission, we can say that ideally the encoding of sensory input by neurons should maximize the mutual information between input and output. However, as with most systems, there is usually some cost associated with maximizing a quantity. For neural systems, a cost function that has received recent attention is that of energy efficiency. Hence, we consider the problem of maximizing mutual information in the extended SSR model subject to a maximum energy constraint. Such a problem is like the classic information theory problem of finding channel capacity subject to a power constraint on the source, except that we are free to optimize the channel by changing the threshold values.

The second problem we discuss can also be related to the neural coding motivation for the SSR model, but is also of relevance to lossy source coding. Recall that in Chapter 6 we discuss methods for decoding the SSR model's output signal to obtain a new output signal that approximately reconstructs the input signal. We use the mean square error (MSE) distortion as a measure for the performance of such a reconstruction, and find that, for the SSR model, the distortion can be decreased by increasing the number of threshold devices, N, and therefore the number of output states, which is $N + 1$. Suppose we define the *rate* of a quantizer as the log of the number of output states, $\log(N + 1)$. Then this result is a very simple illustration of *rate–distortion* theory (Berger and Gibson 1998). In general, the distortion of a quantizer can be reduced by increasing the rate, or, conversely, the rate can be reduced by allowing the distortion to increase. Hence, we consider for the SSR model, and its extension to the arbitrary threshold model, the problem of minimizing the rate subject to a distortion constraint.

Before discussing these problems in more detail, we briefly state relevant results from previous chapters, and the theory of solving constrained optimization problems using the method of Lagrange multipliers.

Review of relevant material

The model we use is the array of threshold elements shown in Fig. 4.1. As in Chapters 4–8, we assume that the input signal is a sequence of samples drawn from a continuously valued probability distribution with probability density function (PDF), $f_x(x)$. The output signal, y, is the sum of the individual outputs of each threshold device, and is a discretely valued signal with $N + 1$ states between zero and N.

As first introduced in Chapter 8, for arbitrary threshold values we let $P_{1|x,i}$ be the probability of threshold device i being 'on', given signal value, x. Then, if the ith threshold value is θ_i, we have

$$P_{1|x,i} = \int_{\theta_i - x}^{\infty} f_\eta(\eta)d\eta = 1 - F_\eta(\theta_i - x), \qquad (9.1)$$

where $f_\eta(\cdot)$ is the noise PDF, $F_\eta(\cdot)$ is the noise cumulative distribution function (CDF), and $i = 1, \ldots, N$.

As first discussed in Chapter 4, the mutual information between the input and output signals in the model is given by

$$I(x, y) = - \sum_{n=0}^{N} P_y(n) \log_2 P_y(n)$$

$$- \int_{-\infty}^{\infty} f_x(x) \sum_{n=0}^{N} P_{y|x}(n|x) \log_2 P_{y|x}(n|x)dx, \qquad (9.2)$$

where $P_y(n) = \int_{-\infty}^{\infty} P_{y|x}(n|x)f_x(x)dx$ is the probability mass function of the output signal, and $P_{y|x}(n|x)$ are the transition probabilities giving the probability that the output is in state $y = n$, given input value x. We saw in Chapter 4 that, for SSR, the mutual information is a function of the noise intensity, σ, that is the ratio of noise standard deviation to signal standard deviation.

For any arbitrary threshold value, θ_i, and noise PDF, $P_{1|x,i}$ can be calculated exactly for any value of x from Eq. (9.1). Assuming $P_{1|x,i}$ has been calculated for desired values of x, each $P_{y|x}(n|x)$ can be calculated from the recursive formulation given in Chapter 8 by Eq. (8.19).

As first discussed in Chapter 6, the optimal mean square error (MSE) distortion decoding is the nonlinear decoding, $\hat{x}$, with values given by $\hat{x}_n = E[x|n], n = 0, \ldots, N$, and an error signal given by

$$\epsilon = x - \hat{x}. \qquad (9.3)$$

This decoding results in the minimum possible MSE distortion for given transition probabilities, which is given by

$$\text{MMSE} = E[\epsilon^2] = E[x^2] - E[\hat{x}^2]. \qquad (9.4)$$

Like the mutual information, the MMSE depends on the transition probabilities, since

$$E[\hat{x}^2] = \sum_{n=0}^{N} \hat{x}_n^2 P_y(n)$$

$$= \sum_{n=0}^{N} E[x|n]^2 P_y(n)$$

$$= \sum_{n=0}^{N} \left(\int_x x P_{x|y}(x|n) dx \right)^2 P_y(n)$$

$$= \sum_{n=0}^{N} \frac{\left(\int_x x f_x(x) P_{y|x}(n|x) dx \right)^2}{P_y(n)}. \tag{9.5}$$

We shall also in this chapter use the expected value of the output signal, y, which is

$$E[y] = \sum_{n=0}^{N} n P_y(n)$$

$$= \sum_{n=0}^{N} n \int_{-\infty}^{\infty} P_{y|x}(n|x) f_x(x) dx$$

$$= \int_{-\infty}^{\infty} f_x(x) \left(\sum_{n=0}^{N} n P_{y|x}(n|x) \right) dx$$

$$= \int_{-\infty}^{\infty} f_x(x) E[y|x] dx. \tag{9.6}$$

Substituting from Eq. (8.16) from Chapter 8 into Eq. (9.6) gives

$$E[y] = \int_{-\infty}^{\infty} f_x(x) \sum_{i=1}^{N} P_{1|x,i} dx$$

$$= \sum_{i=1}^{N} \int_{-\infty}^{\infty} f_x(x) P_{1|x,i} dx$$

$$= \sum_{i=1}^{N} \int_{-\infty}^{\infty} f_x(x) F_\eta(\theta_i - x) dx, \tag{9.7}$$

which holds for arbitrary threshold values. As we saw in Chapters 4 and 6, for the SSR case we have $P_{y|x}(n|x)$ given by the binomial formula, as in Eq. (4.9), and therefore $E[y|x] = N P_{1|x}$. For an even noise PDF, $f_\eta(\eta)$, we have also $E[y] = N/2$.

Constrained optimization

Suppose we wish to fix the number of threshold devices, N, and for a given value of noise intensity, σ, find the threshold settings that either:

(i) maximize the mutual information, subject to a constraint that specifies that E[y] is less than some value, A, or

(ii) minimize the mutual information, subject to a constraint that the MSE distortion is less than some value, B.

Since $I(x, y)$, E[y], and the MSE distortion are all functions of the transition probabilities, $P_{y|x}(n|x)$, it is possible to formulate such an optimization as a variational problem, where the aim is to find the optimal set of transition probabilities. This set will consist of $N + 1$ functions of the continuous variable, x, that is $\{P_{y|x}(n|x)\}$, $n = 0, \ldots, N$. However, as we saw in Chapter 8, each $P_{y|x}(n|x)$ depends entirely on the vector of thresholds, $\vec{\theta}$, and we can therefore solve optimal quantization problems by finding the optimal N-dimensional vector, $\vec{\theta}^*$. We are able to take the same approach for constrained optimization problems, as we now discuss.

Suppose for any given noise intensity, σ, we label the quantity we wish to optimize as $f(\vec{\theta})$, and the variable we wish to constrain as $m(\vec{\theta})$. Then the problem of maximizing the *cost function*, $f(\vec{\theta})$, subject to the constraint that $m(\vec{\theta}) \leq A$, can be expressed as the nonlinear optimization problem

Find:
$$\max_{\vec{\theta}} \; f(\vec{\theta}),$$

subject to:
$$m(\vec{\theta}) \leq A, \vec{\theta} \in \mathbb{R}^N. \tag{9.8}$$

The method of Lagrange multipliers (Gershenfeld 1999) can be used to solve such constrained optimization problems. Using the standard approach to this method, we begin by incorporating the constraint, $m(\vec{\theta}) \leq A$, in a new cost function, $g(\vec{\theta})$, as

Find:
$$\max_{\vec{\theta}} \; g(\vec{\theta}) = f(\vec{\theta}) - \lambda m(\vec{\theta}),$$

subject to:
$$\vec{\theta} \in \mathbb{R}^N, \lambda > 0. \tag{9.9}$$

It can be shown that solving Problem (9.8), for some constraint value A, is equivalent to solving Problem (9.9), for some corresponding value of λ. The simplest way of ensuring that the constraint is met is to begin with a guess for a value of λ, and then solve Problem (9.9). If for this value of λ the constraint is not met, then the *Lagrange multiplier*, λ, can be varied and Problem (9.9) solved with this new value. This process can be repeated until λ is such that the constraint, $m(\vec{\theta}) \leq A$, is satisfied.

As in Chapter 8, we shall find that there are many local optima for our constrained optimization problems. However, we can make use of the same techniques as described in Chapter 8 to find the global optimum.

Chapter structure

The remainder of this chapter is separated into two main sections. Section 9.2 considers neurally motivated energy constraint problems. Section 9.3 considers the problem of rate–distortion tradeoff, which is conventionally part of the domain of lossy source coding theory, but which is also applicable in neural coding situations.

9.2 Information theory and neural coding

There is increasing interest in applying the techniques of electronic engineering and signal processing to neuroscience research. This research field is known as *computational neuroscience* (Rieke *et al.* 1997, Eliasmith and Anderson 2003). The motivation for such studies is obvious: the brain uses electrical signals – as well as chemical signals – to propagate, store, and process information, and must employ some sort of coding and modulation mechanism as part of this process. The fields of information theory and signal processing have many mature techniques for dealing with signal propagation and coding, and these techniques can be employed to gain new insights into the ways the brain encodes, propagates, stores, and processes information.

Of particular relevance to this book is the fact that the brain is capable of performing very well when required to obtain information via the senses in very noisy conditions. Often, the signal-to-noise ratio (SNR) of the sensory neuron is orders of magnitude lower than those usually encountered in electronic systems (Bialek *et al.* 1993). As discussed in Chapter 2, many studies have shown that stochastic resonance (SR) can occur in neurons, so that it appears possible that certain tasks required in the nervous system have evolved to be optimally adapted to operating in such noisy conditions, or, alternatively, have evolved to generate noise to enable its neurons to perform optimally (Longtin 1993, Douglass *et al.* 1993).

There is some debate in the literature regarding which measures are appropriate to use to quantify information transmission in neural systems. Although some authors such as Johnson (2004) argue against the use of mutual information, it has, however, been the preferred measure in numerous papers (Levin and Miller 1996, Rieke *et al.* 1997, Borst and Theunissen 1999, Stocks and Mannella 2001, Abarbanel and Rabinovich 2001). One reason for this is that mutual information provides a measure that is independent of any decoding. While lossy source coding theory tends to use mean square error to measure distortion, it is debatable whether such a distortion measure is relevant for describing the quality of information perception by the brain. Consider, for example, a stochastically quantized signal; such a signal is a discrete random variable. A known and invertible – that is, not stochastic or lossy – transformation of this signal will cause changes in the MSE distortion, but not the mutual information. If the important feature of a signal is its shape when considered after some known transformation, then it does not make sense to consider MSE distortion, as the same information is available even if the distortion becomes huge.

However, given that mutual information has been previously used elsewhere in neural coding research, we also use it here to illustrate the main point, that is to examine how the optimal thresholds in the extended SSR model change when subject to energy constraints. Hence, we now briefly comment on the existing literature on energy constrained neural coding, and define a measure of energy for the extended SSR model.

Energy constraints

Many recent studies in the field of biophysics have shown that energy is an important constraint in neural operation, and have investigated the role of this constraint in neural information processing (Balasubramanian *et al.* 2001, Laughlin 2001, Wilke and Eurich 2001, Bethge *et al.* 2002, Levy and Baxter 2002, Schreiber *et al.* 2002, Hoch *et al.* 2003b).

Recall that in the SSR model, and its extension to arbitrary thresholds, when *iid* additive noise is present at the input to each threshold element, the overall output becomes a randomly quantized version of the input signal, which for the right level of noise is highly correlated with the input signal. In this section, we view the extended SSR model as a population of simple neuron models, and present an investigation of the optimal noisy encoding of a random input signal, subject to constraints on the available energy expenditure. Our model is extremely simplified when compared to real neurons, or realistic models; however, it does encapsulate the basic aspects of how stochastic quantization may occur in a population of neurons.

The only previous work on SSR of relevance to such a goal is work that imposes an energy constraint on the SSR model, and calculates the optimal input signal distribution given that constraint (Hoch *et al.* 2003b).

Previously in this book, Chapter 8 deals with optimizing the thresholds with no constraints. By contrast, here we look at two energy constrained problems. In the first problem, we fix the input distribution and 'neural population size', N, and aim to find the threshold settings that maximize the mutual information, subject to a maximum average output energy constraint. The second problem we tackle is the perhaps more biologically relevant problem, of minimizing the population size, N – and therefore the energy expenditure – given a minimum mutual information constraint and fixed thresholds.

Before we discuss these problems, however, we first define our measures of energy and energy efficiency, and discuss how these measures vary for the SSR model, where all thresholds have the same value.

Average output energy

The simplest approach is to assume that the energy expended is the same constant amount every time a neuron emits a 'spike'. Therefore, minimizing the average output energy consumption requires minimization of the mean output value, which for our model is $E[y]$, as given by Eq. (9.6). Note that for arbitrary thresholds, the threshold values control the output probabilities, $P_y(n)$. Therefore by ensuring some thresholds are set high enough, larger values of y can be made less probable in order to reduce the average output energy expended.

Information efficiency

Previously, Schreiber *et al.* (2002) makes use of an efficiency measure defined as the ratio of mutual information to metabolic energy required in a neural coding. We use a similar metric, the ratio of mutual information to average output energy, which we shall call *information efficiency* and denote by

$$\xi = \frac{I(x, y)}{E[y]}. \tag{9.10}$$

This quantity is a measure of bits per unit energy. Clearly this information efficiency measure is increased if either the mutual information increases, or average output energy decreases. We shall see that this measure is useful only if some constraint is placed on either mutual information or energy, since the efficiency can be made near infinite by ensuring that all thresholds are so large that the output of the model is almost always zero. Hence, for practical purposes, maximizing the information efficiency requires either maximizing the mutual information subject to a maximum output energy constraint, or minimizing the average output energy subject to a minimum mutual information constraint.

However, first we shall consider the situation where all thresholds are not necessarily zero, but are equal to the same value, θ. We compare how the mutual information and average output energy varies with noise intensity, σ and θ, as well as how the mutual information varies for specified energy constraints. Examining the behaviour of the information efficiency given by Eq. (9.10) for these situations will provide a useful benchmark for our later constrained optimal quantization problems.

All thresholds equal

If all thresholds are equal to the signal mean of zero, then the model is in the SSR configuration, and if $f_x(x)$ is an even function, the average output energy is $N/2$. If, however, the value of all thresholds, θ, is changed to be nonzero, then both the mutual information and average output energy will change. We shall consider only the situation of a Gaussian signal and independent Gaussian noise, since we are able to state an exact result for the output energy in this case.

Exact output energy for Gaussian signal and noise

Consider the total probability that the ith device is 'on'. This is a function of that threshold's actual value, θ_i, which we can write as

$$f(\theta_i) = \int_{-\infty}^{\infty} f_x(x) P_{1|x,i} dx$$

$$= 1 - \int_{-\infty}^{\infty} f_x(x) F_\eta(\theta_i - x) dx. \tag{9.11}$$

For even noise distributions, Eq. (9.11) can be written as

$$f(\theta_i) = \int_{-\infty}^{\infty} f_x(x) F_\eta(x - \theta_i) dx. \tag{9.12}$$

However, for fully general $f_\eta(\cdot)$, taking the derivative of both sides of Eq. (9.11) with respect to θ_i gives

$$\frac{df(\theta_i)}{d\theta_i} = -\int_{-\infty}^{\infty} f_x(x) f_\eta(\theta_i - x) dx. \tag{9.13}$$

Note that $-\frac{df(\theta_i)}{d\theta_i}$ is the convolution of $f_x(x)$ and $f_\eta(x)$, which are both Gaussians. The convolution of two Gaussians is another Gaussian (Rényi 1970) and we therefore have

$$\frac{df(\theta_i)}{d\theta_i} = -\mathcal{N}(\theta_i, \sigma_x^2 + \sigma_\eta^2), \tag{9.14}$$

where $\mathcal{N}(\mu, s^2)$ is a Gaussian PDF with mean μ and variance s^2. Thus, we must have

$$f(\theta_i) = \int_{-\infty}^{-\theta_i} \frac{1}{\sqrt{2\pi(\sigma_x^2 + \sigma_\eta^2)}} \exp\left(-\frac{\tau^2}{2(\sigma_x^2 + \sigma_\eta^2)}\right) d\tau$$

$$= 0.5 + 0.5\mathrm{erf}\left(\frac{-\theta_i}{\sqrt{2(\sigma_x^2 + \sigma_\eta^2)}}\right)$$

$$= 0.5 + 0.5\mathrm{erf}\left(\frac{-\theta_i}{\sigma_x\sqrt{2(1 + \sigma^2)}}\right)$$

$$= 0.5 - 0.5\mathrm{erf}\left(\frac{\theta_i}{\sigma_x}\sqrt{\frac{\rho}{2}}\right), \tag{9.15}$$

where ρ is the correlation coefficient between the input to any two different threshold devices, as discussed in Appendix 6. Thus, $f(\theta_i)$ depends on both σ_η and σ_x, not just their ratio, σ. However, for SSR we have $\theta_i = 0$, and $f(0) = 0.5$, regardless of the actual magnitudes of σ_x and σ_η.

The expected value of y is

$$E[y] = \sum_{i=1}^{N} f(\theta_i)$$

$$= \frac{N}{2} - \frac{1}{2}\sum_{i=1}^{N} \mathrm{erf}\left(\frac{\theta_i}{\sigma_x\sqrt{2(1 + \sigma^2)}}\right). \tag{9.16}$$

If all thresholds are equal to the same value, θ, then

$$E[y] = \frac{N}{2} - \frac{N}{2}\mathrm{erf}\left(\frac{\theta}{\sigma_x\sqrt{2(1 + \sigma^2)}}\right), \tag{9.17}$$

and, clearly, the average energy is a function of the ratio of θ to σ_x, as well as σ.

We are now able to present results showing how the mutual information and average output energy vary with θ and σ.

Results

Figure 9.1(a) shows numerical calculations of the mutual information against increasing σ for a Gaussian signal with $\sigma_x = 1$, and Gaussian noise. We have $N = 5$, and consider various threshold values. If the value of σ_x were to be changed, then the thresholds shown would need to be adjusted to keep θ/σ_x constant to give the same results. Figure 9.2 shows the average output energy for the same conditions, calculated from the exact formula of Eq. (9.17) and validated numerically.

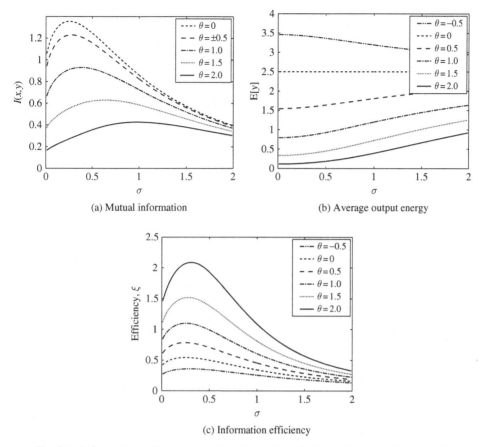

(a) Mutual information

(b) Average output energy

(c) Information efficiency

Fig. 9.1. Information efficiency as a function of σ, all thresholds equal. Figure 9.2 shows the result of numerically calculating the mutual information, $I(x, y)$, as a function of noise intensity, σ, for various threshold values and $\sigma_x = 1$. Note that since both the signal and noise have even PDFs, the mutual information is the same for threshold values with the same magnitude but opposite sign. Hence, as can be seen in Fig. 9.2, the mutual information is identical for $\theta = \pm 0.5$. As the magnitude of the threshold value increases, the mutual information decreases for all σ. However, the value of σ at which the peak value of the mutual information occurs increases with increasing σ. Figure 9.2 shows the average output energy calculated from the exact formula of Eq. (9.17) and validated numerically. The average energy decreases with increasing σ. The information efficiency is shown in Fig. 9.2, which indicates that the efficiency increases with increasing σ, and therefore with increasing output energy.

The information efficiency is plotted in Fig. 9.2. As we expect – given the result in Stocks (2001a) that shows that the optimal threshold value is equal to the signal mean – the mutual information decreases for $\theta \neq 0$. However, as we might also expect, the average output energy is greater than $N/2$ for $\theta < 0$ and less than $N/2$ for $\theta > 0$.

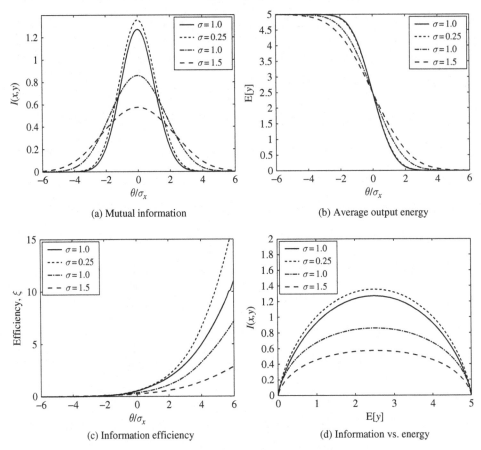

(a) Mutual information

(b) Average output energy

(c) Information efficiency

(d) Information vs. energy

Fig. 9.2. Information efficiency as a function of θ/σ_x, all thresholds equal. Figure 9.2(a) shows the mutual information, $I(x, y)$, calculated numerically for four different values of noise intensity, σ, as a function of θ/σ_x. Of the four values of σ shown, at $\theta = 0$ the mutual information is greatest for $\sigma = 0.25$, as we know it should be, since this is the SSR situation discussed in Chapter 4. However, for θ nonzero, the optimal mutual information can occur for other values of σ, as also illustrated in Figure 9.2. Figure 9.2(b) shows the average output energy calculated from the exact formula of Eq. (9.17) and validated numerically. At $\theta = 0$, the average output energy is $E[y] = N/2$, as we know it always is for SSR. Figure 9.2(c) shows that, for $\theta \geq 0$, the information efficiency is largest for $\sigma = 0.25$, and for each value of σ increases with increasing θ/σ_x, as also shown in Figure 9.2. For $\theta < 0$, the information efficiency is very small in comparison with $\theta > 0$. This is due to the mutual information decreasing with decreasing σ for $\theta < 0$, while the average energy continues to increase. Figure 9.2(d) shows the variation in mutual information with $E[y]$. For a given value of σ, the mutual information is maximized at $N/2$, and decreases as the energy decreases from this value.

More importantly, Fig. 9.2 shows that the information efficiency increases with increasing threshold values, and therefore with decreasing energy. If we wish to maximize the information efficiency, this illustrates the requirement for placing a minimum mutual information or maximum average output energy constraint for nontrivial results.

Figure 9.2 shows the mutual information, average output energy, and informa-tion efficiency for various values of σ as a function of θ/σ_x. Of the four values of σ shown, at $\theta = 0$ the mutual information is greatest for $\sigma = 0.25$, as we know it should be, since this is the SSR situation discussed in Chapter 4. However, for θ nonzero, the optimal mutual information can occur for other values of σ, as also illustrated in Fig. 9.2. At $\theta = 0$, the average output energy is $E[y] = N/2$, as we know it always is for SSR. Figure 9.2(c) shows that for $\theta \geq 0$, the infor-mation efficiency is largest for $\sigma = 0.25$, and for each value of σ increases with increasing θ/σ_x, as also shown in Fig. 9.2. For $\theta < 0$, the information efficiency is very small in comparison with $\theta > 0$. This is due to the mutual information decreasing with decreasing σ for $\theta < 0$, while the average energy continues to increase. Figure 9.2(d) shows the variation in mutual information with $E[y]$. For a given value of σ, the mutual information is maximized at $N/2$, and decreases as the energy decreases from this value.

Suppose we require that $E[y] = A$. Then rearranging Eq. (9.17) gives

$$\frac{\theta}{\sigma_x} = \sqrt{2(1+\sigma^2)}\text{erf}^{-1}\left(1 - \frac{2A}{N}\right). \tag{9.18}$$

Figure 9.3 shows the mutual information and information efficiency for various fixed values of $E[y]$. For each value of $E[y]$, the corresponding value of θ was found from Eq. (9.18).

The main conclusion from these results is that the information efficiency can be made arbitrarily large by decreasing the output energy. However, as the output energy decreases, the mutual information increases. Hence, for any value of σ, there is a tradeoff between mutual information and energy. We now progress to considering two constrained optimal quantization problems.

Energy constrained optimal quantization

The first constrained problem we aim to solve is the problem of maximizing the mutual information subject to the constraint that the average output energy must be no larger than some value, A. This can be expressed as

Find: $\qquad \max_{\vec{\theta}} f(\vec{\theta}) = I(x, y),$

subject to: $\qquad m(\vec{\theta}) = E[y] \leq A, \vec{\theta} \in \mathbb{R}^N. \tag{9.19}$

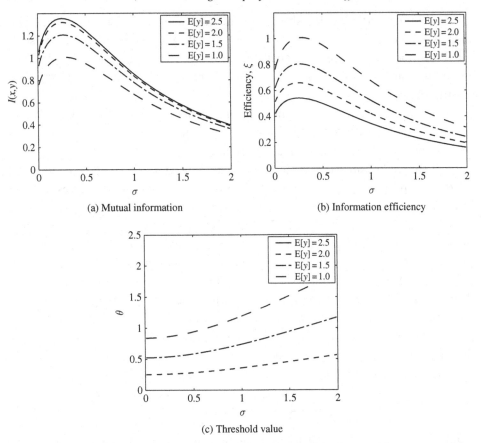

(a) Mutual information   (b) Information efficiency

(c) Threshold value

Fig. 9.3. Numerical calculations of the mutual information, $I(x, y)$, and information efficiency, ξ, for four specified values of output energy, as a function of noise intensity, σ. The threshold value, θ, that achieves the specified value of the average output energy, $E[y]$, is shown in Fig. 9.3(c), and was calculated from the exact formula of Eq. (9.18).

If we now write this in terms of a Lagrange multiplier, λ, we have

Find: $$\max_{\vec{\theta}} \; g(\vec{\theta}) = I(x, y) - \lambda E[y^2],$$

subject to: $$\vec{\theta} \in \mathbb{R}^N, \lambda \geq 0. \tag{9.20}$$

As in Chapter 8, we set $\sigma_x = 1$, and expect that the optimal thresholds will scale in inverse proportion to σ_x.

The results of numerical solutions of Problem (9.20) for $N = 5$, Gaussian signal and noise, and various values of A are shown in Figures 9.4(a) and 9.4(b), which show the mutual information and energy efficiency, for several values of A, and in Fig. 9.5, which shows the optimal thresholds. The case of no constraint is also

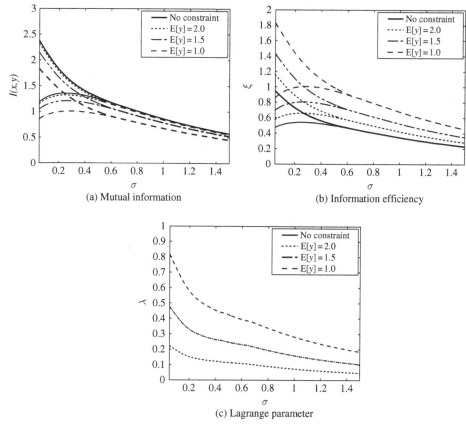

Fig. 9.4. Information efficiency for optimized thresholds. Mutual information, $I(x, y)$, is shown in Fig. 9.4(a) with thick solid lines, against increasing noise intensity, σ, obtained by numerically solving Problem (9.20) for three different values of the constraint on the average output energy. Also shown with thin solid lines is the mutual information obtained for the same constraint when all thresholds have the same value. Figure 9.4(b) shows the corresponding energy efficiency, ξ, against σ. Figure 9.4(c) shows the value of the Lagrange parameter, λ, required to achieve each constraint as a function of σ.

plotted for comparison. Figures 9.4(a) and 9.4(b) also show with thin solid lines the mutual information and energy efficiency obtained for each constraint if all thresholds are set to the same value, as considered in Section 9.2. Note that the energy constraint is always met with equality, since we find that mutual information always decreases if the energy is made smaller than the constraint value. The value of λ that satisfies each constraint also varies with σ. These values are shown for each energy constraint in Fig. 9.4(c).

Figure 9.4(a) shows that as the maximum output energy decreases, the mutual information also decreases, but only by a fraction of a bit per sample. From

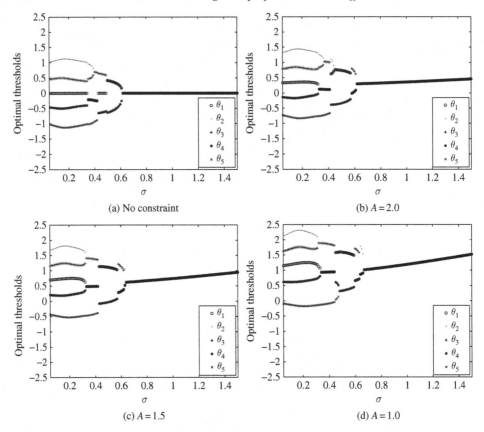

Fig. 9.5. The optimal threshold values obtained when Problem (9.20) is solved for three different values of an energy constraint, for the case of Gaussian signal and noise, and $N = 5$. Also shown are the optimal thresholds for the corresponding unconstrained problem, as solved in Chapter 8. As the maximum average energy constraint decreases, we see an upward shift in the optimal thresholds. Thus, the optimal threshold values are larger than the unconstrained optimal thresholds. This is to be expected, since larger threshold values mean fewer threshold crossings, and therefore a lower average output energy. Otherwise the same quali- tative behaviour as the unconstrained problem occurs. For sufficiently large σ, the optimal solution is for all thresholds to have the same value.

Fig. 9.4(b), the information efficiency increases as the maximum output energy decreases. From this result, it is clear that information efficiency can be substan- tially increased with only a small decrease in mutual information.

Figure 9.5 shows that, as the maximum average energy constraint decreases, there is an upward shift in the optimal thresholds. This shift increases with increas- ing σ. Thus, the optimal threshold values are larger than the unconstrained optimal thresholds. This is to be expected, since larger threshold values mean fewer thresh- old crossings, and therefore a lower average output energy. Otherwise the same qualitative behaviour as the corresponding unconstrained problem occurs. For sufficiently large σ, the optimal solution is for all thresholds to have the same value. This value is obtained by substituting the desired value of A into Eq. (9.18) and solving for θ. For small σ we have the same bifurcational behaviour in the optimal threshold diagram as described in Chapter 8.

Fixed thresholds and a minimum information constraint

In a biological context, increasing the energy efficiency of information encoding by optimizing threshold values may not be particularly realistic. To be useful to an organism, a sensory system must convey a certain amount of information, $\hat{I}$, to the brain. Any less than this and the sensory function is not useful. For example, to avoid becoming prey, a herbivore must be able to distinguish a sound at, say, 100 m, whereas 50 m may not be sufficient. To develop such a sensory ability, the animal must use more neurons, and therefore use more energy – in the form of generated spike trains, as well as energy to grow them in the first place – to convey the information to the brain. For maximum evolutionary advantage the animal must do this in the most energy efficient manner – otherwise it may have to eat more than is feasible – so the problem that evolution solves in the animal is not to fix N and maximize $I(x, y)$, but to fix $I(x, y)$ and minimize N (R. P. Morse, personal communication, 2003). If the energy is an increasing function of N, then this problem is equivalent to minimizing the energy. This is slightly different from the problem of maximizing information subject to an energy constraint, although the most efficient way of transmitting $I(x, y)$ bits of information must be equivalent to finding the value of energy E that gives rise to the optimal amount of information $I(x, y)$.

Thus, for this problem we wish to fix the mutual information and threshold settings, and find the population size, N, that minimizes the energy for a range of σ. For the SSR situation where all thresholds are equal to the signal mean, and $f_x(x)$ and $f_\eta(\eta)$ are zero mean even functions, the average output energy is simply $N/2$ and hence increases linearly with increasing N. Therefore the solution to this problem reduces to finding the minimum N that satisfies $I(x, y) \geq B$, which can be expressed as

$$\text{Find:} \qquad \min N,$$
$$\text{subject to:} \qquad I(x, y) \geq B, \vec{\theta} = \vec{0}, N \in \mathbb{Z}^+. \qquad (9.21)$$

The result of solving Problem (9.21) for Gaussian signal and noise and three values of $I(x, y)$ is shown in Fig. 9.6, as a function of σ. It is clear that N is minimized for a nonzero value of σ, which we might expect, given that we have the SSR situation.

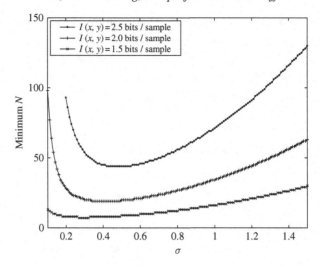

Fig. 9.6. The result of minimizing N for three constraints on $I(x, y)$, for Gaussian signal and noise, as a function of σ. It is clear that N is minimized for a nonzero value of σ for each constraint.

Discussion and open questions

In this section we have formulated and solved two different problems of energy efficient information transmission in the extended SSR model.

For the problem of finding the threshold settings that maximize the information efficiency for fixed N, our results clearly indicate that the information efficiency always increases for decreasing average output energy and always decreases for increasing mutual information. However, a fairly large information efficiency increase can occur for only a small decrease in mutual information. We show also that the maximum information efficiency for optimized thresholds is always strictly decreasing for increasing noise intensity, σ, that is no SR effect is seen. However, since populations of real neurons are not known for having widely distributed thresholds, the information efficiency results shown in Fig. 9.2 are the most biologically relevant. This figure shows that the maximum information efficiency occurs for nonzero noise. This is due to the same mechanism as SSR, where the noise acts to randomly distribute the effective threshold values.

We also briefly discuss the problem of minimizing the number of 'neurons' required to achieve a certain minimum mutual information. For the SSR model, the minimum N occurs for nonzero σ. In contrast, we expect that if the same problem is solved for arbitrary thresholds, N will increase with increasing σ. However, we also expect that, for small σ, the optimal thresholds will be widely distributed, and, for large σ, the optimal thresholds will be the SSR situation of all thresholds equal to the signal mean.

There is much further work to be completed on these problems, and the results presented are intended to convey 'proof of principle' rather than a comprehensive investigation. Future research might, for example, introduce energy constraints and variable thresholds to Problem (9.21). Insight gained from the study of these highly simplified neural population model problems may be of benefit to the understanding of neural coding in the presence of noise in more realistic neuron models and real neurons.

We now, in Section 9.3, consider the rate–distortion tradeoff issue in the SSR model and its extension to arbitrary thresholds.

9.3 Rate–distortion tradeoff

From a quantizer design point of view, it is desirable to minimize both the *rate* and the *distortion*. Although the exact definitions of both rate and distortion can vary, in quantization, the term 'rate' loosely corresponds to the amount of compression obtained by a quantizer, and hence rate-distortion theory falls into the category of *lossy source coding* or *lossy compression* theory. However, since rate and distortion are dependent on the same variables, they cannot both be simultaneously minimized, and there must be a tradeoff between rate and distortion. Thus, the rate-distortion problem is usually formulated as the problem of minimizing the rate, subject to a specified constraint that the distortion can be no larger than some fixed value, D. The minimum possible rate that achieves distortion D for a given source distribution is known as the *rate-distortion function*, denoted as $R(D)$. In general, $R(D)$ is a theoretical limit that cannot be achieved in practice.

In such rate-distortion theory, rate is generally defined as the mutual information between the input and output of a quantizer. Therefore, if we consider the SSR model – or its extension to arbitrary thresholds – from such a viewpoint, we must perform the opposite task to that considered in Section 9.2, and minimize the mutual information rather than maximize it. The reason that we can consider both goals without contradiction is that in Section 9.2 we seek to optimize information transmission – that is, we address the question of 'what is the largest amount of information, on average, that can be transmitted in a channel, subject to some constraints?' By contrast, this section addresses a compression problem, by aiming to answer the question 'What is the smallest amount of information required to represent a signal with, on average, a distortion no larger than some value, D?'

Of particular relevance to this discussion is a simple formula known as the *information transmission inequality* (Berger and Gibson 1998), which relates rate, distortion, and channel capacity, C. This formula is

$$D \geq R^{-1}(C). \tag{9.22}$$

Note that in this formula, channel capacity is the maximum mutual information for a given channel, and the rate-distortion function, $R(D)$, is the minimum mutual information to achieve distortion D. This formula says that if one is trying to transmit data from a source with rate-distortion function $R(D)$, over a channel with capacity C, the average distortion achieved will be greater than the inverse of the rate-distortion function evaluated at C (Berger 1971).

Looking at this another way, suppose a communications system designer is required to transmit information through a channel with capacity C, and is aiming for fidelity, or average distortion, D. Suppose also that the information source has a known rate–distortion function, $R(D)$. Then the designer knows immediately whether there is any chance of achieving their requirements, because success will only be possible if $C \geq R(D)$. This illustrates the utility of knowing the theoretical optimal rate–distortion tradeoff, and why there are different reasons for both maximizing and minimizing mutual information. We now discuss in more detail the basic theory of the rate–distortion function.

Rate–distortion theory

The optimal tradeoff between rate and distortion is measured using the rate–distortion function (Berger and Gibson 1998), often expressed as $R(D)$, where R is the rate – defined as the mutual information – and D is some arbitrary distortion measure, often taken to be MSE distortion. The rate–distortion function is defined as the solution to the following constrained optimization problem

Find: $\min I(x, y)$,

subject to: Distortion $\leq D$.   (9.23)

Thus, $R(D)$ is defined as being the minimum possible rate to achieve distortion D. This implies that, in practice, the actual rate achieved in a quantizer will always be $I(x, y) \geq R(D)$.

Exact $R(D)$ for a Gaussian source

Shannon derived an exact analytic result for the rate–distortion function for a Gaussian source with variance σ_x^2, and the MSE distortion measure (Berger and Gibson 1998, Cover and Thomas 1991). This exact result states that

$$R(D) = \begin{cases} 0.5 \log_2 \left(\frac{\sigma_x^2}{D} \right) & 0 \leq D \leq \sigma_x^2, \\ 0 & D \geq \sigma_x^2 \end{cases} \tag{9.24}$$

and says that no quantization scheme can achieve a distortion less than D with a rate smaller than $R(D)$, for a Gaussian source with variance σ_x^2. In other words, a quantization scheme with rate R will provide a mean square distortion no smaller than D. Thus, we have a lower bound for the mutual information for quantization of a Gaussian source as

$$I(x, y) \geq 0.5 \log_2 \left(\frac{\sigma_x^2}{D} \right). \tag{9.25}$$

We briefly explain why $R(D)$ is zero for $D \geq \sigma_x^2$. Consider the region $D > \sigma_x^2$; here, distortion larger than σ_x^2 can be achieved without transmitting any information. For a zero mean Gaussian signal, the critical distortion, $D = \sigma_x^2$, is achievable by simply always guessing that the input signal is zero. In this case, the MSE distortion is simply the variance of the signal, σ_x^2. For larger distortions, if we always guess that the input is some nonzero value, say a, then the MSE distortion is $\sigma_x^2 + a^2$. Therefore for distortion $D > \sigma_x^2$, guessing $a = \sqrt{(D - \sigma_x^2)}$ gives distortion D with a rate of zero. However, in the region $D < \sigma_x^2$, to achieve some distortion, $D < \sigma_x^2$, it is necessary to transmit some information, $R(D) > 0$.

The rate–distortion function can be inverted to obtain the *distortion–rate function*, $D(R)$. Carrying this out for a Gaussian source using Eq. (9.24) gives

$$D(R) = \sigma_x^2 2^{-2R}. \tag{9.26}$$

Equation (9.26) says that no quantization of a Gaussian source can achieve a distortion smaller than D, if the rate is to be no larger than R.

Equation (9.26) can be expressed as the signal-to-quantization-noise ratio (SQNR)

$$\text{SQNR} = \frac{\sigma_x^2}{D(R)} = 2^{2R}. \tag{9.27}$$

Taking the base ten logarithm of both sides of Eq. (9.26) and multiplying by 10 gives

$$10 \log_{10}(D) = 10 \log_{10} \sigma_x^2 - 20 \log_{10}(2) R$$
$$\simeq 20 \log_{10} \sigma_x - 6.02R. \tag{9.28}$$

Rearranging Eq. (9.28) gives the maximum possible SQNR of a Gaussian source with rate R in decibels as

$$10 \log_{10} \left(\frac{\sigma_x^2}{D} \right) = \left(20 \log_{10}(2) \right) R \simeq 6.02R \quad \text{dB}. \tag{9.29}$$

This corresponds with a well-known rule of thumb in quantizer design that states that a one-bit increase in rate gives about a 6 dB increase in SNR (Gray and Neuhoff 1998). Furthermore, it shows that the maximum possible SQNR for a given R is proportional to R.

R(D) for non-Gaussian source distributions

Exact expressions for $R(D)$ such as that given by Eq. (9.24) are quite rare. For most source distributions, numerical techniques are required to find the rate–distortion function, and, fortunately, the Arimoto–Blahut algorithm can be used to achieve this (Blahut 1972, Arimoto 1972).[1]

However, although analytical formulas for the exact rate–distortion function are difficult to find, much attention has been focused on finding upper or lower bounds to $R(D)$ for non-Gaussian sources. In particular, consider the following reasoning, where y is the output of the encoding operation of a quantizer, $\hat{x}$ is the MMSE distortion decoding of y, and $\epsilon = x - \hat{x}$ is the error signal

$$
\begin{aligned}
I(x, y) &= H(x) - H(x|y) \\
        &= H(x) - H(x|\hat{x}) \\
        &= H(x) - H(x - \hat{x}|\hat{x}) \\
        &= H(x) - H(\epsilon|\hat{x}) \\
        &\geq H(x) - H(\epsilon).
\end{aligned}
\tag{9.30}
$$

The above steps hold since, first, the average conditional entropy of x given y is identical for a decoding of y. Second, the conditional entropy of ϵ given $\hat{x}$ is the same as the conditional entropy of x. Third, the conditional entropy of ϵ is always smaller than the unconditioned entropy (Cover and Thomas 1991).

We also know that the entropy of a continuous random variable, v, with variance, σ_v^2, is maximized by the Gaussian distribution, and is given by

$$
H(v) = 0.5 \log_2 (2\pi e \sigma_v^2).
\tag{9.31}
$$

Therefore, since for optimal decoding the mean of the error is zero, then the variance of the error signal, ϵ, is the MSE distortion, D, and from Inequality (9.30)

$$
I(x, y) \geq H(x) - 0.5 \log_2 (2\pi e D).
\tag{9.32}
$$

If we let

$$
R_L(D) = H(x) - 0.5 \log_2 (2\pi e D),
\tag{9.33}
$$

it can be shown (Berger and Gibson 1998, Linder and Zamir 1999) that

$$
I(x, y) \geq R(D) \geq R_L(D).
\tag{9.34}
$$

Eq. (9.33) is therefore known as the *Shannon lower bound* for the rate–distortion function. It applies to any source PDF, $f_x(x)$, and gives a lower bound on the rate–distortion function.

---

[1] The Arimoto–Blahut algorithm can also be used to calculate channel capacity – see Chapter 4.

For the specific case of a Gaussian source, we also have $H(x) = 0.5 \log_2 (2\pi e \sigma_x^2)$, and therefore

$$R_L(D) = 0.5 \log_2 \left(\frac{\sigma_x^2}{D} \right). \tag{9.35}$$

In fact, the reasoning given to arrive at this result forms the first part of the proof of Eq. (9.24) for a Gaussian source. The second part of the proof is to find a condition for which the bound of Inequality (9.32) is achievable, and is given in Cover and Thomas (1991). This also shows that $R_L(D)$ becomes tight with the actual $R(D)$ as given by Eq. (9.24), so that $R_L(D) = R(D)$ for a Gaussian source.

For non-Gaussian sources, the Shannon lower bound will always be smaller than that for a Gaussian source with the same variance as the non-Gaussian source. Equation (9.33) can also be simplified to be written in a manner similar to Eq. (9.35) using the concept of *entropy power*. Consider a random variable with PDF $f_x(x)$ and differential entropy, $H(x)$. The entropy power, Q_o, of this random variable is defined as being the variance of a Gaussian random variable that has differential entropy equal to $H(x)$ (Berger and Gibson 1998). Since the differential entropy of a Gaussian distribution is given by Eq. (9.31), this means that

$$Q_o = \frac{2^{2H(x)}}{2\pi e}, \tag{9.36}$$

and therefore

$$R_L(D) = 0.5 \log_2 \left(\frac{Q_o}{D} \right). \tag{9.37}$$

We have also the SQNR in dB corresponding to $R_L(D)$ as

$$10 \log_{10} \left(\frac{\mathrm{E}[x^2]}{D} \right) = 10 \log_{10} \left(\frac{\mathrm{E}[x^2]}{Q_o} \right) + (20 \log_{10}(2)) R_L(D). \tag{9.38}$$

The entropy and Shannon lower bound corresponding to three different distributions are shown in Table 9.1. Note that there is no simple analytical expression available for the entropy and Shannon lower bound of the logistic distribution for arbitrary variance, although these quantities can easily be calculated numerically.

R(D) and its relationship to correlation coefficient

Recall the formula in Chapter 6 that expresses the SQNR in terms of the correlation coefficient between the input and decoded output of the SSR model, ρ, as

$$\mathrm{SQNR} = \frac{\mathrm{E}[x^2]}{D} = \frac{1}{1 - \rho^2}. \tag{9.39}$$

Table 9.1. *The PDF, variance, entropy, and Shannon lower bound for various distributions*

Distribution	$f_X(x)$	Variance	$H(x)$	$R_L(D)$
Gaussian	$\dfrac{1}{\sqrt{2\pi\sigma_x^2}}\exp\left(-\dfrac{x^2}{2\sigma_x^2}\right)$	σ_x^2	$0.5\log_2(2\pi e\sigma_x^2)$	$0.5\log_2\left(\dfrac{\sigma_x^2}{D}\right)$
Uniform	$\begin{cases}\dfrac{1}{\sigma_x}, & x \in \left[-\dfrac{\sigma_x}{2}, \dfrac{\sigma_x}{2}\right], \\ 0, & \text{otherwise}\end{cases}$	$\dfrac{\sigma_x^2}{12}$	$\log_2(\sigma_x)$	$0.5\log_2\left(\dfrac{\sigma_x^2}{2\pi eD}\right)$
Laplacian	$\dfrac{1}{\sqrt{2}\sigma_x}\exp\left(\dfrac{-\sqrt{2}|x|}{\sigma_x}\right)$	σ_x^2	$\log_2(\sqrt{2}e\sigma_x)$	$0.5\log_2\left(\dfrac{e\sigma_x^2}{\pi D}\right)$

In terms of decibels this is

$$10\log_{10}(\text{SQNR}) = -10\log_{10}(1 - \rho^2). \tag{9.40}$$

This formula holds both for optimal linear (Wiener) decoding, and for MMSE decoding, although the correlation coefficient is smaller for MMSE decoding, for which we achieve the largest possible correlation coefficient, $\rho_{x\hat{x}}$. Substituting from Eq. (9.39) into Eq. (9.35) then gives the rate–distortion function for a Gaussian source – which is also the Shannon lower bound – in terms of the correlation coefficient as

$$R(D) = R_L(D) = -0.5\log_2\left(1 - \rho_{x\hat{x}}^2\right). \tag{9.41}$$

Recall also that the correlation coefficient for any linear decoding is independent of the actual decoding, and is the same as the correlation coefficient for no decoding. Hence, for the SSR model we have, for a Gaussian source

$$R(D) = R_L(D) > -0.5\log_2\left(1 - \rho_{xy}^2\right). \tag{9.42}$$

Bounds for the mutual information in terms of correlation coefficient, or coherence function, are well known in the literature (Pinsker 1964, Nikitin and Stocks 2004).

We shall now consider numerically the relationship between the Shannon lower bound and the actual mutual information for SSR and various signal and noise distributions.

Rate–distortion tradeoff for SSR

We aim to examine the *operational* rate–distortion performance of the SSR model, and compare it to the theoretical Shannon lower bound for the sources given in

Table 9.1, as well as the numerically calculated Shannon lower bound for the logistic source.

Operational rate–distortion

Recall that Chapter 4 provides figures showing numerical calculations of the mutual information in the SSR model as a function of noise intensity, σ, for various matched signal and noise distributions. Chapter 6 provides similar figures for the MMSE distortion. Therefore, plotting the value of $I(x, y)$ against the corresponding value of MMSE distortion for each value of σ gives provides a plot of the operational rate–distortion tradeoff for each value of σ in the SSR model. The result of carrying this out for $0 \leq \sigma \leq 5$ is shown in Figs 9.7 and 9.8 for various values of N, and the cases of matched Gaussian, Laplacian, and logistic signal and noise. Fig. 9.7 shows the mutual information plotted against the MMSE distortion, while Fig. 9.8 shows the mutual information plotted against the output SQNR. In quantization theory, it is quite often the case that plots of rate against distortion are shown in terms of SQNR in dB, due to the loglinear nature of the Shannon lower bound. In each figure, points on the curve for each value of N correspond to a particular value of σ. The thick black lines show the theoretical Shannon lower bound, $R_L(D)$.

For each distribution, the curves for each value of $N > 1$ start at a rate of one bit per sample, then increase with rate and decrease with MSE distortion, as σ increases. The rate then reaches its maximum before the MMSE distortion reaches its minimum. Then with continuing increasing σ, the curves reach the MMSE distortion minimum, before curling back down towards the $R_L(D)$ curve. Note that this means that – except for large σ – there are two values of σ for which the same distortion can occur, corresponding to two different rates. This is due to the SR behaviour seen in both the mutual information and distortion. If the main goal of a quantizer is to operate with minimum distortion, this observation indicates that the optimal value of input SNR to use is the one that achieves the minimum distortion, rather than the maximum rate.

A further observation is the fact that, for very large σ – that is, low input SNR – the SSR operational rate–distortion is very close to the Shannon lower bound.

Comparison of $I(x, y)$ with $R_L(D)$

Figure 9.9 shows (with thick lines) the mutual information for each distribution plotted against increasing σ, as in Chapter 4. It also shows (with thin lines) the Shannon lower bound for each value of σ, where the distortion value used for each σ, say $D(\sigma)$, is the operational distortion, obtained with optimal MMSE distortion decoding. The difference between the two curves for each value of N is therefore the difference in rate between the actual rate that achieves $D(\sigma)$ and the Shannon

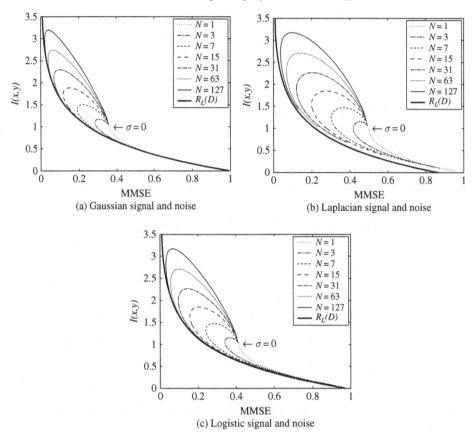

Fig. 9.7. Mutual information against MMSE distortion for three matched signal and noise distributions, and a number of values of N. The thick black line shows the Shannon lower bound, $R_L(D)$. Points on the curve for each value of N correspond to different values of noise intensity, σ, where σ starts at zero at the indicated point. Note that there are in general two values of mutual information that achieve the same MMSE distortion and two values of MMSE distortion that achieve the same mutual information. This is due to the SR behaviour of the mutual information and MMSE distortion for increasing σ.

lower bound for the rate that achieves $D(\sigma)$. As is also seen in Figs 9.7 and 9.8, it is clear that, for large σ, the actual mutual information gets closer and closer to the Shannon lower bound.

Exact result for uniform signal and noise and $\sigma \leq 1$

In Chapter 4, we stated an exact result for the SSR mutual information and $\sigma \leq 1$, in terms of N and σ. This result, given in Eq. (4.58), is first derived in Stocks (2001c). In Chapter 6, we derived a new result for the MMSE distortion under the same conditions, as given by Eq. (6.100). Therefore, we have enough information

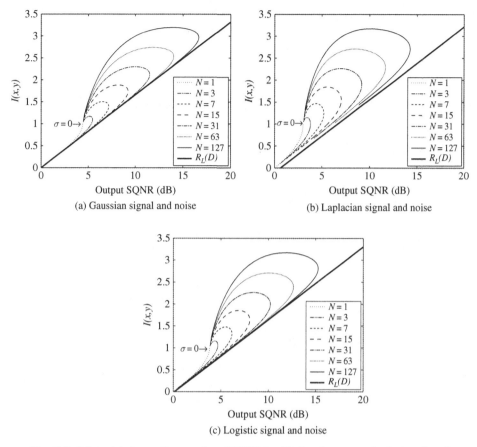

Fig. 9.8. Mutual information against SQNR in dB for three matched signal and noise distributions, and a number of values of N. The thick black line shows the Shannon lower bound, $R_L(D)$, converted to dB. Points on the curve for each value of N correspond to different values of noise intensity, σ, where σ starts at zero at the indicated point.

to calculate analytical operational rate–distortion points in these circumstances. However, there is no simple way to combine these two expressions to obtain an expression for the mutual information in terms of MMSE distortion. The best we can do is calculate from each equation the mutual information and distortion for a range of values of σ, and then plot the resultant operational rate–distortion curve, as shown for other distributions in Fig. 9.7. Since we can also numerically calculate the rate and distortion for $\sigma > 1$, we shall not restrict our attention to the exact formulas, but plot in Fig. 9.10 the mutual information against rate for $0 \leq \sigma \leq 5$.

For uniform signal and noise and $\sigma = 1$, we saw in Chapter 6 that MMSE decoding has linearly spaced reconstruction points and is therefore a linear decoding

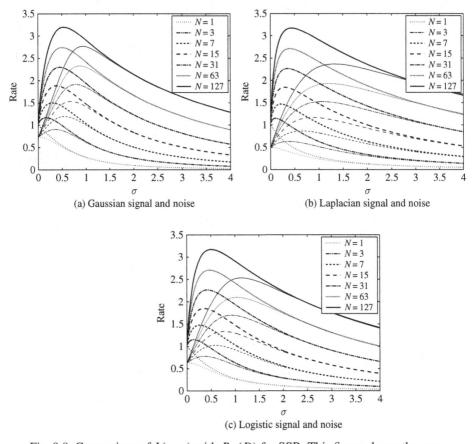

Fig. 9.9. Comparison of $I(x, y)$ with $R_L(D)$ for SSR. This figure shows the operational mutual information for SSR (thick lines), as well as the Shannon lower bound (thin lines) corresponding to the minimum achievable distortion for each value of σ. Clearly, the actual mutual information is far larger than the lower bound for small σ, but gets closer to the bound as σ increases.

scheme. This means that the optimal correlation coefficient is the linear correlation coefficient of Eq. (6.64) evaluated at $\sigma = 1$, which is

$$\rho_{xy} = \sqrt{\left(\frac{N}{N+2}\right)}. \tag{9.43}$$

Substituting from Eq. (9.43) into Eq. (9.39) gives

$$\text{SQNR} = \frac{\sigma_x^2}{12D} = \frac{N+2}{2}, \tag{9.44}$$

and therefore

$$\frac{\sigma_x^2}{D} = 6(N+2), \tag{9.45}$$

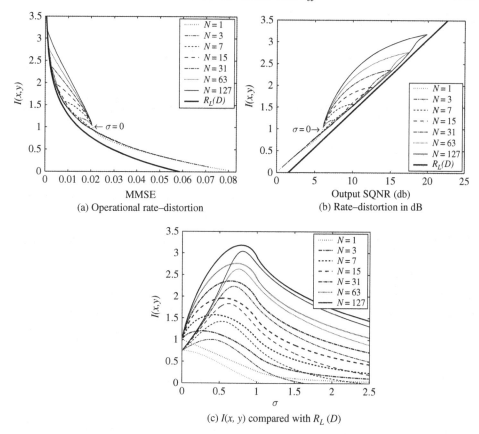

(a) Operational rate–distortion

(b) Rate–distortion in dB

(c) $I(x, y)$ compared with $R_L (D)$

Fig. 9.10. Operational rate–distortion tradeoff for SSR for uniform signal and noise and various N. Figure 9.10(a) shows the mutual information plotted against MMSE distortion, while Fig. 9.10(b) shows the mutual information plotted against output SQNR. As with Figs 9.7 and 9.8, points on the curve for each value of N correspond to different values of noise intensity, σ, where σ starts at zero at the indicated point, and the thick black line shows the Shannon lower bound, $R_L(D)$. Figure 9.10(c) shows both the operational mutual information, $I(x, y)$, (thick lines) and the Shannon lower bound, $R_L(D)$, (thin lines) plotted against increasing noise intensity, σ.

and we have the Shannon lower bound on rate in terms of the minimum possible distortion at $\sigma = 1$ as

$$R_L(D) = 0.5 \log_2 \left(\frac{3(N + 2)}{\pi e} \right). \tag{9.46}$$

We have also from Chapter 5 a large N approximation to the mutual information at $\sigma = 1$ as

$$I(x, y) \simeq 0.5 \log_2 \left(\frac{(N + 2)e}{2\pi} \right). \tag{9.47}$$

Therefore, for large N and uniform signal and noise with $\sigma = 1$ we have

$$I(x, y) - R_L(D) \simeq 0.5 \log_2 \left(\frac{\exp(2)}{6} \right) \simeq 0.15. \tag{9.48}$$

This result says that the mutual information for SSR is always at least approximately 0.15 bits per sample larger than the Shannon lower bound. This difference is clearly visible in Fig. 9.10(c).

Rate–distortion tradeoff for optimized thresholds

The operational rate–distortion tradeoff for the case of optimized thresholds, $N = 5$ and Gaussian signal and noise – as discussed in Chapter 8 – is shown in Fig. 9.11. Both the cases of maximized mutual information and minimized MMSE distortion are shown, as well as the SSR case. This plot clearly shows that, when the MMSE distortion is minimized, the corresponding mutual information is smaller than that obtained by maximizing the mutual information. Conversely, maximizing the mutual information results in a larger distortion. When the distortion is minimized, the resultant curve is relatively close to the Shannon lower bound,

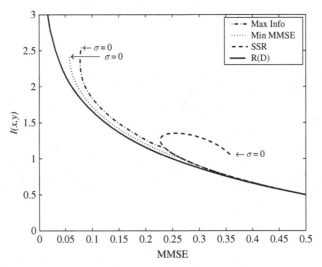

Fig. 9.11. Operational rate–distortion tradeoff for $N = 5$ and Gaussian signal and noise, for optimized thresholds. Both the cases of maximized mutual information and minimum MMSE distortion are shown, with the SSR case also plotted for comparison. Points on the curve for each situation correspond to different values of noise intensity, σ, where σ starts at zero at the indicated point. The thick solid line shows the Shannon lower bound, $R_L(D)$. It is clear that optimizing the thresholds provides a much larger mutual information and smaller MMSE distortion than is achievable with SSR.

$R(D)$, but, as we might expect, given the SSR results shown previously, does not reach it. This plot also clearly illustrates that the performance of SSR at best reaches about half the optimal mutual information, but only about five times the minimum possible distortion.

9.4 Chapter summary

The introductory section of this chapter briefly discusses the need for trade-offs between performance and cost in the design of engineered systems. It also reviews the theory from previous chapters relevant to this chapter, and discusses the Lagrange multiplier method for constrained optimization.

Section 9.2 introduces the problem of energy efficient information transfer to the SSR model and its extension to arbitrary thresholds. We define measures of energy and energy efficiency, and discuss how these measures vary with the threshold value, θ, and noise intensity, σ, in the SSR model. We then formulate and solve a constrained extension to the optimal quantization problem first discussed in Chapter 8, and show how the same qualitative behaviour occurs in optimal thresholds, but with an increase in the mean threshold value with increasing σ. For sufficiently large σ, the situation of all thresholds equal to the same value is optimal. We also briefly discuss a related problem of minimizing the number of threshold devices required to achieve a certain mutual information constraint in the SSR model.

Section 9.3 begins with a discussion of the tradeoff between rate and distortion in a quantizer, and then formalizing this discussion by stating relevant theory. We then examine the operational rate–distortion tradeoff in the SSR model, and compare the performance achieved with known lower bounds. We find that for large σ, the operational rate for a given distortion is relatively close to the Shannon lower bound, when compared with small σ.

Chapter 9 in a nutshell

This chapter includes the following highlights:

- The introduction of energy constraints to the SSR model and the arbitrary threshold extension of the SSR model, and formulation of these constraints into constrained optimization problems.
- Solutions of an energy constrained stochastic quantization problem.
- Discussion of the operational rate–distortion tradeoff for the SSR model.

Open questions

Possible future work and open questions arising from this chapter might include:

- Incorporation of energy constraints into optimal stochastic quantization of arrays of more realistic neuron models, such as the FitzHugh–Nagumo model considered in Stocks and Mannella (2001).
- Consideration of arbitrary thresholds for the problem of minimizing N for a given minimum mutual information, or distortion constraint.
- Comparison of the operational rate–distortion tradeoff for the SSR model, and optimally stochastic quantized model, with the operational tradeoff found in conventional scalar quantizers. The conventional quantizer can be considered both with and without independent threshold noise, and with and without companding.
- Introduction of a constraint on output entropy for the optimal stochastic quantization model. Such a problem has often been considered in quantization theory (György and Linder 2000).

This concludes Chapter 9, which considers various tradeoffs between cost and performance in the SSR model, and its extension to arbitrary thresholds. Chapter 10 now examines a further extension of the SSR model to incorporate biologically relevant features. The emphasis is on modelling stochastic resonance in the auditory system, and a proposition to improve cochlear implant performance using SSR effects.

10

Stochastic resonance in the auditory system

In this chapter we illustrate the relevance of stochastic resonance to auditory neural coding. This relates to natural auditory coding and also to coding by cochlear implant devices. Cochlear implants restore partial hearing to profoundly deaf people by trying to mimic, using direct electrical stimulation of the auditory nerve, the effect of acoustic stimulation.

10.1 Introduction

It is not surprising that the study of auditory neural coding involves stochastic phenomena, due to one simple fact – signal transduction in the ear is naturally a very noisy process. The noise arises from a number of sources but principally it is the Brownian motion of the stereocilia (hairs) of the inner hair cells that has the largest effect (Hudspeth 1989). Although the Brownian motion of the stereocilia appears small – typically causing displacements at the tips of the stereocilia of 2–3 nm (Denk and Webb 1992) – this is not small compared with the deflection of the tips at the threshold of hearing. Evidence exists that suggests, at threshold, the tip displacement is of the order of 0.3 nm (Sellick *et al.* 1982, Crawford and Fettiplace 1986), thus yielding a signal-to-noise ratio (SNR) of about -20 dB at threshold. Of course, under normal operating conditions the SNR will be greater than this, but, at the neural level, is typically of the order of 0 dB (DeWeese and Bialek 1995).

Given the level of noise and the fact that neurons are highly nonlinear makes the auditory system a prime candidate for observing stochastic resonance (SR) type behaviour. Indeed, many studies have now demonstrated that Brownian motion of stereocilia is approximately of the right magnitude to enhance signal transduction via SR (Jaramillo and Wiesenfeld 1998, Rattay *et al.* 1998, Jaramillo and Wiesenfeld 2000, Gebeshuber 2000). However, all of the cited papers focused on sinusoidal signals.

Clearly, stimuli of this nature do not commonly arise in the context of the mammalian ear. Consequently, the question as to whether SR occurs for more natural stimuli, such as speech, still remains. This is the central question reviewed in Section 10.3. As a precursor to this question, in Section 10.2 the effect of signal distribution on SR is shown in a simple threshold (comparator) model. Here it can be seen that the strength of the SR effect is strongly dependent on the form of the signal distribution. Finally, in Section 10.4 the application of suprathreshold stochastic resonance (SSR) to cochlear implant coding is discussed.

10.2 The effect of signal distribution on stochastic resonance

Despite the large literature that now exists on SR, one question that has received relatively little attention is whether SR can be observed for any type of signal and, if so, how does the magnitude of the SR effect depend on the form of the distribution? Clearly, for signal processing applications this is an important question – not least in the context of biological sensory systems, where signals are rarely periodic and usually take on a wide variety of forms.

In this section we consider in what forms of aperiodic signals SR can occur. It will be observed that the magnitude of the SR effect is strongly dependent on the form of the signal distribution – signal distributions with long tails (such as speech) tend to show a reduced SR effect.

Similar to earlier studies in Chapter 2, a simple comparator model – such as shown in Fig. 4.1 if $N = 1$, or in Fig. 2.6 if $T[.]$ is a binary threshold – is considered. The input signal is $x(t)$ and the random noise is $\eta(t)$. We consider only Gaussian noise, with a standard deviation of σ_η. The output signal, $y(t)$, is given by the response function

$$y(t) = \begin{cases} 1 & \text{if } x(t) + \eta(t) > x_{\text{th}} \\ 0 & \text{otherwise}, \end{cases} \qquad (10.1)$$

where x_{th} is the threshold level.

The aim is to show whether or not SR effects occur for general types of continuously distributed aperiodic signals and how the strength of the effect depends on signal distribution. The input signal is generated as a sequence of independently random samples from either a Gaussian or Laplacian distribution and therefore simulates discrete time samples taken from an ideal white-noise process.

Gaussian signals have been used in many other studies, where SR effects have been shown to occur, for example Collins *et al.* (1995a). It is therefore anticipated that they will be observed in this system also. The Laplacian distribution, which has a sharper peak and more slowly decaying tails than a Gaussian distribution – see Chapter 4 – has been chosen for comparison because it is commonly used to model music and speech (Dunlop and Smith 1984).

The PDFs of the Gaussian and Laplacian distributions are given in Table 4.1 in Chapter 4. In both cases we assume a signal mean of zero and a standard deviation of σ_x.

Two different measures are used to quantify the performance of the system given by Eq. (10.1): (i) the correlation coefficient, ρ_{xy}, between the input signal, $x(t)$, and the response $y(t)$ and (ii) the mutual information between $x(t)$ and $y(t)$, that is, $I(x, y)$.

Gaussian and Laplacian signal

Figure 10.1 shows the correlation coefficient and mutual information plotted against noise intensity for various threshold values for a Gaussian input signal. For both measures, stochastic resonance (that is, a noise induced maximum in performance) is observed but only when x_{th} is greater than $1.7\sigma_x$ or $2\sigma_x$ for correlation coefficient and mutual information respectively. This is consistent with the literature that shows SR only occurs if the signal is predominantly below the threshold value, that is the signal is 'subthreshold' (Bulsara and Zador 1996, Stocks 2001a).

It is also observed that the SR effect is more pronounced in the correlation coefficient curve than in the mutual information curve. Indeed, the noise-induced enhancement in the mutual information is relatively weak. From Fig. 10.1(b) it can

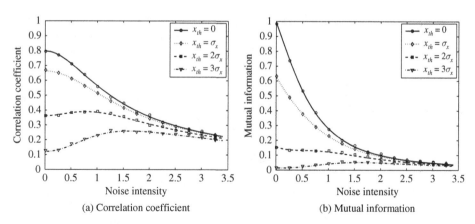

(a) Correlation coefficient (b) Mutual information

Fig. 10.1. Correlation coefficient and mutual information (in bits per sample) against noise intensity ($\sigma = \sigma_\eta/\sigma_x$) for various values of x_{th} for a Gaussian input signal. The data points are the results of digital simulation and the solid lines were obtained theoretically. Stochastic resonance is observed for sufficiently large threshold values.

be seen that, in the absence of noise, one-bit of information is transmitted when the threshold is zero; this is the maximum amount of information that this system can transmit (that is, the channel capacity), and hence gives a benchmark to gauge the size of the SR effect. In comparison to channel capacity, the noise-induced information gains are relatively weak. For threshold values above $2\sigma_x$ (that is, where SR effects are observed), the mutual information never attains a value greater than 0.1 bits – less than a tenth of the channel capacity.

Consequently, we have to conclude that SR results in a rather modest improvement in performance – in general it is preferable to operate this system with smaller thresholds, where SR is not observed.

Figure 10.2 shows the correlation coefficient and mutual information plotted against noise intensity for various threshold values for a Laplacian input signal.

This figure shows a number of interesting effects compared to the Gaussian case. First, larger threshold values are required to observe the SR effect compared with Gaussian signals – SR is only observed if the threshold is greater than approximately $3\sigma_x$ compared to $2\sigma_x$ for the Gaussian case; second, the SR effect is further reduced in magnitude – this is particularly the case for the mutual information, where extremely small noise-induced information gains are observed.

Indeed, for Laplacian signals the information gains are so small that they are difficult to observe due to the scale of the plots. For this reason, the size of the noise-induced gains for both Gaussian and Laplacian signals and for correlation and mutual information have been plotted in Fig. 10.3. This figure was obtained by calculating (from simulation) the value of the maximum noise-induced correlation (or information) and subtracting off the correlation (or information) at zero

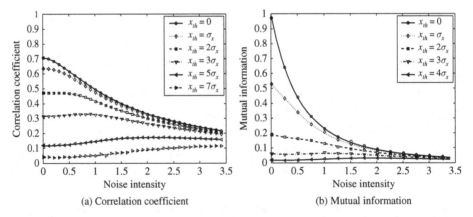

(a) Correlation coefficient   (b) Mutual information

Fig. 10.2. Correlation coefficient and mutual information (in bits per sample) against noise intensity ($\sigma = \sigma_\eta/\sigma_x$) for various values of x_{th} for a Laplacian input signal x. The data points are the results of digital simulation and the solid lines were obtained theoretically.

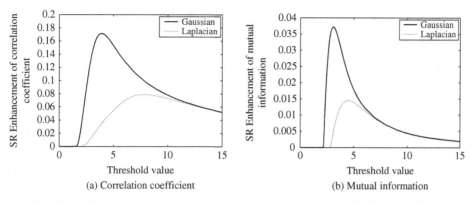

Fig. 10.3. Noise-induced performance gains against threshold, for correlation coefficient and mutual information (in bits per sample). The results were obtained from digital simulation. The solid lines are the results for a Gaussian signal and the dashed lines for a Laplacian signal.

noise; this exercise was then repeated as a function of threshold value. The solid curves show the case for Gaussian signal and the dashed lines for Laplacian. The most striking feature is that the SR effect is clearly much stronger in the correlation coefficient than the mutual information. For the Gaussian signal the maximum noise-induced correlation gain is approximately 0.17, to be compared with a maximum correlation (obtained at zero noise and zero threshold) of approximately 0.80. For the Laplacian signal the corresponding figures are a correlation gain of 0.08 and a maximum correlation of 0.70. Although the correlation gain is lower for the Laplacian signal, it still amounts to an 11% increase. In contrast the noise-induced gains in the mutual information are very weak. For Gaussian signals the maximum noise-induced gain is approximately 3.7 percent of channel capacity and for the Laplacian signal it is less than 1.5 percent. It is questionable whether such small gains would prove useful in practical applications. Clearly, these results bring into doubt the utility of SR as a mechanism for improving system performance when aperiodic signals are used.

An obvious question is why are the information gains so small for Laplacian signals? A possible explanation is because the maximum information gain is observed at relatively large threshold values relative to standard deviation – indeed, at a value of just less than five (relative to signal standard deviation) – for Laplacian signals. It is necessary to recall that the strength of the classical SR effect with a sinusoidal signal is strongly dependent on the distance between the peak signal level – that is, the maximum level the signal attains – and the threshold value. This statement holds for the case when the amplitude is not small and hence the dynamics are highly nonlinear (Stocks 1995). If the amplitude of the signal is set such that the peak signal level is just below threshold, then the addition of small amounts of noise will lead to a large improvement in the output SNR. Lowering the signal amplitude increases the distance between the peak signal and the threshold and therefore necessitates the addition of larger levels of noise to maximize the SNR. This leads to a smaller maximum SNR. Alternatively, increasing the amplitude above threshold – that is, so the signal is suprathreshold in the absence of noise – removes the SR effect entirely.

The question then arises as to how this picture carries over to aperiodic signals whose distribution does not have a finite bounded support – that is, signal distribution with 'tails' in their probability density functions (PDFs) – and hence for which a peak signal to threshold distance does not exist. First, it is clear that because of the tails the signal will not always be subthreshold – excursions of the signal above threshold will always occur and, at least for these portions of the signal, no SR effect is possible. In contrast, for those portions of the signal that lay below threshold, noise can in principle aid transmission. However, the value of noise that optimally achieves this depends on the distance between the signal and the threshold. Given that this distance is now a random quantity – due to the random nature of the signal – it is clear that, for a fixed level of noise, performance will only be improved for some – possibly small – portions of the signal; that is, for those parts of the signal that are the appropriate distance – which is governed by the noise – away from the threshold. For other parts of the signal the noise will either be too strong or too weak to result in improved performance.

From the above discussion it would be expected that the strength of the SR effect will depend crucially on the shape of the PDF of the signal. For signals with long tails only small portions of the signal will be an appropriate distance below threshold to be enhanced by the SR mechanism. The remainder of the signal will be either suprathreshold or too far below threshold for the noise to be useful. Consequently, SR is likely to occur only if the tails of the distribution decay at a quick enough rate so that, by an appropriate choice of threshold, sufficient portions of the signal are the appropriate distance below threshold for noise to be of benefit. From the results presented above we can see that the tails of a Gaussian signal decay quicker than the Laplacian and hence a stronger SR effect is observed for Gaussian signals. Following this line of reasoning it is to be expected that signal distributions that decay slower than a Laplacian (that is, slower than exponentially) will show a further reduced SR effect.

Implications for biological sensory coding

The implication of these results could be quite profound. Stochastic resonance has been forwarded as a possible mechanism by which sensory neurons can enhance the transmission of weak sensory stimuli (Longtin *et al.* 1991, Douglass *et al.* 1993,

Wiesenfeld and Moss 1995, Collins *et al.* 1996b, Levin and Miller 1996). In general, these stimuli are anticipated to be aperiodic. Although the comparator system we use in this example is a somewhat oversimplified model of a biological neuron, nevertheless we expect the results will generalize to these systems – neurons are essentially threshold devices and their dynamics reflect this. Our discussion, so far, would therefore imply that, for a broad class of aperiodic stimuli, SR does not lead to significant enhancement of the transmitted information. Indeed, many real-world signals would fall into this category. For example, the long-term distribution of speech or music is close to Laplacian (Dunlop and Smith 1984) – the implication, therefore, is that SR would not lead to significant improvements in performance for these types of signals. Of course, this does not mean that the signal could not be pre-processed – for example, by a compressive nonlinearity – to modify the probability distribution into a form suitable for SR to occur. Indeed, we describe in Section 10.3 that at least for the auditory system this appears to be the case.

However, for the role of SR to be established in biological sensory systems there is a need to consider biologically plausible stimuli in biologically realistic models. Only then can more definite statements regarding the applicability of SR to neural coding be made. In Section 10.3 this is discussed, where the coding of speech information is considered in a complex auditory model.

Finally, it can be noted that even though for some aperiodic signals SR is very weak, it is still possible to get large SR effects via SSR in parallel arrays of such elements – see Chapter 4 and Stocks (2001a). Indeed, SSR effects have been shown for Laplacian signals in Chapter 4 and McDonnell and Abbott (2004b). In Fig. 10.4 we show that SSR occurs in an array of 15 elements for various values of the

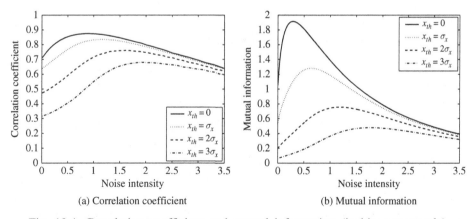

(a) Correlation coefficient   (b) Mutual information

Fig. 10.4. Correlation coefficient and mutual information (in bits per sample) against noise intensity, $\sigma = \sigma_\eta/\sigma_x$, for various values of x_{th} for an array of 15 devices, and Laplacian signal and noise. A pronounced SSR effect is evident in both figures.

threshold value. Clearly, comparison of Figs 10.2(a) and 10.4(a) and Figs 10.2(b) and 10.4(b) demonstrates that SSR can be observed for Laplacian signals even if conventional SR effects are very weak. In this respect, SSR can be viewed as a more general SR effect that is not limited by signal amplitude, threshold setting, or signal distribution – SSR is observed at all values of the threshold and signal amplitudes.

10.3 Stochastic resonance in an auditory model

In the previous section we illustrated that SR effects are very weak for certain types of signal distribution. Signals that have slowly decaying tails are not expected to exhibit strong SR effects and this is highlighted explicitly for Laplacian signals. Natural stimuli, such as speech and music, are known to possess slowly decaying tails and, indeed, are often modelled using Laplacian distributions (Dunlop and Smith 1984). Consequently, this brings into question whether or not SR can really be taken seriously as an auditory coding mechanism. Although many studies have reported SR effects relating to the auditory system, all of these studies have utilized periodic stimulation. Therefore, the question as to whether SR effects do occur with realistic stimuli remains, which we now discuss following the approach of Aruneema Das as part of her doctoral work (Das 2006). An auditory model, based on the Meddis hair cell model (Meddis 1986, Sumner *et al.* 2002), is extended to take into account the Brownian motion of the stereocilia. Coding at the auditory nerve level is studied as a function of the added noise for both sinusoidal and speech signals.

The Meddis hair cell model

The heart of the auditory model is the inner hair cell (IHC) model originally proposed by Meddis (1986) and then updated (Sumner *et al.* 2002). The model describes the transduction of mechanical acoustic stimuli into neural action potentials. The model was developed to mimic the processes of the auditory pathway so that it can be used in research in cochlear functioning, speech perception, and other areas of auditory physiology. The main components of the Meddis model are shown in Fig. 10.5.

The input to the model is the basilar membrane (BM) velocity and its output is a stream of neural spike trains. Figure 10.5 shows the complete processing path for the model (Sumner *et al.* 2002) from the input to the stereocilia of the IHC model to spiking on the auditory nerve. The dynamics of the model are divided into four broad sections as shown in Fig. 10.5. These sections will now be discussed in detail.

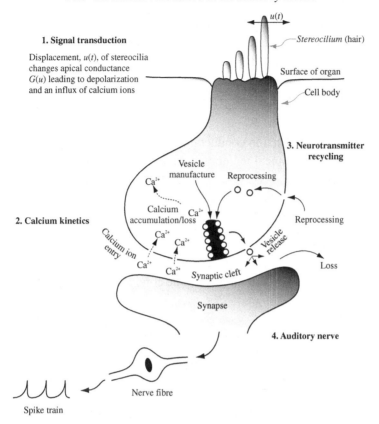

1. **Signal transduction**

Displacement, $u(t)$, of stereocilia
changes apical conductance
$G(u)$ leading to depolarization
and an influx of calcium ions

Stereocilium (hair)

Surface of organ

Cell body

3. **Neurotransmitter
recycling**

Vesicle
manufacture   Reprocessing

2. **Calcium kinetics**

Ca^{2+}

Calcium
accumulation/loss Ca^{2+}

Reprocessing

Calcium ion
entry

Ca^{2+}

Ca^{2+}

Ca^{2+} Ca^{2+}

Ca^{2+} Ca^{2+} Synaptic cleft

Vesicle
release

Loss

Synapse

4. **Auditory nerve**

Nerve fibre

Spike train

$u(t)$

Fig. 10.5. Schematic diagram of the Meddis hair cell model (modified from Sumner *et al.* (2002)). The model comprises four main stages: (1) conversion of the basilar membrane velocity into changes in intracellular potential; (2) the change in intracellular potential causes calcium ion channels to open; (3) the influx of calcium gives rise to the release of neurotransmitter; (4) the propagation of neurotransmitter across the synaptic cleft generates action potentials in an auditory neuron.

The model dynamics are based on the description of Sumner *et al.* (2002) where more details can be found.

Signal transduction, generation of IHC receptor potential

The transduction of acoustic energy into an electrical potential is accomplished by the stereocilia – the small hairs that emanate from the top of the hair cell. Displacement of the stereocilia results in the opening of ion channels that in turn leads to a depolarization of the intracellular potential. The displacement of the IHC stereocilia, $u(t)$, is represented as a function of BM velocity, $v(t)$, by

$$\tau_c \frac{du(t)}{dt} + u(t) = \tau_c C_{\text{cilia}} v(t), \qquad (10.2)$$

where C_{cilia} is a gain factor and τ_c is a time constant. The number of open ion channels and the apical conductance, $G(u)$, change with the stereocilia displacement. The total apical conductance is given by

$$G(u) = G_{cilia}^{max}\left[1 + \exp\left(-\frac{u(t) - u_o}{s_0}\right)\left(1 + \exp\left(-\frac{u(t) - u_1}{s_1}\right)\right)\right]^{-1} + G_a,$$

(10.3)

where G_{cilia}^{max} is the transduction conductance with all channels open, G_a is the passive conductance in the apical membrane, and s_0, u_0, s_1, and u_1 are constants. The membrane potential of the cell body is modelled with a passive electrical circuit, described by

$$C_m\frac{dV(t)}{dt} + G(u)(V(t) - E_t) + G_k(V(t) - E_k') = 0,$$

(10.4)

where $V(t)$ is the intracellular hair cell potential, C_m is the cell capacitance, G_k is the voltage-invariant basolateral membrane conductance, E_t is the endocochlear potential, and

$$E_k' = E_k + \frac{E_t R_p}{R_t + R_p}$$

(10.5)

is the reversal potential of the basal current E_k, corrected for the resistance (R_t, R_p) of the supporting cells.

Calcium controlled neurotransmitter release

The depolarization of the IHC leads to the opening of calcium ion channels in the basal part of the cell. The calcium current ($I_{ca}(t)$) as a result is

$$I_{ca}(t) = G_{ca}^{max}m_{I_{ca}}^3(t)(V(t) - E_{ca}),$$

(10.6)

where E_{ca} is the reversal potential for calcium, G_{ca}^{max} is the calcium conductance near the synapse, with all the channels open, and $m_{I_{ca}}^3$ is the fraction of calcium channels that are open. With the opening of calcium ion channels, calcium ions enter the cell and accumulate near the synapse. Calcium concentration $[Ca^{2+}](t)$ in the IHC is modelled as a function of calcium current, $I_{ca}(t)$

$$\tau_{ca}\frac{d[Ca^{2+}](t)}{dt} + [Ca^{2+}](t) = I_{ca}(t),$$

(10.7)

where τ_{ca} is a time constant. The probability, $k(t)$, of the release of neurotransmitter is proportional to the calcium ion concentration, and can be written as

$$k(t) = \max\left[z\left([Ca^{2+}]^3(t) - [Ca^{2+}]_{th}^3\right), 0\right],$$

(10.8)

where $[Ca^{2+}]_{th}$ is a threshold constant and z is a scalar for converting calcium concentration levels into release rate.

Neurotransmitter flow process

The neurotransmitter flow process, denoted by module 3 in Fig. 10.5, is easier to explain with the aid of the flow diagram; this is shown in Fig. 10.6 and has been adapted from Meddis (1986).

The entire process is quantal and stochastic and is described by the equations

$$\frac{dq(t)}{dt} = N(w(t), x) + N([M - q(t)], y) - N(q(t), k(t)), \tag{10.9}$$

$$\frac{dc(t)}{dt} = N(q(t), k(t)) - lc(t) - rc(t), \tag{10.10}$$

$$\frac{dw(t)}{dt} = rc(t) - N(w(t), x). \tag{10.11}$$

The rate of release of individual vesicles of neurotransmitter $k(t)$ from the free transmitter pool (q) store into the cleft (c) is dependent on the calcium concentration. In the cleft some of the transmitter is lost at a rate l. The remaining transmitter in the cleft is taken back into the cell into a reprocessing (w) store at a rate r. The reprocessing store converts them back into original vesicles and they are returned to the free transmitter pool at a rate x. Equation (10.9) describes the rate of change of neurotransmitter level concentration in the hair cell. Equation (10.10) describes the rate of change of the amount of transmitter in the synaptic cleft. Equation (10.11) describes the rate of change of the amount of transmitter in the reprocessing store. The neurotransmitter in (q) is quantal and (q) is continuously supplied with new transmitter vesicles at a rate, y, where M represents the maximum number of

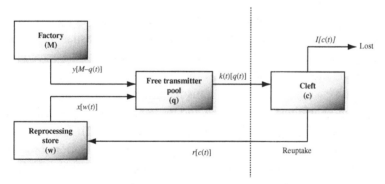

Fig. 10.6. Neurotransmitter dynamics. The dashed vertical line denotes the hair cell membrane, everything to the left of it is in the hair cell and everything to the right of it is outside the hair cell (in the synaptic cleft). The basic cycle is as follows: the factory (M) generates neurotransmitter that resides in the free transmitter pool (q). The neurotransmitter is released into the synaptic cleft (c) at a rate that depends on the calcium concentration. Some neurotransmitter is lost from the hair cell but some is retaken up and reprocessed by the reprocessing store (w).

transmitter quanta that can be held in (q). All the transport processes of neurotransmitter are stochastic and are described by the function $N(n, \rho)$, in which each of n
quanta has an equal probability of release, ρdt, in a single simulation epoch. For
example, for the function $N(q(t), k(t))$ the number of quanta released in a short
period of time dt is given by the function $N(q, kdt)$ that decides randomly how
many of the $q(t)$ independent quanta will be released assuming that each has an
equal probability of being released. The final response of the system is a sequence
of vesicle release events into the synaptic cleft.

Auditory nerve response

An action potential (AP) or spike is generated in the auditory nerve (AN) fibre if a
neurotransmitter vesicle is released and $p(t)$ exceeds a random number between 0
and 1, that is

$$p(t) = \begin{cases} 0 & t - t_1 < R_A \\ 1 - c_r \exp\left(-\frac{t - t_1 - R_A}{s_r}\right) & t - t_1 \geq R_A, \end{cases} \qquad (10.12)$$

where c_r determines the maximum contribution of the relative refractory period in
$p(t)$, s_r is the time constant of refraction, t is time now, t_l is the time of the last
spike, and R_A is the absolute refractory period. The process is denoted by module
4 in Fig. 10.5.

The complete auditory model

The Meddis hair cell model describes only the signal transduction process – it does
not include the preprocessing of the middle ear or of the basilar membrane itself.
To produce a complete auditory model, the preprocessing has to be accounted for –
again we use the models proposed in Sumner *et al.* (2002).

In the preprocessing before signal transduction, the motion of the BM is represented as a nonlinear filtering operation and the middle ear response is represented
as low- and high-frequency attenuation (Sumner *et al.* 2002). The response of the
middle ear is modelled by a second-order linear bandpass Butterworth filter. The
filtering of the BM is modelled with a dual resonance nonlinear (DRNL) filter
architecture (Meddis *et al.* 2001) as shown in Fig. 10.7.

A single DRNL filter consists of two parallel pathways, one linear and the other
nonlinear, whose outputs are summed to produce the filter output. The linear path
consists of a bandpass function, a low pass function, and a gain factor in a cascade.
The nonlinear path is also a cascade consisting of a bandpass function, a compression function, a second bandpass function, and a low pass function. The output
of the system is the sum of the outputs of the linear and the nonlinear paths. The

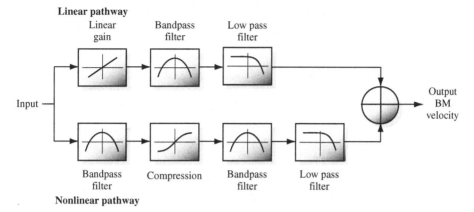

Fig. 10.7. Basic architecture of a dual resonance non-linear (DRNL) filter. The filter consists of two pathways, one linear, the other nonlinear (representing the active gain mechanism), that are summed to produce the overall response.

three bandpass functions each consist of a cascade of two or more gammatone filters (Slaney 1993). The low pass filters consist of a cascade of second-order low pass filters. The compression in the nonlinear pathway is described by

$$y[t] = \text{sign}(x[t])\min[a|x[t]|, b|x[t]|^v]. \tag{10.13}$$

Here a, b, and v are parameters of the model used to control the gain of the nonlinear path.

The DRNL filter representing the BM and the IHC model and the AN response together form the complete model of the inner ear (cochlea). All that remains is to specify the preprocessing of the middle ear, which is simply taken to be a bandpass filter. The complete auditory model is shown in Fig. 10.8.

The complete auditory model has been validated (Sumner *et al.* 2002) against physiological data and is, therefore, thought to represent a plausible model of the real auditory system. However, the model does have deficiencies. For example, some noise sources that are known to occur in the inner ear have been omitted from the model – the most notable of these is the Brownian motion of the stereocilia. Although the updated model (Sumner *et al.* 2002) does take into account the stochastic and quantal nature of neurotransmitter release, it has not included the fluctuations of the stereocilia, which, in turn, lead to the presence of a significant noise floor superimposed on the intracellular potential. In practice, therefore, spontaneous neurotransmitter release can occur through Brownian motion of the stereocilia in addition to the stochastically driven synaptic release of vesicles.

In a series of numerical experiments Das (2006) has added a Gaussian fluctuating component directly to the BM velocity of Eq. (10.2), to mimic the effect of

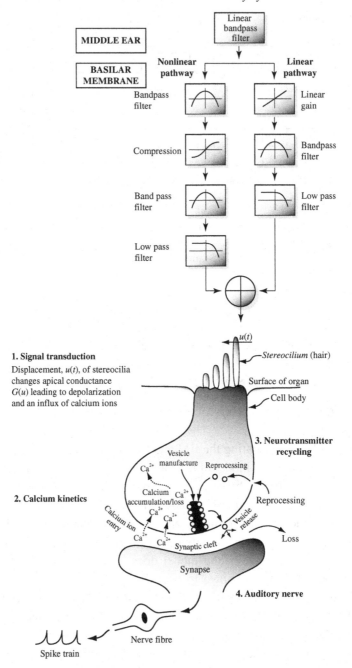

Fig. 10.8. Complete auditory model, after Sumner *et. al.* (2002).

Brownian motion. The effect of noise has then been studied using a range of stimuli to investigate whether SR can occur.

Response for sinusoidal signals

The coherence function is used to quantify the response of the system to noise. This is a function related to cross-correlation, and is defined in terms of power spectral densities and a cross-spectral density term as

$$C_{xy}(f) = \frac{|P_{xy}(f)|^2}{P_{xx}(f)P_{yy}(f)}. \tag{10.14}$$

The coherence $C_{xy}(f)$ is a real dimensionless function of frequency having values in the range between zero and unity. It gives a measure of correlation between x and y at each frequency, f. It is a function of the power spectral densities of x and y – that is, $P_{xx}(f)$ and $P_{yy}(f)$ respectively – and the cross power spectral density between x and y, $P_{xy}(f)$. The total coherence estimate is obtained as $\frac{1}{f_{max}} \int_0^{f_{max}} C_{xy}(f)df$, which also ranges between zero and unity.

Comparison with results from physiological experiments

In a physiological experiment, Jaramillo and Wiesenfeld (1998) presented results that suggested that the root mean square (rms) displacements of the stereocilia due to Brownian motion are at a level appropriate for the enhancement of signal transduction via SR. Specifically, they demonstrated that optimal signal transmission was observed for stereocilia displacements of approximately 2 nm rms – a value known to occur in real hair cells (Denk *et al.* 1989, Denk and Webb 1992).

Using an equivalent experimental paradigm, Das (2006) carried out simulations based on the complete auditory model introduced above. However, for direct comparison, only Eqs (10.3)–(10.8) were implemented, and the neurotransmitter dynamics and the auditory neuron were not included. The reason for this is that in the physiological experiments the hair cells – from the *maculae sacculi* of leopard frogs – were removed before a stimulus was applied using a stiff glass probe. Consequently, the response was measured presynaptically. In principle the model could also have been truncated at Eq. (10.4).

The stimulus Das (2006) used was a sinusoid of frequency of 500 Hz, and amplitude 2 nm (subthreshold), as used in the physiological hair cell experiments of Jaramillo and Wiesenfeld (1998). The signal length was 10 seconds and Gaussian white noise was added to the stimulus. The displacement noise amplitude is the rms value of the displacement of the stereocilia ranging from 0 to 16 nm. The stimulus was added to Gaussian noise, which was filtered with a 1 kHz cutoff low pass Butterworth filter before being input to the IHC model. The values of

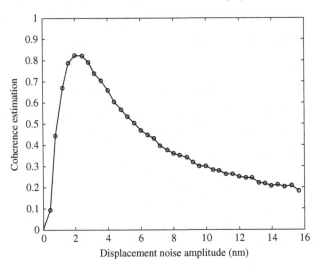

Fig. 10.9. Coherence against displacement noise amplitude for a 500 Hz sinusoidal signal, as obtained from digital simulation. The optimal displacement noise amplitude of 2.2 nm rms is consistent with the physiological results of Jaramillo and Wiesenfeld (1998) and the known level of Brownian motion in hair cells (Denk *et al.* 1989, Denk and Webb 1992).

G_{ca}^{max}, $[Ca^{2+}]_{th}$, and M are adjusted to get the best fit with the physiological results. Further details can be found in Das (2006).

The results of the simulation are shown in Fig. 10.9. This figure clearly shows an SR effect with the optimal displacement noise amplitude around 2.2 nm rms. This result is consistent with the experiments of Jaramillo and Wiesenfeld (1998) and yields optimal noise in the known range of displacements induced by Brownian motion (Denk *et al.* 1989, Denk and Webb 1992). It also confirms that the model is able to capture the results of physiological experiment and therefore gives confidence that the inclusion of the noise in Eq. (10.2) is reasonable from a modelling perspective. The result also provides additional confirmation that Brownian motion of the stereocilia enhances subthreshold sinusoidal stimuli in the auditory system.

Simulations of the full auditory model

Here we present results using the full auditory model shown in Fig. 10.8. Figure 10.10 shows plots of coherence against noise intensities for various frequencies of the sinusoidal stimulus. The hair cells used to generate these figures had best frequencies (BF) of 15 kHz and 500 Hz, respectively. Owing to the tuning characteristics of the hair cells, the stimulus threshold varies with BF. However, for these results the amplitude was fixed at 20 dB – this is subthreshold for most hair cell BFs.

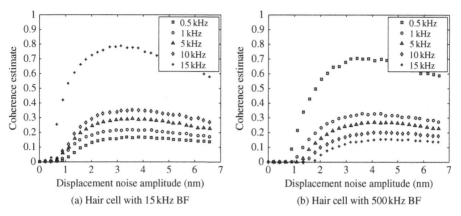

Fig. 10.10. Coherence estimation against displacement noise amplitude for the complete auditory model for a hair cell with (a) a 15 kHz BF and (b) a 500 Hz BF. The stimulus was a 20 dB sinusoid of varying frequency; values are shown in the legend.

The figures show the occurrence of SR. Hence these results indicate that SR is plausible in the human auditory system – at least for subthreshold periodic stimuli. Maximum coherence is observed for stimulus frequencies equal to the BFs of the hair cells. This is to be expected because the hair cells are tuned to the BFs and hence their responses are maximal at these frequencies. As the stimulus frequency moves away from the BF, the hair cells become detuned and the coherence is reduced. It is also noted that at BF the optimal displacement noise amplitude is in the range 3–3.5 nm – again this is consistent with known stereocilia displacements (Denk *et al.* 1989, Denk and Webb 1992) due to Brownian motion.

Another interesting observation is that the optimal noise level is reduced as the stimulus frequency approaches the BF. Consequently, the noise reinforces the tuning characteristics and preferentially enhances signals that are close to BF. The increase in the hair cell response as the BF is approached arises due to the tuning characteristics of the basilar membrane; at BF the BM and hence the stereocilia displacement is maximized. Consequently, the hair cell stimulus increases as the BF is approached and becomes closer to threshold. This means that smaller noise levels are required to induce crossings of the threshold thus lowering the optimal noise level.

Figure 10.11 shows the coherence plotted against displacement noise amplitude for various input signal intensities. The hair cell used for this figure had a BF of 5 kHz, but similar results were obtained with other BFs (Das 2006). The input stimulus frequency was set to the same value as the BF. The figure shows that as the stimulus level increases, the coherence also improves, but eventually the SR effect disappears at sufficiently high stimulus levels. This picture is consistent

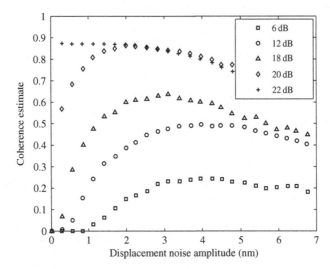

Fig. 10.11. Coherence against displacement noise amplitude for a 5 kHz sinusoidal stimulus for various stimulus levels. The hair cell BF was fixed at the stimulus frequency of 5 kHz.

with the known behaviour of SR. In a single element – for example, neuron, hair cell, comparator etc. – SR is normally only observed if the signal is subthreshold; for suprathreshold signal levels, the SR effect does not persist. This is precisely the behaviour observed in Fig. 10.11. For stimulus levels below 18 dB the coherence is zero in the absence of noise, implying that threshold crossing is not possible. However, at 20 dB the coherence is approximately 0.45 at zero noise. This can occur only if the signal can cross the threshold in the absence of noise. Consequently, the threshold for these parameters is close to 20 dB. For signal levels above this value the signal is suprathreshold and hence SR is not anticipated to occur – as indeed is observed.

Response for speech signals

As discussed in the introduction to Section 10.3, one of the central motivations for studying the role of noise in a complete auditory model is the fact that SR effects are very weak when signals have slowly decaying tails in their distributions. In particular, we demonstrated that for Laplacian signal distributions the SR effect led to noise-induced enhancements of the information of less than 1.5% of channel capacity. The question remains, therefore, can SR be observed with real speech signals? We now address this issue by using real speech signals as the input to the auditory model. The stimulus is obtained from an audio file consisting of the word 'phonetician' spoken by a male voice. Plotting a histogram of this signal, as shown

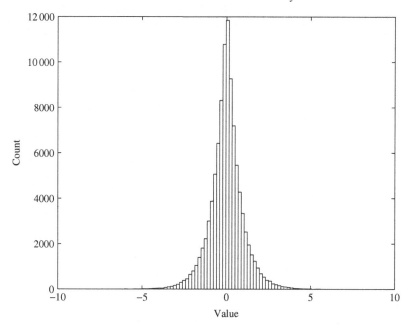

Fig. 10.12. Histogram of the sample amplitudes of the speech signal of the word 'phonetician'.

in Fig. 10.12, finds that it possesses slowly decaying tails, and that therefore the underlying PDF is similar in form to a Laplacian distribution (Das 2006).

The signal was passed through the model at BF values of 500 Hz, 5 kHz, and 15 kHz. The signal intensities used were between 20 and 40 dB, because the threshold values of the BFs used lie within this range. The amplitude of the signals was based on the rms value.

Figure 10.13 shows the coherence computed between the speech signal and the resultant spike train for stimulus levels of (a) 20 dB and (b) 40 dB. Considering Fig. 10.13(a), for a 5 kHz BF, the coherence does not show any SR type behaviour and the monotonically decreasing curve suggests that at this frequency the signal is effectively suprathreshold. A very weak SR effect is observed at BFs of 15 kHz and 500 Hz and clearly the signal is subthreshold at these frequencies. However, the coherence values are very low and the small noise-induced gains are probably not significant for auditory processing.

Figure 10.13(b) shows results for a 40 dB signal level. The increase in signal level has resulted in an increase in the coherence values, but the signal is now suprathreshold at all frequencies and hence there are no SR effects observed.

There are two possible reasons why SR effects are very weak and the coherence values are low. First it is possible that strong SR effects are not observed because of the slowly decaying tails in the signal distribution, that is, because of

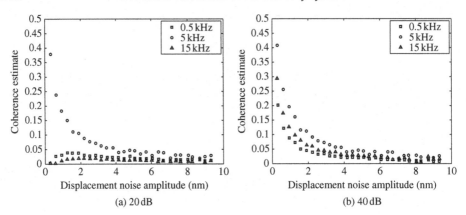

Fig. 10.13. Coherence against displacement noise amplitude for a speech signal of (a) 20 dB and (b) 40 dB intensity. The legends show the BFs used. The coherence is calculated between the original speech signal and the resultant spike train obtained from the complete auditory model.

the Laplacian shape of the signal distribution. However, most likely it is due to the way the coherence estimate is computed. The coherence is calculated between the original speech signal and the resultant spike train obtained from a single hair cell with a fixed BF. Given the frequency selectivity of the BM and hair cell, the spike train is coding only information in a narrow range of frequencies around the BF. Consequently, the spike train is not strongly correlated to the signal frequencies outside of the BF range and hence, for these frequencies, the response of the hair cell is dominated by noise. This results in a low overall coherence estimate.

To clarify this point the simulations were repeated and the coherence calculated between the BM velocity (output of the DRNL filter) and the spike train. This should be a more reliable indicator of the role of noise in auditory coding because the coherence is now computed between the signal to be transduced – that is, the input to the hair cell – and the resultant coded spike train.

Figure 10.14 shows the results. Figure 10.14(a) displays dominant SR effects for the 0.5 kHz and the 15 kHz BFs (where clearly the signal is subthreshold) and a slight SR effect for BF 5 kHz. Figure 10.14(b) also shows SR behaviour. However, the stimulus is clearly making excursions across the threshold in the absence of noise and hence the SR effect is reduced. Nevertheless, these results clearly show that the auditory system has the potential for using Brownian motion (noise) at the stereocilia level to enhance signal coding at the neural level.

An interesting observation is that there is no evidence that the Laplacian nature of the speech signal limits the SR effect. This seems to be in contrast to the results presented in Section 10.2. However, the ear is highly compressive and therefore it is to be anticipated that the Laplacian nature of the signal is removed by the compression. More specifically, the compression should remove the tails in the signal distribution. To confirm that this is indeed the case, a final set of simulations is undertaken. The preprocessing of the BM was bypassed – and so also the signal compression – by bypassing the DRNL filters. In this scenario the signal applied to the hair cells would be expected to be of a Laplacian nature and hence strong SR effects are not anticipated. The results are shown in Fig. 10.15. No SR effects are observed and hence this appears to confirm the hypothesis.

The compression is clearly an important part of the signal processing chain from the perspective of SR. This is perhaps not surprising. There is no evidence

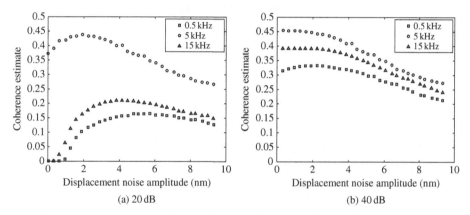

Fig. 10.14. Coherence against displacement noise amplitude for a speech signal of (a) 20 dB and (b) 40 dB intensity. The legends show the BFs used. The coherence is calculated between the output of the DRNL filters (that is, the input to the hair cell) and the resultant spike train.

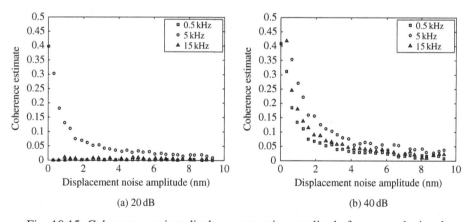

Fig. 10.15. Coherence against displacement noise amplitude for a speech signal of (a) 20 dB and (b) 40 dB intensity. The legends show the BFs used. The input stimulus has not been preprocessed using the DRNL filters before reaching the IHC stereocilia.

to suggest that the level of noise known to occur in the auditory system is tunable (adjustable) and, therefore, it should probably be viewed as fixed. It is well understood that, in principle, fixed noise could be a barrier to the use of SR as a neural coding mechanism (Collins *et al.* 1995b). In principle, to maximize information flow the input noise has to be adjusted to keep it at its optimal value as the signal level varies. However, if strong signal compression is employed, the input SNR can be maintained between set bounds and hence be close to optimal. For this reason there is a reduced necessity to adjust the noise level, instead the signal level is compressed (adjusted) to maintain the system close to its optimal operation point.

10.4 Stochastic resonance in cochlear implants

In this section we discuss a proposal to use noise, and in particular suprathreshold stochastic resonance (SSR), in cochlear implants to improve performance, for example improvement of speech comprehension and music appreciation.

Cochlear implants are devices that can be used to restore partial hearing to the profoundly deaf. In many profoundly deaf people the hair cells of the inner ear that transduce mechanical sound vibration into electrical signals are damaged or missing (Kiang *et al.* 1970, Hinojosa and Lindsay 1980, Nadol, Jr. 1981). A cochlear implant can often be surgically implanted into the cochlea to replace the function of the hair cells (Wilson 1997, Loizou 1999, Rosen 1996). Cochlear implants work by direct electrical stimulation of the cochlear nerve. The purpose of the implant is to evoke, with electrical stimulation, a similar pattern of nerve activity to that expected in a healthy ear when stimulated acoustically. A schematic diagram of a cochlear implant is shown in Fig. 10.16.

Cochlear implants consist of a number of electrodes (typically 16–22) that each stimulate a sub-population of the total population of nerve fibres. The signal is first passed through a bank of passband filters, or channels – each channel having a different centre frequency. The output from each channel is then connected to just one electrode. The signal processing is designed to reflect the different tuning characteristics of the hair cells and hence mimic the effect of 'place coding' in the ear.

So how can noise be used to improve cochlear implants – surely there is enough noise already without adding more? In fact, the noise is much reduced, indeed almost removed entirely, in cochlear implants. In a healthy ear it is the Brownian motion of the hair cells and the synaptic transmission of neurotransmitter to the postsynaptic afferents that are the major sources of noise. Therefore, in cochlear implant patients, the absence of the hair cells leads to a significantly lower than normal stochastic nerve activity. Studies have demonstrated that the healthy mammalian cochlea has a significant amount of spontaneous activity (neural firing) 'in quiet' conditions (Kiang 1966) that is largely not observed if the hair cells

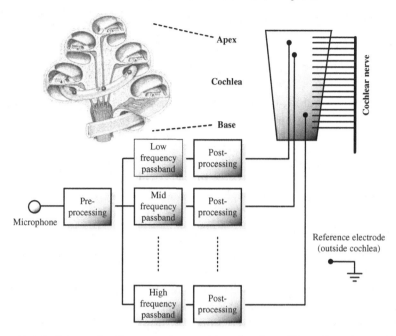

Fig. 10.16. Schematic diagram of a cochlear implant. The microphone picks up the acoustic stimulus and converts it into an electrical signal. After preprocessing – typically automatic gain control and compression – the signal is split into a series of channels by an array of bandpass filters – lowest frequency at the top, highest frequency at the bottom. After postprocessing – typically envelope extraction and carrier modulation – the signal is transmitted across the skull to the implanted electrodes. The snail-shaped structure shows an interior view of the cochlea and is adapted from Loeb (1985).

and presynaptic connections are damaged or absent (Kiang *et al.* 1970, Liberman and Dodds 1984). This led to the suggestion by Robert Morse and Edward Evans in 1996 that noise should be re-injected back into the deafened ear (Morse and Evans 1996). They demonstrated that the addition of noise to cochlear implant signals enhanced vowel coding. This seminal work subsequently stimulated a wide range of other studies on the use of stochastic resonance effects to improve cochlear implant coding (Rubinstein *et al.* 1999, Morse and Evans 1999a, Morse and Evans 1999b, Morse and Roper 2000, Morse and Meyer 2000, Matsuoka *et al.* 2000b, Matsuoka *et al.* 2000a, Zeng *et al.* 2000, Chatterjee and Robert 2001, Hohn 2001, Stocks *et al.* 2002, Morse *et al.* 2002, Rubinstein and Hong 2003, Behnam and Zeng 2003, Xu and Collins 2003, Hong and Rubinstein 2003, Allingham *et al.* 2004, Xu and Collins 2004, Xu and Collins 2005, Morse and Stocks 2005, Chatterjee and Oba 2005, Hong and Rubinstein 2006, Morse *et al.* 2007).

The application of suprathreshold stochastic resonance to cochlear implants

Under normal operating conditions the signals used in cochlear implants are suprathreshold relative to the neural threshold – otherwise they would not transmit any information. Consequently, if we are trying to use an SR paradigm to enhance information flow, it does not make sense to use subthreshold signals and add noise as this is known to reduce information flow relative to suprathreshold signals without noise (Levin and Miller 1996, Stocks 2000c). In short, trying to utilize standard subthreshold SR is likely to lead to a reduction in performance. This also means that adding noise directly to the cochlear implant signal is unlikely to enhance performance – although noise induced linearization effects may be beneficial (Dykman *et al.* 1994). The only SR paradigms that suggest that noise could be useful in cochlear implants are: (i) aperiodic stochastic resonance (ASR) in neural arrays, as introduced by Collins *et al.* (1995b); and (ii) suprathreshold stochastic resonance (SSR). Note that ASR and SSR are very similar, indeed SSR is the suprathreshold variant of ASR – ASR being introduced in the context of subthreshold signals only. However, SSR has one major advantage over ASR, in that it results in a greater transmission of signal information. Consequently, the use of suprathreshold signals, and hence SSR, is advocated.

To implement SSR at the neural level it is necessary to give every neuron its own noise source. This will ensure that the spontaneous firing across the neural population is incoherent in the absence of a stimulus. In a real cochlear implant we cannot expect to electrically stimulate each nerve fibre independently and hence it is not possible to introduce a large number of independent noise sources. In some of the simulations presented here, each neuron has been given its own noise source. To this extent, we regard these results as establishing the principle that noise can enhance the rate of information transmission through a cochlear implant. Further work is required to find optimal methods for increasing the independence between nerve fibres. To this end, preliminary experiments have shown that some independence can be achieved by the addition of noise to cochlear implant electrodes – the noise intensity on each electrode being carefully chosen to enhance the degree of independence (Morse and Meyer 2000). An alternative strategy using high-frequency 'conditioning' pulse trains has also been proposed as a means of generating stochasticity at the individual neural level (Rubinstein *et al.* 1999, Matsuoka *et al.* 2000b, Matsuoka *et al.* 2000a, Zeng *et al.* 2000, Rubinstein and Hong 2003, Hong and Rubinstein 2006).

Here some results are presented on the use of SSR to improve cochlear implant performance (Allingham *et al.* 2004, Morse and Stocks 2005). This work extends a study presented in Stocks *et al.* (2002) – which considered a single-channel (electrode) – to the case of multiple channels.

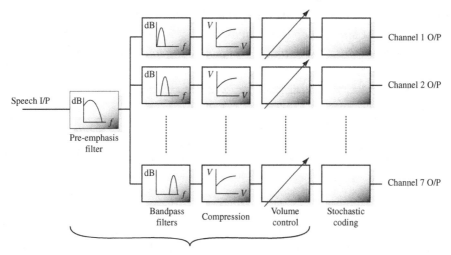

Fig. 10.17. Simulated cochlear implant coding strategy. The figure shows two parts to the coding. The first part is a standard coding strategy for analogue cochlear implants; the input signal is pre-emphasized, split into seven frequency bands, and the output of each bandpass filter is compressed. The second part is a stochastic coding stage in which the input is either added to or multiplied by the output of an independent noise source.

Cochlear implant model

Two different cochlear implant models have been simulated, a basic model in which no pre-emphasis and compression stages were implemented and a more complete, and thus realistic model, in which all processing stages have been included. A schematic of the complete model is shown in Fig. 10.17.

The model is based around the simultaneous analogue strategy (SAS) used in the Clarion cochlear implant (Advanced Bionics Ltd). The Clarion implant has 16 electrodes, which can either be stimulated individually relative to a distant ground electrode (monopolar stimulation), or stimulated as seven bipolar pairs, with two electrodes unused. The purpose of the multiple electrodes inside the cochlea is to mimic, albeit crudely, the coding of frequency that occurs with normal acoustic stimulation. In the normal ear, the frequency of an acoustic signal can potentially be coded both in terms of which fibres are most active (place coding) and the temporal pattern of activity of each fibre (time coding) (Evans 1978, Hartmann 1996). The place coding of information is mimicked in a cochlear implant by filtering the input signal by a bank of bandpass filters, each with a different passband. In accord with the normal tonotopic arrangement of the cochlea, bandpass filters with low-frequency passbands are used to stimulate electrodes in the apex of the cochlea and those with a high-frequency passband are used to stimulate electrodes in the base. The SAS strategy is normally used with seven channels of bipolar stimulation

Table 10.1. *Passbands of the 6th order Butterworth filters used to mimic sound processing by the simultaneous analogue strategy of the Clarion cochlear implant (Loizou 1998). F_{lower} and F_{upper} denote the lower and upper cutoff frequencies in Hz.*

Channel	1	2	3	4	5	6	7
F_{lower}	250	500	875	1150	1450	2000	2600
F_{upper}	500	875	1150	1450	2000	2600	6800

and therefore has seven bandpass filters. Here, filters with the same characteristics (Table 10.1) have been used, but with monopolar stimulation. Monopolar stimulation has been used because bipolar stimulation can evoke nerve discharges during both phases of stimulation (van den Honert and Stypulkowski 1987), whereas nerve fibres in the normal ear respond to only one phase (Kiang 1966). The temporal responses of cochlear nerve fibres to bipolar electrical stimulation are therefore unlike those to acoustic stimulation in the normal ear.

In the complete cochlear implant model, the simulated cochlear implant had a pre-emphasis filter before the bank of bandpass filters (Fig. 10.17), which was identical to that used for SAS; the filter was a first-order highpass filter (cutoff 1500 Hz) cascaded with a third-order lowpass filter (cutoff 7800 Hz). Also in common with the SAS strategy, each bandpass filter was proceeded by logarithmic compression, to match the dynamic range of speech, which is 40 to 50 dB over short time intervals (Cox *et al.* 1988), to the smaller psychophysical dynamic range for electrical stimulation, which from 'threshold' to 'uncomfortably loud' is only 8–30 dB (Michelson 1971). In contrast, however, to the standard SAS strategy, the logarithmic compression was followed by a stochastic coding stage. Two different stochastic coding schemes have been implemented: an additive scheme, where Gaussian white-noise was added to each electrode prior to neural stimulation, and a multiplicative scheme, where the signal is half wave rectified and then multiplied by a Gaussian white-noise 'carrier'. In addition to these studies, the case when the noise is added at the individual neural level (not at the electrodes) has been considered. The precise model used is indicated on the figure captions.

Model of the electrically stimulated ear

The output of the cochlear implant simulation was used as the input to a model of the electrically stimulated ear. For simplicity, the spiral structure of the cochlea was modelled as an uncoiled cylinder and the array of nerve fibres was modelled as being in a plane orthogonal to the electrodes. The length of the uncoiled cochlea was taken to be 34 mm (Nadol Jr. 1988) and the maximum electrode insertion depth from the round window to be 25 mm (Loizou 1998). Furthermore, to match the Clarion cochlear implant, the 16 electrodes in the model were spaced 1 mm apart (from 10–25 mm from the round window). In typical cochlear implant patients, each electrode is about 0.5–1.0 mm away from the nearest afferent cell body (Shepherd *et al.* 1993, Finley *et al.* 1990), which has been taken to be the region of initial excitation. In this study, this distance was taken to be 0.5 mm.

The electrodes in a real cochlea are surrounded by perilymphatic fluid, which is fairly conductive. Currents from individual electrodes therefore tend to spread in the conductive medium, and distant cochlear nerve fibres may be excited by the combined current from several electrodes (Simmons and Glattke 1972, Merzenich *et al.* 1974). Here, current spread from an electrode was modelled by an exponential decay of current with distance using a space-constant of 3.6 dB/mm (Wilson *et al.* 1994, Bruce *et al.* 1999). The nerve fibres were simulated using a phenomenological model based on the leaky integrate-and-fire model (Tuckwell 1988). Each cochlear nerve fibre was modelled by

$$\tau_m \dot{V}_i = -V_i + s(t) + \sqrt{2D}\xi_i(t), \qquad (10.15)$$

where $\langle \xi_i(t)\xi_j(t)\rangle = \delta_{ij}(t - t')$, $\langle \xi_i \rangle = 0$, V_i is the membrane potential of the ith neuron, τ_m is the membrane time constant, D is the noise intensity, $s(t)$ is the input stimulus, δ_{ij} is the Kronecker delta function, and $\xi_i(t)$ is a Gaussian noise source independent of the noise sources of the other neurons. In contrast to the standard leaky integrate-and-fire model, the model used in this study had a dynamic threshold derived from a study of the sodium inactivation variable, h, in the Frankenhauser–Huxley model, which is given by

$$\tau_h \dot{h} = -h + h_\infty \qquad (10.16)$$

where τ_h is the threshold recovery time constant and h_∞ is a sigmoidal function defined by

$$h_\infty = \frac{1}{1 + \exp\left(\frac{v(t) - \mu_\infty}{\sigma_\infty}\right)}. \qquad (10.17)$$

Here, μ_∞ is the mean of the sigmoid and σ_∞ is its standard deviation. Given h, the threshold at time t is defined by

$$\theta(t) = \frac{\theta_M}{h^P} + \theta_0, \qquad (10.18)$$

where θ_M, P, and θ_0 are model parameters.

At each simulation step t, the membrane potential $V(t)$ is calculated. If it exceeds the dynamic threshold, $\theta(t)$, then an action potential is deemed to have been initiated and the membrane potential and h are reset to zero. Each action potential is followed by an absolute refractory period, τ_{abs}, during which no further action potential can be evoked, irrespective of signal amplitude. During this period the membrane potential and h remain at zero.

For an ensemble of N fibres, the output of the model is a set of binary spike trains such that the output of neuron i at time t is given by

$$x_i(t) = \begin{cases} 1 & \text{if an action potential is generated at time } t, \\ 0 & \text{otherwise,} \end{cases} \tag{10.19}$$

where $i = 1, 2, ..., N$. The output ensemble of the neural array, $x(t)$, is taken to be the sum of the individual spike trains such that

$$x(t) = \sum_{i=1}^{N} x_i(t). \tag{10.20}$$

Finally, it is assumed that the N fibres are distributed evenly across the cochlea.

The equations were simulated using a two-step Heun algorithm for numerical integration with a time-step of 45.35 μs to match the sampling frequency of the WAV files containing the real speech sentences used in the simulation results presented below. Further modelling details and parameters can be found in Allingham *et al.* (2004) for the additive noise studies and Morse and Stocks (2005) for the multiplicative noise.

Information transmission rates

Following the method of Pinsker (1964) and Stein *et al.* (1972), the information rate between the input stimulus, $s(t)$, and the ensemble output spike train, $x(t)$, is calculated in bits per second according to

$$I = -\frac{1}{2\pi} \int \log_2 (1 - \rho(\omega)^2) d\omega, \tag{10.21}$$

where the coherence function, $\rho(\omega)$, is defined by

$$\rho(\omega)^2 = \frac{|S_{sx}(\omega)|^2}{|S_{ss}(\omega)||S_{xx}(\omega)|}. \tag{10.22}$$

This approach, based on the application of Gaussian channel theory, has previously been shown to give an accurate measure of the information rate even when the system dynamics are highly nonlinear (Nikitin and Stocks 2004). For non-Gaussian channels Eq. (10.21) is expected to yield a lower bound estimate of the transmitted information.

Additive noise

As a first step, initial studies focused on the basic cochlear implant model, that is excluding pre-emphasis and compression. Furthermore, noise was added to each neuron and not at the electrodes. To make the study relevant to previous studies of speech intelligibility, real speech stimuli selected from the BKB (Bamford–Kowal–Bench) sentence set were used (Bench *et al.* 1979). This is a set of standard sentences used to test speech intelligibility of cochlear implant users.

The rate of information transmission is first studied using ensembles of various sizes. Figure 10.18 shows these results for four BKB sentences and using 100, 1000, and 5000 nerve fibres. All the curves show a maximum at a nonzero level of noise and thus demonstrate an SR effect. At low levels of noise it can be seen

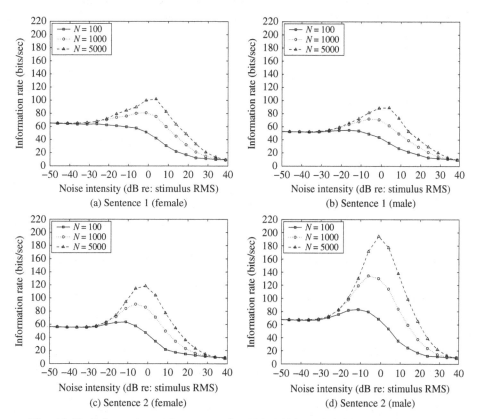

Fig. 10.18. Information-noise curves for 100, 1000, and 5000 nerve fibres. The results were obtained using two sentences: 'The kettle boiled quickly' and 'The man's painting a sign'. These will be referred to as sentence 1 and sentence 2. The transmitted information rate rises by around one-third for a large number of nerve fibres at the optimal noise intensity when compared with the low noise case, although a much bigger gain is observed in (d).

that the information transmission is still quite high due to the suprathreshold nature of the speech signal – this confirms that it is indeed SSR that is being observed. For large numbers of nerve fibres the rate of information transmission increases by approximately one-third at the optimal noise intensity when compared with the low noise case – except in Fig. 10.18(d), where the increase is much larger. In many respects these information gains appear modest. However, it should be noted that the system parameters (for example, channel gains) have not been optimized to maximize the transmitted information; it is anticipated that greater information gains are possible for optimized system parameters. This statement is supported by the results in Fig. 10.18(d), which displays a threefold increase in information. This preferential information gain is probably due to a chance matching of the spectral characteristics of the signal to the individual channel gains. A better overall matching of the system parameters to speech should increase the noise-induced gain in information. Nevertheless, it can be concluded that, in principle, SSR can occur in the auditory nerve and hence improved information transmission is possible. Consequently, SSR may provide a method of achieving genuine functional benefits in the performance of cochlear implants.

As already discussed each electrode stimulates a sub-population of fibres through the mechanism of current spread. Generally, current spread is perceived to be a negative aspect of electrical stimulation because it leads to cross-channel interaction. This interaction leads to a misrepresentation of the tonotopic structure of the ear because a given frequency can stimulate a wide population of nerve fibres. Current spread can be controlled by using current steering techniques to enhance localization of currents and hence give better pitch selectivity (Donaldson *et al.* 2005). However, here it is shown there is a potential positive side to current spread – it can enhance information flow. Figure 10.19 shows the effect of increasing the current spread by reducing the attenuation from 3.6 dB/m to 0.5 dB/m. At low noise the current spread reduces the information, while at noise intensities close to the optimal values the information is enhanced by about 10 percent. This demonstrates that current spread does not have an adverse effect on information transmission and can potentially be beneficial.

Finally, to demonstrate that the SSR effect occurs for speech signals other than those used in Figs. 10.18 and 10.19, the simulations are repeated using 16 different BKB sentences. The results are shown in Fig. 10.20. Both female and male voices are used and the total population of nerve fibres was fixed at 1000. The SSR effect is present in all cases, indicating that the results presented earlier are general. The enhancement of the rate of information transmission through a cochlear implant is expected to occur for almost all speech signals.

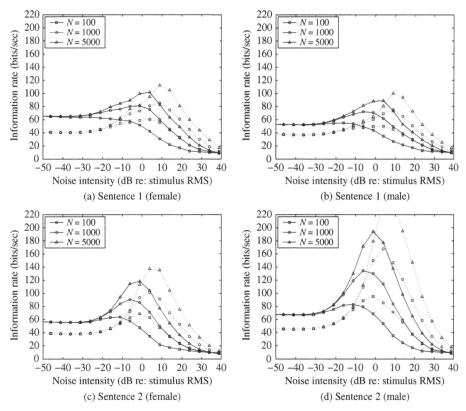

Fig. 10.19. Altering the current spread space-constant. This figure shows the effect of altering the current spread space-constant, which determines the exponential attenuation of current away from the electrodes. All results were obtained under identical conditions to those shown in Fig. 10.18. Two values of current spread have been used, 3.6 dB/mm (solid curves) and 0.5 dB/mm (dotted curves). The SSR effect is strengthened by this change. At low noise intensities the information transmission rate is reduced for the larger current spread, but the rate at the optimal noise intensity is increased by around 10 percent compared with the low current spread (high attenuation) case.

Multiplicative noise

The results for additive noise are encouraging. However, the dominant source of noise in normal hearing may be multiplicative, that is signal dependent. The multiplicative character originates from the quantal nature of synaptic transmission between the inner hair-cells and the cochlear nerve. Crucially, auditory models that exclude these noise sources fail to predict the correct physiological response to certain stimuli, such as high-level clicks and low-frequency sinusoidal

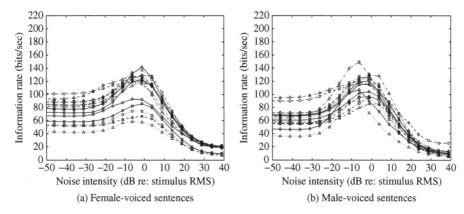

Fig. 10.20. Information-noise curves for 1000 nerve fibres for 16 (a) female-voiced and (b) male-voiced sentences. The SSR effect is present in all cases.

stimulation (Evans 1975). Furthermore, at high stimulus levels, these models predict a reduction in spike-time variability that is not observed in physiological data (Johnson 1980).

Because profound deafness is associated with loss of the inner hair-cells, the multiplicative (synaptic) noise sources are absent in the deafened ear. It is therefore possible that the response of the electrically stimulated cochlear nerve would more closely resemble the response of the acoustically stimulated nerve if the stimulus were the product of noise and some compressive function of the stimulus.

For this reason studies using multiplicative noise have been carried out and the transmitted information compared to those obtained using additive noise (Morse and Stocks 2005). Here results of a preliminary computational investigation on the use of multiplicative noise to enhance cochlear implant coding are reported. The full cochlear implant model shown in Fig. 10.17 was used for this study.

Figure 10.21(a) shows the results for 1000 simulated nerve fibres when the additive noise is internal to the fibres. The stimulus is the sentence, 'The clown had a funny face', presented at a level 40 dB above threshold, which is defined here to be the level that caused a spike to be evoked in 50 percent of trials when the noise intensity, D, was zero. With internal additive noise, standard suprathreshold stochastic resonance (SSR) is observed; the information transmitted by the array of fibres was optimized for a nonzero level of noise. If, however, independent noise is applied to each of the 16 electrodes (Fig. 10.21(a): solid circles), then the SSR effect is at best marginal and the array of fibres behaves in a manner that would be expected for a single nonlinear device with internal noise.

Figure 10.21(b) shows results for 1000 fibres using the same stimulus, but with multiplicative noise. When the multiplicative noise is internal to each nerve fibre

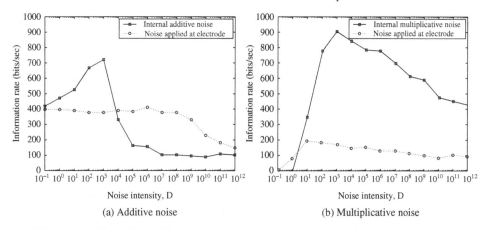

(a) Additive noise

(b) Multiplicative noise

Fig. 10.21. The effect of noise intensity, D, on information rate for (a) additive noise and (b) multiplicative noise. Results are shown for 1000 simulated cochlear nerve fibres for the BKB sentence 'The clown had a funny face'. The square symbols show the results when the noise was internal to each nerve fibre and the circles show results when the noise was applied at the electrodes.

(square symbols), the information rate can be optimized by a nonzero noise level. The effect is similar to the SSR effect observed with internal additive noise. The optimum information rate, however, is about 25 percent greater with multiplicative noise than with additive noise. Also, compared with internal additive noise, the information rate with internal multiplicative noise is less sensitive to noise intensities around the optimal level. At low noise intensities, the information rate with multiplicative noise is zero because the product of the speech signal and the noise becomes subthreshold. In contrast, for SSR with independent additive noise, the information rate for suprathreshold signals asymptotes to a constant nonzero level (Stocks 2000c, Stocks 2001a).

For multiplicative noise applied at the electrodes, the effect of noise on the information rate is similar to the effect for internal multiplicative noise, in that low noise intensities result in zero information transmission and the information rate is optimized by a nonzero noise level. The optimum information rate, however, is much lower than the optimum rate for internal multiplicative noise and lower than the information rate for the deterministic system – shown to be about 400 bits/s in Fig. 10.21(a).

With 1000 simulated fibres, additive noise applied to the 16 electrodes does not seem to increase information transmission. This suggests that the effective noise stimulating each nerve fibre is largely coherent, as might be expected because of the current spread. The results are quite different if noise is applied at the neural level. These demonstrate the potential benefit of using noise for cochlear implant stimulation: at the optimum level of internal noise, the information rate is about twice as large as for stimulation without noise. To exploit additive noise for cochlear implant coding it is clearly crucial that methods are required to increase the independence of the effective noise stimulus at each neuron.

It has been shown that internal multiplicative noise can lead to greater information rates than internal additive noise. Given that multiplicative noise is characteristic of synaptic transmission between the hair cells in the inner ear and the innervated cochlear nerve fibres, this suggests that multiplicative noise might have a functional role in normal auditory coding. If so, then it should motivate the use of multiplicative noise in cochlear implant coding as it would not only code the signals in a form usual to the auditory system, but also potentially lead to increased information transmission and therefore speech comprehension. However, the efficacy of this method is yet to be demonstrated when the noise is applied to the simulated intracochlear electrodes rather than to the fibres directly. Therefore, although these results demonstrate that multiplicative noise might be exploited for cochlear implants, further investigation is required to determine how this stochastic resonance type effect can be implemented.

10.5 Chapter summary

This chapter commences in Section 10.2 by studying the question of whether SR in a single comparator model depends strongly on the form of the signal distribution. It is demonstrated that, for Laplacian signals, the SR effect is very weak. In particular, the mutual information is observed to show a maximum noise-induced increase of less than 1.5 percent of channel capacity. Gaussian signals also showed modest performance gains with, at most, a 3.7 percent gain being observed. In general, it is suggested that signal distributions that have long – that is, slowly decaying – tails in their PDFs are not likely to show strong SR effects.

For this reason in Section 10.3 it was argued that studies that try and establish the role of SR in neural coding, in particular the coding of auditory signals by the ear, should use biologically realistic signals. Previous studies have used sinusoidal signals with fixed amplitudes; such signals are not representative of typical acoustic stimuli and may well show significantly enhanced SR effects compared to realistic stimuli. This background paved the way for the studies presented here that used speech signals in a biologically plausible auditory model. The results demonstrate that, despite the speech signals having slowly decaying distributions, strong SR effects are possible. The compressive nature of the ear is sufficient to remove the tails of the signal distribution. Furthermore, the level of noise introduced by the Brownian motion of the stereocilia of the hair cells is found to be at an appropriate level for SR effects to occur – this is in agreement with previous studies.

Finally, in Section 10.4, the role of SSR in the context of cochlear implants is discussed. Simulations show that in principle the introduction of noise at the neural level can improve the transmission of information via the SSR effect. Additionally, it is shown that noise introduced in a multiplicative manner – that is, the noise is multiplied by the signal instead of added to it – can lead to large SSR effects. The possibility of using noise to enhance the performance of cochlear implants is therefore promising.

Chapter 10 in a nutshell

This chapter includes the following highlights:

- The magnitude of the SR effect is shown to strongly depend on the form of the signal distribution.
- Signals with slowly decaying distributions display a much reduced SR effect.
- SR effects are observed in biologically plausible models of auditory sensory coding and hence may well form part of a natural auditory coding mechanism.
- Models of cochlear implants display an enhanced performance when noise is added at the neural level.

Open questions

Interesting open questions arising from this chapter might include:

- Does the SSR mechanism naturally arise in the auditory system or in other biological sensory systems?
- Does synaptic noise in the auditory system strongly affect the transmission of signal information?
- Is it possible to improve cochlear implants by adding noise to the electrodes?
- In cochlear implants in practice, is it preferable to use multiplicative (that is, signal-dependent) noise rather than additive noise?

11

The future of stochastic resonance and suprathreshold stochastic resonance

To conclude this book, we summarize our main results and conclusions, before briefly speculating on the most promising areas for future research.

11.1 Putting it all together

Stochastic resonance

Chapter 2 presents a historical review and elucidation of the major epochs in the history of stochastic resonance (SR) research, and discussion of the evolution of the term 'stochastic resonance'. A list of the main controversies and debates associated with the field is given.

Chapter 2 also demonstrates qualitatively that SR can actually occur in a single threshold device, where the threshold is set to the signal mean. Although SR cannot occur in the conventional signal-to-noise ratio (SNR) measure in this situation, if ensemble averaging is allowed, then the presence of an optimal noise level can decrease distortion.

Furthermore, Chapter 2 contains a discussion and critique of the use of SNR measures to quantify SR, the debate about SNR gains due to SR, and the relationship between SNRs and information theory.

Suprathreshold stochastic resonance

Chapter 4 provides an up-to-date literature review of previous work on suprathreshold stochastic resonance (SSR). It also gives numerical results, showing SSR occurring for a number of matched and mixed signal and noise distributions not previously considered. A generic change of variable in the equations used to determine the mutual information through the SSR model is introduced. This change of variable results in a probability density function (PDF) that describes the average transfer function of the SSR model. This PDF is derived for several specific cases, for which it is proved that the mutual information is a function of the noise intensity parameter, σ, rather than a function of both the noise variance and the signal variance independently.

Chapter 5 both improves on previous results and derives several new large N approximations to the mutual information, output entropy, average conditional output entropy, output distribution, and channel capacity in the SSR model. An expression for a channel capacity achieving input PDF for any given noise PDF is found, which holds under the conditions for which the large N mutual approximation formula holds. This formula gives an upper bound for the achievable channel capacity for SSR.

Stochastic quantization with identical thresholds

Chapter 6 introduces and applies the concept of decoding to the SSR model, in order to provide a signal that approximates, or reconstructs, the input signal. The noise in such a reconstruction can be measured with the mean square error (MSE) distortion, or, equivalently, the signal-to-quantization-noise ratio. Plots are presented of the MSE distortion and reconstruction points for a number of linear and nonlinear decoding schemes applied to the SSR model, and it is shown that SR occurs in the MSE distortion measure.

Chapter 6 also gives an analytical derivation of an expression for the minimum mean square error (MMSE) distortion and optimal reconstruction points for the SSR model for the case of uniform signal and noise, with noise intensities, $\sigma \le 1$.

The *information bound* is applied to find a lower bound, named here as the *average information bound*, on the MSE distortion for decoding of the SSR model. It is shown that a linear decoding scheme for the SSR model is a biased efficient estimator, and that MMSE decoding is biased and does not meet the average information bound with equality. Analysis of the average information bound confirms that the value of noise intensity, σ, which minimizes the MSE distortion means finding the best tradeoff between average error variance and mean square bias. This tradeoff is analogous to the tradeoff between output entropy and average conditional output entropy required to maximize the mutual information.

Chapter 7 derives large N expressions for the optimal linear MSE distortion, reconstruction points and linear correlation coefficient. It also demonstrates that the MSE distortion cannot be made asymptotically small for small noise intensities in the SSR model, even for infinite N.

Optimal stochastic quantization

Chapter 8 extends the SSR model to allow all threshold devices to have arbitrary threshold values. In order to calculate the transition probabilities for this extension, we derive a very general recursive formula, with $O(N^2)$ computational complexity, that makes it straightforward to calculate the transition probabilities for any given noise distribution, and threshold values.

We then formulate mathematically the problem of optimal stochastic quantization of the array of threshold devices, in terms of the vector of optimal thresholds, $\vec{\theta}$. The solution to this problem resulted in the discovery of an unexpected bifurcation pattern in the optimal threshold values, as the noise intensity increases. At the same time, this solution numerically validates that the SSR situation of all thresholds equal to the signal mean is in fact optimal for sufficiently large noise intensity, for a range of signal and noise distributions, and both the mutual information and MSE distortion measures. Furthermore, we validate that SSR remains optimal for sufficiently large noise intensity, even if N becomes very large.

Also in Chapter 8, we derive an approximation to the Fisher information for arbitrary thresholds, and apply this approximation to find a lower bound on the MSE distortion. Optimally setting the thresholds was shown to be related to finding the optimal tradeoff between the two components of this lower bound, the mean square bias, and the average error variance.

Stochastic quantizer performance tradeoffs

In Chapter 9, we introduce energy constraints to the SSR model and the arbitrary threshold extension of the SSR model, and formulate these constraints into constrained optimization problems. Solutions of such an energy constrained stochastic quantization problem show similar bifurcations to the results of Chapter 8.

Chapter 9 also discusses, for the first time, the operational rate–distortion tradeoff for the SSR model.

Suprathreshold stochastic resonance in cochlear implants

Chapter 10 contains discussion of the possible role of stochastic resonance or suprathreshold stochastic resonance in the auditory system. It is shown that in principle the introduction of random noise at the neural level can improve information transmission in cochlear implants via SSR.

11.2 Closing remarks

Specific suggestions for further research that may be of interest are given at the end of each chapter. However, in general there are two directions in which there are many challenges and potential for new results based on this research.

First, for accomplished mathematicians, there is potential for many of the mathematical results presented to be placed in a more rigorous and general mathematical framework. Furthermore, many of the results given here were obtained by numerical methods alone, and it is our belief that the application of advanced mathematical approaches will find analytical explanations or predictions in some cases. In particular, it may be possible to further explain the complex bifurcation patterns plotted in Chapters 8–9, and it may be possible to predict the location of the bifurcations.

Secondly, in the other direction, the challenge for engineers is to design circuits or quantization schemes based on the theoretical results contained in this book. Such work may also require extensions to theory, such as consideration of non-ideal threshold devices (Martorell *et al.* 2004), and utilization of adaptive methods to control the noise present in the system (Mitaim and Kosko 1998). We already know that DIMUS sonar arrays made use of stochastic quantization in the 1960s. Furthermore, SSR has been proposed as a means of improving cochlear implant encoding (Stocks *et al.* 2002). Given the current trend towards arrays of small, low-power, and low-cost networks of sensors (Pradhan *et al.* 2002), and for experimental observations of SR in nanoscale devices (Lee *et al.* 2003, Badzey and Mohanty 2005), it is feasible that a form of distributed quantization that utilizes ambient noise will find new applications in the near future.

Appendix 1

Suprathreshold stochastic resonance

A1.1 Maximum values and modes of $P_{y|x}(n|x)$

This section derives Eqs (4.11) and (4.12), which give the values of the modes of $P_{y|x}(n|x)$ for SSR, in terms of the inverse cumulative distribution function (ICDF) of the noise, and are stated in Section 4.3 of Chapter 4.

To begin, the derivative of the transition probabilities, $P_{y|x}(n|x)$, can be expressed in terms of $P_{y|x}(n|x)$ as

$$\frac{dP_{y|x}(n|x)}{dx} = P_{y|x}(n|x) \left(\frac{n - NP_{1|x}}{P_{1|x}(1 - P_{1|x})} \right) \frac{dP_{1|x}}{dx}$$

$$= P_{y|x}(n|x) \left(\frac{n - NP_{1|x}}{P_{1|x}(1 - P_{1|x})} \right) R_n(x - \theta), \quad n = 0, \ldots, N. \quad (A1.1)$$

Note that for $n = 0$ and $n = N$, $P_{y|x}(0|x) = (1 - P_{1|x})^N$ and $P_{y|x}(N|x) = P_{1|x}^N$.

Setting the derivative to zero in Eq. (A1.1) gives

$$P_{1|x} = \frac{n}{N}, \quad n = 1, \ldots, N - 1. \quad (A1.2)$$

For $n = 0$, $P_{y|x}(0|x)$ is maximized when $P_{1|x} = 0$, which means at the minimum possible value of x; and for $n = N$, $P_{y|x}(N|x)$ is maximized when $P_{1|x} = 1$, which means at the maximum possible value of x.

Therefore, for all n, the maximum of $P_{y|x}(n|x)$ occurs when

$$1 - F_\eta(\theta - x) = \frac{n}{N}, \quad n = 0, \ldots, N, \quad (A1.3)$$

which implies that

$$x = \theta - F_\eta^{-1}\left(1 - \frac{n}{N}\right), \quad n = 0, \ldots, N, \quad (A1.4)$$

where $F_\eta^{-1}(\cdot)$ is the ICDF of the noise. For even noise probability density functions (PDFs)

$$x = \theta + F_\eta^{-1}\left(\frac{n}{N}\right), \quad n = 0, \ldots, N. \tag{A1.5}$$

Note that differentiating the natural logarithm of $P_{y|x}(n|x)$ – that is, the log-likelihood function – rather than $P_{y|x}(n|x)$ by itself, gives the same result.

The value of x at which the maximum of $P_{y|x}(n|x)$ occurs is known as the *mode* of $P_{y|x}(n|x)$ for each n. Note that if the noise distribution has a PDF with infinite support and tails, then the mode of $P_{y|x}(0|x)$ and $P_{y|x}(N|x)$ is at $x = \pm\infty$.

A1.2 A proof of Equation (4.38)

This section proves the identity given by Eq. (4.38) in Section 4.3 of Chapter 4

$$-\sum_{n=0}^{N} \log_2\binom{N}{n} = \sum_{n=1}^{N}(N+1-2n)\log_2 n. \tag{A1.6}$$

The proof is

$$\text{lhs} = -\sum_{n=0}^{N} \log_2\binom{N}{n}$$

$$= \sum_{n=0}^{N} \log_2\frac{n!(N-n)!}{N!}$$

$$= \sum_{n=1}^{N} \log_2 n! + \sum_{n=1}^{N} \log_2(N-n)! - \sum_{n=1}^{N} \log_2 N!$$

$$= \sum_{n=1}^{N}\sum_{j=1}^{n} \log_2 j + \sum_{n=1}^{N}\sum_{j=1}^{N-n} \log_2 j - \sum_{n=1}^{N}\sum_{j=1}^{N} \log_2 j$$

$$= \sum_{k=1}^{N}(N-k+1)\log_2 k + \sum_{k=1}^{N}(N-k)\log_2 k - N\sum_{k=1}^{N} \log_2 k$$

$$= \sum_{n=1}^{N}(N-2n+1)\log_2 n$$

$$= \text{rhs}.$$

A1.3 Distributions

This section states the PDF, CDF, and ICDF for each of seven different continuously valued probability distributions. It also uses the CDF to state the probability that a single threshold device is 'on', $P_{1|x}$, for threshold value, θ, and uses the ICDF to state the modes of $P_{y|x}(n|x)$. Some of these distributions are used throughout this book, but are first used in Section 4.3 in Chapter 4.

Gaussian signal and noise

If the input signal has a Gaussian distribution with zero mean and variance σ_x^2, then

$$f_x(x) = \frac{1}{\sqrt{2\pi\sigma_x^2}} \exp\left(-\frac{x^2}{2\sigma_x^2}\right). \tag{A1.7}$$

If the independent noise in each device is Gaussian with zero mean and variance σ_η^2, then

$$f_\eta(\eta) = \frac{1}{\sqrt{2\pi\sigma_\eta^2}} \exp\left(-\frac{\eta^2}{2\sigma_\eta^2}\right). \tag{A1.8}$$

The cumulative distribution function (CDF) of the noise evaluated at $\eta = z$ is

$$F_\eta(z) = 0.5 + 0.5\,\mathrm{erf}\left(\frac{z}{\sqrt{2}\sigma_\eta}\right), \tag{A1.9}$$

where erf$(\cdot)$ is the error function (Spiegel and Liu 1999). Therefore

$$P_{1|x} = 0.5 + 0.5\,\mathrm{erf}\left(\frac{x-\theta}{\sqrt{2\sigma_\eta^2}}\right). \tag{A1.10}$$

The ICDF of the Gaussian noise is

$$F_\eta^{-1}(w) = \sqrt{2}\sigma_\eta\,\mathrm{erf}^{-1}(2w-1), \tag{A1.11}$$

where $w \in [0, 1]$, and erf$^{-1}(\cdot)$ is the inverse error function. From Eq. (4.12), the values of x at which the maximum of each $P_{y|x}(n|x)$ occurs – that is, the mode of $P_{y|x}(n|x)$, for each n – are

$$x = \theta + \sqrt{2}\sigma_\eta\,\mathrm{erf}^{-1}\left(\frac{2n}{N}-1\right), \quad n = 1, \ldots, N-1. \tag{A1.12}$$

For $n = 0$ and $n = N$, the mode is at $x = \mp\infty$, respectively.

Uniform signal and noise

If the input signal x is uniformly distributed between $-\sigma_x/2$ and $\sigma_x/2$, with zero mean, then

$$f_x(x) = \begin{cases} 1/\sigma_x & \text{for } -\sigma_x/2 \le x \le \sigma_x/2, \\ 0 & \text{otherwise.} \end{cases} \tag{A1.13}$$

If the independent noise η in each device is uniformly distributed between $-\sigma_\eta/2$ and $\sigma_\eta/2$, with zero mean, then

$$f_\eta(\eta) = \begin{cases} 1/\sigma_\eta & \text{for } -\sigma_\eta/2 \le \eta \le \sigma_\eta/2, \\ 0 & \text{otherwise.} \end{cases} \tag{A1.14}$$

The CDF of the noise evaluated at $\eta = z$ is

$$F_\eta(z) = \begin{cases} 0 & \text{for } z < -\sigma_\eta/2, \\ z/\sigma_\eta + 1/2 & \text{for } -\sigma_\eta/2 \le z \le \sigma_\eta/2, \\ 1 & \text{for } z > \sigma_\eta/2. \end{cases} \tag{A1.15}$$

Therefore

$$P_{1|x} = \begin{cases} 0 & \text{for } x < \theta - \sigma_\eta/2, \\ x/\sigma_\eta + 1/2 - \theta/\sigma_\eta & \text{for } \theta - \sigma_\eta/2 \le x \le \theta + \sigma_\eta/2, \\ 1 & \text{for } x > \theta + \sigma_\eta/2. \end{cases} \tag{A1.16}$$

The ICDF of the uniform noise is

$$F_\eta^{-1}(w) = \sigma_\eta(w - 0.5), \tag{A1.17}$$

where $w \in [0, 1]$. From Eq. (4.12), the mode of each $P_{y|x}(n|x)$ is

$$x = \theta + \sigma_\eta \left(\frac{n}{N} - \frac{1}{2} \right), \quad n = 0, \ldots, N. \tag{A1.18}$$

Laplacian signal and noise

If the input signal x has a Laplacian distribution with zero mean and variance σ_x^2, then

$$f_x(x) = \frac{1}{\sqrt{2}\sigma_x} \exp\left(\frac{-\sqrt{2}|x|}{\sigma_x} \right). \tag{A1.19}$$

If the independent noise η in each device has a Laplacian distribution with zero mean and variance σ_η^2, then

$$f_\eta(\eta) = \frac{1}{\sqrt{2}\sigma_\eta} \exp\left(\frac{-\sqrt{2}|\eta|}{\sigma_\eta} \right). \tag{A1.20}$$

The CDF of the noise evaluated at $\eta = z$ is

$$F_\eta(z) = 0.5 \left(1 + \text{sign}(z) \left(1 - \exp \left(\frac{-\sqrt{2}|z|}{\sigma_\eta} \right) \right) \right),$$

(A1.21)

where $\text{sign}(\cdot)$ indicates the signum (sign) function. Therefore we have

$$P_{1|x} = 0.5 \left(1 + \text{sign}(x - \theta) \left(1 - \exp \left(\frac{-\sqrt{2}|x - \theta|}{\sigma_\eta} \right) \right) \right).$$

(A1.22)

It is clearer to write this as

$$P_{1|x} = \begin{cases} 0.5 \exp \left(\frac{-\sqrt{2}\theta}{\sigma_\eta} \right) \exp \left(\frac{\sqrt{2}x}{\sigma_\eta} \right) & \text{for } x \le \theta, \\ 1 - 0.5 \exp \left(\frac{\sqrt{2}\theta}{\sigma_\eta} \right) \exp \left(\frac{-\sqrt{2}x}{\sigma_\eta} \right) & \text{for } x \ge \theta. \end{cases}$$

(A1.23)

The ICDF of the Laplacian noise is

$$F_\eta^{-1}(w) = \begin{cases} \frac{\sigma_\eta}{\sqrt{2}} \ln (2w) & \text{for } w \in [0, 0.5], \\ -\frac{\sigma_\eta}{\sqrt{2}} \ln (2(1 - w)) & \text{for } w \in [0.5, 1). \end{cases}$$

(A1.24)

From Eq. (4.12), the mode of each $P_{y|x}(n|x)$ is

$$x = \begin{cases} \theta + \frac{\sigma_\eta}{\sqrt{2}} \ln \left(\frac{2n}{N} \right) & \text{for } 0 < \frac{n}{N} \le 0.5, \\ \theta - \frac{\sigma_\eta}{\sqrt{2}} \ln \left(2(1 - \frac{n}{N}) \right) & \text{for } 0.5 \le \frac{n}{N} < 1. \end{cases}$$

(A1.25)

For $n = 0$ and $n = N$, the mode is at $x = \mp\infty$ respectively.

Logistic signal and noise

If the input signal x has a logistic distribution with zero mean and variance σ_x^2, then

$$f_x(x) = \frac{\exp \left(-\frac{x}{b_x} \right)}{b_x \left(1 + \exp \left(-\frac{x}{b_x} \right) \right)^2},$$

(A1.26)

where $\sigma_x^2 = \frac{\pi^2 b_x^2}{3}$.

Note that this distribution can also be written in terms of the hyperbolic cosine function as

$$f_x(x) = \frac{1}{4b_x \cosh^2 \left(\frac{x}{2b} \right)}.$$

(A1.27)

If the independent noise η in each device has a logistic distribution with zero mean and variance σ_η^2, then

$$f_\eta(\eta) = \frac{\exp\left(-\frac{\eta}{b_\eta}\right)}{b_\eta\left(1 + \exp\left(-\frac{\eta}{b_\eta}\right)\right)^2},$$ (A1.28)

where $\sigma_\eta^2 = \frac{\pi^2 b_\eta^2}{3}$.

The CDF of the noise evaluated at $\eta = z$ is

$$F_\eta(z) = \frac{1}{1 + \exp\left(-\frac{z}{b_\eta}\right)}.$$ (A1.29)

Note that the logistic CDF is the solution to the logistic equation, that is the differential equation

$$\frac{dF_\eta(z)}{dz} = \frac{1}{b_\eta} F_\eta(z)(1 - F_\eta(z)).$$ (A1.30)

Now

$$P_{1|x} = \frac{1}{1 + \exp\left(-\frac{(x-\theta)}{b_\eta}\right)},$$ (A1.31)

and hence

$$\frac{dP_{1|x}}{dx} = \frac{1}{b_\eta} P_{1|x}(1 - P_{1|x}).$$ (A1.32)

The ICDF of the logistic noise is

$$F_\eta^{-1}(w) = -b_\eta \ln\left(\frac{1 - w}{w}\right),$$ (A1.33)

where $w \in [0, 1)$. From Eq. (4.12), the mode of each $P_{y|x}(n|x)$ is

$$x = \theta + \frac{\sqrt{3}\sigma_\eta}{\pi} \ln\left(\frac{n}{N - n}\right), \qquad n = 1, \ldots, N - 1.$$ (A1.34)

For $n = 0$ and $n = N$, the mode is at $x = \mp\infty$ respectively.

Cauchy signal and noise

If the input signal x has a Cauchy (or Lorentzian) distribution with location parameter equal to zero and dispersion parameter $\lambda_x > 0$, then

$$f_x(x) = \frac{\lambda_x}{\pi} \frac{1}{\lambda_x^2 + x^2}.$$ (A1.35)

Unlike the other distributions above, the Cauchy distribution has undefined even-order moments and infinite odd-order moments and therefore cannot be characterized by its variance. Instead, it is characterized by the dispersion parameter λ_x, which is the PDF's *full width at half-maximum* (FWHM). The FWHM of a function is the distance between points on its curve at which the function reaches half its maximum value.

If the independent noise η in each device has a Cauchy distribution with location parameter equal to zero and FWHM λ_η, then

$$f_\eta(\eta) = \frac{\lambda_\eta}{\pi} \frac{1}{\lambda_\eta^2 + x^2}. \tag{A1.36}$$

The CDF of the noise evaluated at $\eta = z$ is

$$F_\eta(z) = \frac{1}{2} + \frac{1}{\pi} \arctan\left(\frac{z}{\lambda_\eta}\right). \tag{A1.37}$$

Therefore we have

$$P_{1|x} = \frac{1}{2} + \frac{1}{\pi} \arctan\left(\frac{x - \theta}{\lambda_\eta}\right). \tag{A1.38}$$

The ICDF of the Cauchy noise is

$$F_\eta^{-1}(w) = \lambda_\eta \tan\left(\pi(w - 0.5)\right), \tag{A1.39}$$

where $w \in [0, 1]$. From Eq. (4.12), the mode of each $P_{y|x}(n|x)$ is

$$x = \theta + \lambda_\eta \tan\left(\pi\left(\frac{n}{N} - 0.5\right)\right), \quad n = 1, \ldots, N - 1. \tag{A1.40}$$

For $n = 0$ and $n = N$, the mode is at $x = \mp\infty$ respectively.

Exponential signal and noise

If the input signal $x \ (\geq 0)$ has an exponential distribution with mean σ_x, then

$$f_x(x) = \frac{1}{\sigma_x} \exp\left(-\frac{x}{\sigma_x}\right). \tag{A1.41}$$

Unlike the previous distributions, the exponential distribution is not an even function, or symmetric about its mean. It is defined only for values greater than or equal to zero.

If the independent noise $\eta \ (\geq 0)$ in each device has an exponential distribution with mean σ_η, then

$$f_\eta(\eta) = \frac{1}{\sigma_\eta} \exp\left(-\frac{\eta}{\sigma_\eta}\right). \tag{A1.42}$$

The CDF of the noise evaluated at $\eta = z$ is

$$F_\eta(z) = 1 - \exp\left(-z/\sigma_\eta\right).\qquad(\text{A1.43})$$

Therefore

$$P_{1|x} = \begin{cases} \exp\left(-\frac{(\theta-x)}{\sigma_\eta}\right) & \text{for} \quad x < \theta, \\ 1 & \text{for} \quad x \geq \theta. \end{cases}\qquad(\text{A1.44})$$

The ICDF of the exponential noise is

$$F_\eta^{-1}(w) = \sigma_\eta \ln\left(\frac{1}{1-w}\right),\qquad(\text{A1.45})$$

where $w \in [0, 1)$.

Rayleigh signal and noise

If the input signal x (≥ 0) has a Rayleigh distribution with mean $\sigma_x\sqrt{\pi/2}$, then

$$f_x(x) = \frac{x}{\sigma_x^2} \exp\left(-\frac{x^2}{2\sigma_x^2}\right).\qquad(\text{A1.46})$$

As in the exponential distribution, the Rayleigh distribution is not an even function or symmetric about its mean. It is defined only for values greater than or equal to zero.

If the independent noise η (≥ 0) in each device has a Rayleigh distribution with mean $\sigma_\eta\sqrt{\pi/2}$, then

$$f_\eta(\eta) = \frac{\eta}{\sigma_\eta^2} \exp\left(-\frac{\eta^2}{2\sigma_\eta^2}\right).\qquad(\text{A1.47})$$

The CDF of the noise evaluated at $\eta = z$ is

$$F_\eta(z) = 1 - \exp\left(-\frac{z^2}{2\sigma_\eta^2}\right).\qquad(\text{A1.48})$$

Therefore

$$P_{1|x} = \begin{cases} \exp\left(-\frac{(\theta-x)^2}{2\sigma_\eta^2}\right) & \text{for} \quad x < \theta, \\ 1 & \text{for} \quad x \geq \theta. \end{cases}\qquad(\text{A1.49})$$

The ICDF of the exponential noise is

$$F_\eta^{-1}(w) = \sigma_\eta\sqrt{2\ln\left(\frac{1}{1-w}\right)},\qquad(\text{A1.50})$$

where $w \in [0, 1)$.

A1.4 Proofs that $f_\varrho(\tau)$ is a PDF, for specific cases

This section provides a proof that $f_\varrho(\tau)$ is a PDF for two of the infinite tail signal and noise distributions given in Table 4.2 in Section 4.3 of Chapter 4. Since each $f_\varrho(\tau)$ is nonnegative, for $f_\varrho(\tau)$ to be a PDF it is necessary only to show that

$$\int_{\tau=0}^{\tau=1} f_\varrho(\tau)d\tau = 1.$$

Gaussian signal and noise

For Gaussian signal and noise, and $\sigma \neq 1$, the change of variable from τ to $u = \mathrm{erf}^{-1}(2\tau - 1)$ is useful. The result is

$$\int_{\tau=0}^{\tau=1} f_\varrho(\tau)d\tau = \int_{u=-\infty}^{u=\infty} \sigma \exp\left((1 - \sigma^2)u^2\right)\frac{1}{\sqrt{\pi}}\exp\left(-u^2\right)du$$

$$= \frac{\sigma}{\sqrt{\pi}}\int_{u=-\infty}^{u=\infty} \exp\left(-\sigma^2 u^2\right)du$$

$$= \frac{\sigma}{\sqrt{\pi}}\sqrt{2\pi\frac{1}{2\sigma^2}}$$

$$= 1.$$

Thus, given this result, $f_\varrho(\tau)$ is a PDF for $\sigma \neq 1$.

For the case of $\sigma = 1$, $f_\varrho(\tau) = 1 \ \forall \ \tau \in [0, 1]$, and is hence a PDF.

Laplacian signal and noise

For Laplacian signal and noise and $\sigma \neq 1$

$$\int_{\tau=0}^{\tau=1} f_\varrho(\tau)d\tau = \int_{\tau=0}^{\tau=0.5} \sigma(2\tau)^{(\sigma-1)}d\tau + \int_{\tau=0.5}^{\tau=1} \sigma(2(1-\tau))^{(\sigma-1)}d\tau$$

$$= \sigma 2^{(\sigma-1)}\left(\int_{\tau=0}^{\tau=0.5} \tau^{(\sigma-1)}d\tau + \int_{\tau=0.5}^{\tau=1}(1-\tau)^{(\sigma-1)}d\tau\right)$$

$$= \sigma 2^{(\sigma-1)}\left(\left[\frac{1}{\sigma}\tau^\sigma\right]_0^{0.5} + \left[\frac{-1}{\sigma}(1-\tau)^\sigma\right]_{0.5}^{1}\right)$$

$$= \sigma 2^{(\sigma-1)}\left(\frac{1}{\sigma}2^{-\sigma} + \frac{1}{\sigma}2^{-\sigma}\right)$$

$$= 1.$$

Thus, given this result, $f_\varrho(\tau)$ is a PDF for $\sigma \neq 1$.

For the case of $\sigma = 1$, $f_\varrho(\tau) = 1 \ \forall \ \tau \in [0, 1]$, and is hence a PDF.

A1.5 Calculating mutual information by numerical integration

This section gives some details regarding calculating the mutual information between the input and output of the SSR model, via numerical integration. This method is required in Chapters 4, 8, and 9, and analogous methods are required for numerical integration of other quantities elsewhere.

Integrating over the input's support

To obtain $I(x, y)$ numerically, it is necessary to perform numerical integrations. The simplest form of numerical integration is to approximate the signal PDF by a discrete version, with resolution $\Delta x \ll 1/N$, where $N + 1$ is the number of output states in the SSR model. Hence, if in the case of a continuously valued PDF $f_x(x)$ we have support $x \in [a, b]$, then discretization with resolution Δx gives discrete values $x = a + i\Delta x$, $i = 0, 1, \ldots, (b - a)/\Delta x$. This simple method of numerical integration is easily justified for calculating mutual information – see Cover and Thomas (1991).

Using this discretization of $f_x(x)$, the mutual information can be written as

$$I(x, y) = -\sum_{n=0}^{N} P_y(n) \log_2 P_y(n)$$

$$-\Delta x \sum_x f_x(x) \sum_{n=0}^{N} P_{y|x}(n|x) \log_2 P_{y|x}(n|x), \quad \text{(A1.51)}$$

where

$$P_y(n) = \Delta x \sum_x P_{y|x}(n|x) f_x(x). \quad \text{(A1.52)}$$

Given these formulas, $f_x(x)$ and $P_{y|x}(n|x)$ need only be calculated for each specified value of x.

For an input distribution with finite support, such as the uniform distribution, only specification of the resolution is required. However for a distribution that has a PDF with infinite support and tails, such as the Gaussian distribution, it is necessary to restrict the upper and lower bounds of the support of x to finite values. This is achieved in a discretization of $f_x(x)$ by setting the maximum and minimum values of x to a multiple, w, of the standard deviation, σ_x. Thus, $x \in [-w\sigma_x, w\sigma_x]$ and $f_x(x)$ is then discretized to a resolution of Δx.

This method of numerical integration has been found to be sufficient. Its accuracy has been verified by calculating the mutual information by more sophisticated numerical integration schemes, such as Simpson quadrature (Press *et al.* 1992).

Numerical integration using $f_Q(\tau)$

Instead of numerically integrating Eq. (4.5) and the integral on the rhs of Eq. (4.17) between $\pm\infty$ (for signal densities with infinite tails), the change of variables from x to τ appears to allow a simpler integration of Eq. (4.24) and the integral on the rhs of Eq. (4.23), between the limits of zero and unity.

Note that although $f_Q(\tau)$ is infinite at $\tau = 0$ and $\tau = 1$ for $\sigma < 1$, the integrand in both integrals is zero at these limits, except for $B_y(0)$ when $\tau = 0$, and $B_y(N)$ when $\tau = 1$. Thus, other than these two cases, both integrals are easily obtainable by standard numerical integration techniques. Using the law of total probability avoids the need to deal with the singularity in the integrals of $B_y(0)$ and $B_y(N)$, since by the evenness of the signal and noise densities, $P_y(0) = P_y(N)$. Once $B_y(n)$ has been calculated for all other n, the corresponding $P_y(n)$s can be derived. Hence $P_y(0) = P_y(N) = 0.5 - 0.5 \sum_{n=1}^{N-1} P_y(n)$.

Comments on numerical integration

The main difficulty arising in numerical calculation of the mutual information arises for larger N. Calculations of the output probabilities can require multiplying a small number, $B_y(n)$, by a very large number, $\binom{N}{n}$. For large N, $\binom{N}{n}$ gets very large, and sufficient precision can soon be lost. Even if 32-bit floating point representation is used, the final calculation of $I(x, y)$ can be highly inaccurate. A way to circumvent this problem is to avoid calculating the set of $\binom{N}{n}$ for a given N. This can be carried out by calculating $P_{y|x}(n|x)$ recursively, using the technique described in Section 8.2 of Chapter 8.

Appendix 2

Large N suprathreshold stochastic resonance

A2.1 Proof of Eq. (5.14)

This section proves the identity given by Eq. (5.14) in Section 5.2 of Chapter 5.

$$-\sum_{n=0}^{N} P_y(n) \log_2 \binom{N}{n}$$

$$= -\frac{1}{N+1} \sum_{n=0}^{N} \log_2 \binom{N}{n}$$

$$= \frac{1}{N+1} \sum_{n=0}^{N} \log_2 \frac{n!(N-n)!}{N!}$$

$$= \frac{1}{N+1} \left(-\sum_{n=0}^{N} \log_2 N! + \sum_{n=0}^{N} \log_2 n! + \sum_{n=0}^{N} \log_2 (N-n)! \right)$$

$$= \frac{1}{N+1} \left(-(N+1) \log_2 N! + \sum_{n=0}^{N} \sum_{j=1}^{n} \log_2 j + \sum_{n=0}^{N} \sum_{j=1}^{N-n} \log_2 j \right)$$

$$= -\log_2 N!$$

$$+ \frac{1}{N+1} \left(\sum_{k=1}^{N} (N-k+1) \log_2 k + \sum_{k=1}^{N} (N-k+1) \log_2 k \right)$$

$$= -\log_2 N! + \frac{2}{N+1} \left(\sum_{n=1}^{N} (N-n+1) \log_2 n \right)$$

$$= -\log_2 N! + 2 \sum_{n=1}^{N} \log_2 n - \frac{2}{N+1} \sum_{n=1}^{N} n \log_2 n$$

$$= -\log_2 N! + 2\log_2 N! - \frac{2}{N+1} \sum_{n=1}^{N} n \log_2 n$$

$$= \log_2 N! - \frac{2}{N+1} \sum_{n=1}^{N} n \log_2 n. \tag{A2.1}$$

A2.2 Derivation of Eq. (5.18)

This section gives the derivation of Eq. (5.18) in Section 5.2 of Chapter 5 that states

$$\frac{2}{N+1} \sum_{n=1}^{N} n \log_2 n \simeq N \log_2 (N+1) - \frac{N(N+2)}{2\ln 2(N+1)} + O\left(\frac{\log N}{N}\right). \tag{A2.2}$$

The Euler–Maclaurin summation formula (Spiegel and Liu 1999) states that a summation over n terms can be replaced by an integral plus correction terms, as

$$\sum_{n=1}^{N-1} F(n) = \int_{n=0}^{N} F(n)dn - 0.5F(0) - 0.5F(N)$$

$$+ \sum_{p=1}^{\infty} (-1)^{p-1} \frac{B_p}{(2p)!} \left(F^{(2p-1)}(N) - F^{(2p-1)}(0)\right), \tag{A2.3}$$

where $F^{(2p-1)}$ indicates the $(2p-1)$th derivative of F, and B_p is the pth Bernoulli number (Spiegel and Liu 1999).

Noting that $\sum_{n=1}^{N} n \log_2 n = \sum_{n=1}^{N-1} (n+1) \log_2 (n+1)$, then $F(n) = (n+1) \log_2 (n+1)$, $F(0) = 0$, and $F(N) = (N+1) \log_2 (N+1)$. Denoting the remainder term of Eq. (A2.3) involving derivatives of $F(n)$ by R, then

$$\sum_{n=1}^{N} n \log_2 n = \sum_{n=1}^{N-1} (n+1) \log_2 (n+1)$$

$$= -0.5F(0) - 0.5F(N) + R + \int_{n=0}^{n=N} (n+1) \log_2 (n+1)dn$$

$$= 0.5(N+1)^2 \log_2 (N+1) - \frac{N(N+2)}{4\ln 2}$$

$$- 0.5(N+1) \log_2 (N+1) + R. \tag{A2.4}$$

Thus, the desired simplification for the second term in Eq. (5.14) is

$$\frac{2}{N+1} \sum_{n=1}^{N} n \log_2 n = N \log_2 (N+1) - \frac{N(N+2)}{2\ln 2(N+1)} + \frac{2R}{N+1}. \tag{A2.5}$$

This is an exact expression that holds for any N, and can be obtained by evaluating R. In order to obtain a large N simplification we need to consider the behaviour of R for large N.

Note that for $p = 1$, $F'(N) - F'(0) = \log_2(N + 1)$ and for all $p > 1$,

$$F^{(2p-1)}(N) - F^{(2p-1)}(0) = \frac{(2p - 3)!}{\ln 2} \left(1 - (N + 1)^{-2p+2}\right). \tag{A2.6}$$

Substituting Eq. (A2.6) into the remainder term of Eq. (A2.3) and multiplying by $\frac{2}{N+1}$ gives

$$\frac{2R}{N + 1} = \frac{1}{6(N + 1)} \log_2(N + 1)$$

$$+ \sum_{p=2}^{\infty} (-1)^{p-1} \frac{B_p}{\ln 2(4p^3 - 7p^2 + 2p)} \left(\frac{1}{N + 1} - (N + 1)^{-2p+1}\right). \tag{A2.7}$$

Hence, the remainder term scales with $\frac{\log N}{N}$, which approaches zero for large N, and our approximation to the second term in Eq. (5.14), for large N, is

$$\frac{2}{N + 1} \sum_{n=1}^{N} n \log_2 n \simeq N \log_2(N + 1) - \frac{N(N + 2)}{2 \ln 2(N + 1)} + O\left(\frac{\log N}{N}\right).$$

A2.3 Proof that $f_S(x)$ is a PDF

This section shows that the function, $f_S(x)$, given by Eq. (5.75) in Section (5.5) is a probability density function (PDF).

As shown in Section A3.6 of Appendix 3, the Fisher information for the SSR model is given by

$$J(x) = \frac{N f_\eta(\theta - x)^2}{P_{1|x}(1 - P_{1|x})}, \tag{A2.8}$$

where $P_{1|x} = 1 - F_\eta(\theta - x)$. Consider the function

$$f_S(x) = \frac{\sqrt{J(x)}}{\pi \sqrt{N}} = \frac{f_\eta(\theta - x)}{\pi \sqrt{F_\eta(\theta - x)(1 - F_\eta(\theta - x))}}. \tag{A2.9}$$

Since $f_\eta(\cdot)$ is a PDF and $F_\eta(\cdot)$ is the CDF of η, we have $f_S(x) \geq 0 \,\forall\, x$.

Letting $h(x) = F_\eta(\theta - x)$, Eq. (A2.9) can be written as

$$f_S(x) = \frac{-h'(x)}{\pi \sqrt{h(x) - h(x)^2}}. \tag{A2.10}$$

Suppose $f_\eta(x)$ has support $x \in [-a, a]$. Integrating $f_S(x)$ over all x gives

$$\int_{x=-a}^{x=a} f_S(x)dx = \int_{x=-a}^{x=a} \frac{-h'(x)}{\pi \sqrt{h(x) - h(x)^2}} dx$$

$$= -\frac{1}{\pi} \left(2 \arcsin \left(\sqrt{h(x)} \right) \big|_{x=-a}^{x=a} \right)$$

$$= -\frac{2}{\pi} (\arcsin(0) - \arcsin(1)) = 1, \qquad (A2.11)$$

which means $f_S(x)$ is a PDF.

Appendix 3

Suprathreshold stochastic resonance decoding

A3.1 Conditional output moments

This section derives expressions for the first and second conditional output moments of the SSR model, as well the conditional output variance. These expressions are used throughout Chapter 6.

The mth moment of the output given x is given by

$$E[y^m|x] = \sum_{n=0}^{N} n^m P_{y|x}(n|x).$$ (A3.1)

Instead of using this definition directly, we shall make use of the fact that all devices have identical thresholds. Using Eq. (4.2) from Chapter 4, the first conditional moment is given by

$$E[y|x] = E\left[\frac{1}{2}\sum_{i=1}^{N} \text{sign}[x + \eta_i] + \frac{N}{2}\bigg|x\right]$$

$$= \frac{N}{2}E[\text{sign}(x + \eta)|x] + \frac{N}{2},$$ (A3.2)

where since all N additive noise components are *iid*, the subscripts in the η_is have been removed. This leads to

$$E[y|x] = \frac{N}{2}(-(1 - P_{1|x}) + P_{1|x}) + \frac{N}{2}$$

$$= N P_{1|x}.$$ (A3.3)

The conditional variance is given by $\text{var}[y^2|x] = E[y^2|x] - E[y|x]^2$. The second conditional moment is

$$E[y^2|x] = E\left[\left(\frac{1}{2}\sum_{i=1}^{N}\text{sign}(x+\eta_i) + \frac{N}{2}\right)^2 |x\right]$$

$$= \frac{1}{4}E\left[\left(\sum_{i=1}^{N}\text{sign}(x+\eta_i)\right)^2 |x\right] + \frac{N^2}{2}E[\text{sign}(x+\eta)|x] + \frac{N^2}{4}.$$

$$\tag{A3.4}$$

We have also

$$E\left[\left(\sum_{i=1}^{N}\text{sign}(x+\eta_i)\right)^2 |x\right]$$

$$= NE[(\text{sign}(x+\eta))^2 |x]$$
$$+ N(N-1)E[\text{sign}(x+\eta_j)\text{sign}(x+\eta_k)|x]|_{\forall\ j\neq k}$$
$$= N + N(N-1)(-2P_{1|x}(1-P_{1|x}) + (1-P_{1|x})^2 + P_{1|x}^2)$$
$$= N + N(N-1)(2P_{1|x}-1)^2.$$

$$\tag{A3.5}$$

Substituting Eq. (A3.5) into Eq. (A3.4) gives

$$E[y^2|x] = \frac{1}{4}\left(N + N(N-1)(2P_{1|x}-1)^2\right) + \frac{N^2}{2}(2P_{1|x}-1) + \frac{N^2}{4}$$

$$= NP_{1|x}(1-P_{1|x}) + N^2 P_{1|x}^2.$$

$$\tag{A3.6}$$

Thus, using Eqs (A3.3) and (A3.6) leaves the conditional variance as

$$\text{var}[y|x] = NP_{1|x}(1-P_{1|x}).$$

$$\tag{A3.7}$$

The correctness of these derivations of the conditional mean and variance can be seen by noting that the probability distribution of the output given the input is the binomial distribution, as given by Eq. (4.9). It is well known that the expected value of such a binomially distributed variable is $\sum_{n=0}^{N} n P_{y|x}(n|x) = NP_{1|x}$, and that the variance is $\sum_{n=0}^{N} n^2 P_{y|x}(n|x) - \left(\sum_{n=0}^{N} n P_{y|x}(n|x)\right)^2 = NP_{1|x}(1-P_{1|x})$ (Kreyszig 1988).

A3.2 Output moments

This section derives expressions for the mean and mean square value of the output of the SSR model, as well the output variance. These expressions are used throughout Chapter 6.

The *m*th moment of the output is given by

$$E[y^m] = E[E[y^m|x]] = \int_x \sum_{n=0}^{N} n^m f_{xy}(x, y)dx \qquad (A3.8)$$

$$= \int_x f_x(x) \sum_{n=0}^{N} n^m P_{y|x}(n|x)dx \qquad (A3.9)$$

$$= \sum_{n=0}^{N} n^m P_y(n). \qquad (A3.10)$$

If the output distribution is known, it is generally easier to apply Eq. (A3.10). However, in the following sections it will most often be the case that using Eq. (A3.9) will be more convenient. In particular, we have the conditional mean given by Eq. (A3.3) and the conditional second moment given by Eq. (A3.6). Thus

$$E[y] = NE[P_{1|x}], \qquad (A3.11)$$

and

$$E[y^2] = NE[P_{1|x}] + N(N-1)E[P_{1|x}^2]. \qquad (A3.12)$$

The output variance cannot be obtained by integration of Eq. (A3.7), but is easily obtained by noting that $\text{var}[y] = E[y^2] - E[y]^2$ as

$$\text{var}[y] = N(N-1)E[P_{1|x}^2] - NE[P_{1|x}](NE[P_{1|x}] - 1). \qquad (A3.13)$$

Even signal and noise PDFs, all thresholds zero

For zero-mean noise PDFs that are even functions, and for all threshold values equal to zero, $P_{1|x}$ is given by Eq. (4.8), that is $P_{1|x} = F_\eta(x)$, where $F_\eta(\cdot)$ is the CDF of the noise. This can be written as

$$P_{1|x} = \frac{1}{2} + \int_{\eta=0}^{\eta=x} f_\eta(\eta)d\eta. \qquad (A3.14)$$

Therefore, the expected value of $P_{1|x}$ over the signal distribution is

$$E[P_{1|x}] = \frac{1}{2} + E\left[\int_{\eta=0}^{\eta=x} f_\eta(\eta)d\eta\right]$$

$$= \frac{1}{2} + \int_{x=-\infty}^{x=\infty} \left(\int_{\eta=0}^{\eta=x} f_\eta(\eta)d\eta\right) f_x(x)dx. \qquad (A3.15)$$

Since $f_\eta(\eta)$ is even, $\int_0^x f_\eta(\eta)d\eta$ is odd with respect to x, and therefore the integral above is zero, as $f_x(x)$ is even. Thus, $E[P_{1|x}] = 0.5$.

Thus, Eqs (A3.11) and (A3.12) can be simplified to

$$E[y] = \frac{N}{2},$$ (A3.16)

and

$$E[y^2] = N(N-1)E[P_{1|x}^2] + \frac{N}{2}.$$ (A3.17)

This means that the variance of y is

$$\text{var}[y] = N(N-1)E[P_{1|x}^2] - \frac{N(N-2)}{4}.$$ (A3.18)

A3.3 Correlation and correlation coefficient expressions

This section derives expressions for the correlation coefficient at various stages of the SSR model. These expressions are used throughout Chapter 6.

Input correlation coefficient at any two thresholds

Although the noise at the input to any given threshold device in the SSR model is uncorrelated with the noise at the input of any other device, due to the presence of the same input signal on each device there is a correlation between the inputs of any given pair of devices. This can be measured using the correlation coefficient, which for zero-meaned inputs to any two comparators, i and j, is given by

$$
\begin{aligned}
\rho_i &= \frac{\text{cov}[x + \eta_i, x + \eta_j]}{\sqrt{\text{var}[x + \eta_i]\text{var}[x + \eta_j]}} \\
&= \frac{E[(x + \eta_i)(x + \eta_j)]}{\sqrt{(\text{var}[x + \eta])^2}} \\
&= \frac{E[x^2]}{\sqrt{(E[x^2] + E[\eta^2])^2}} \\
&= \frac{\sigma_x^2}{\sigma_x^2 + \sigma_\eta^2} \\
&= \frac{1}{1 + \sigma^2},
\end{aligned}
$$ (A3.19)

where we have denoted the variance of the signal as σ_x^2 and the variance of the noise as σ_η^2 and let $\sigma^2 = \sigma_\eta^2/\sigma_x^2$. It is clear that the correlation coefficient is unity only in the absence of noise, and decreases towards zero as the noise variance increases.

Output correlation coefficient at any two thresholds

As in Section A3.3, which derives a formula for the correlation coefficient between the *inputs* to any two threshold devices, we can derive an expression for the correlation coefficient between the *outputs* of any pair of threshold devices, i and j. This is

$$
\begin{aligned}
\rho_o &= \frac{\text{cov}[y_i, y_j]}{\sqrt{\text{var}[y_i]\text{var}[y_j]}} \\
&= \frac{E[y_i y_j] - E[y_i]E[y_j]}{\sqrt{(\text{var}[y_i])^2}} \\
&= \frac{E[y_i y_j] - \frac{1}{4}}{\frac{1}{4}} \\
&= \frac{E[P_{1|x}^2] - \frac{1}{4}}{\frac{1}{4}} \\
&= 4E[P_{1|x}^2] - 1.
\end{aligned}
\tag{A3.20}
$$

Input–output correlation

The output of any given threshold device is correlated with its input. This correlation can be expressed as

$$
\begin{aligned}
E[x y_i] &= \sum_{y_i=0}^{1} \int_{-\infty}^{\infty} x y_i \, f_{x,y_i}(x, y_i) dx \\
&= \int_{-\infty}^{\infty} 0 x (1 - P_{1|x}) f_x(x) + 1 x P_{1|x} f_x(x) dx \\
&= \int_{-\infty}^{\infty} x P_{1|x} f_x(x) dx \\
&= E[x P_{1|x}].
\end{aligned}
\tag{A3.21}
$$

Assuming the input signal has a mean of zero and variance σ_x^2, the correlation coefficient between x and y_i is

$$
\begin{aligned}
\rho_{xy_i} &= \frac{E[x y_i]}{\sigma_x \sqrt{\text{var}[y_i^2]}} \\
&= \frac{E[x P_{1|x}]}{\sigma_x \sqrt{\text{var}[y_i^2]}}.
\end{aligned}
\tag{A3.22}
$$

The overall system output encoding, y, is also correlated with the input signal, x. The input–output correlation is

$$E[xy] = \int_{-\infty}^{\infty} x E[y|x] f_x(x) dx \qquad (A3.23)$$

$$= \sum_{n=0}^{N} n E[x|n] P_y(n). \qquad (A3.24)$$

The fact that the above two identities are equal can easily be shown by substituting for the definition of expected value and the use of Bayes' rule, before changing the order of integration and summation. Substituting Eq. (A3.3) into Eq. (A3.23) gives an expression for the input–output correlation in terms of the correlation between the inputs to any two devices as

$$E[xy] = \int_{-\infty}^{\infty} x N P_{1|x} f_x(x) dx$$

$$= N E[x P_{1|x}] \qquad (A3.25)$$

$$= N E[x y_i]. \qquad (A3.26)$$

Hence, the overall input–output correlation is N times larger than the correlation between the input and output of a single device.

Assuming the input signal has a mean of zero and variance σ_x^2, the correlation coefficient between x and y is

$$\rho_{xy} = \frac{E[xy]}{\sigma_x \sqrt{\mathrm{var}[y]}}$$

$$= N \frac{E[x P_{1|x}]}{\sigma_x \sqrt{E[y^2] - E[y]^2}}$$

$$= N \frac{E[x P_{1|x}]}{\sigma_x \sqrt{N(N-1) E[P_{1|x}^2] - N^2 E[P_{1|x}]^2 + N E[P_{1|x}]}}. \qquad (A3.27)$$

To progress further requires knowledge of the noise PDF, $f_\eta(\eta)$. Once this is specified, $P_{1|x}$ can be derived, and therefore $E[x P_{1|x}]$ and the output moments.

A3.4 A proof of Prudnikov's integral

This section gives a proof of an integral used in Section 6.4 of Chapter 6 for the calculation of the output variance for the SSR model for Gaussian signal and noise. The integral we prove here is listed in Prudnikov *et al.* (1986) as

$$f(a) = \int_{x=-\infty}^{x=\infty} \exp\left(-a^2 x^2\right) \mathrm{erf}^2(x) dx = \frac{2}{a\sqrt{\pi}} \arctan\left(\frac{1}{a\sqrt{a^2+2}}\right). \qquad (A3.28)$$

Observe that $f(a) = f(-a)$, and that the arctan function is an odd function. Eq. (A3.28) is expressed in terms of the arctan function so that the relationship holds for both positive and negative a. If we are only interested in the case of $a > 0$, the equation can be expressed in terms of the arcsin function as

$$f(a) = \int_{x=-\infty}^{x=\infty} \exp(-a^2 x^2) \operatorname{erf}^2(x) dx = \frac{2}{a\sqrt{\pi}} \arcsin\left(\frac{1}{1+a^2}\right). \tag{A3.29}$$

The proof of this result follows. Note that the integrand is an even function of x. Hence

$$f(a) = 2 \int_{x=0}^{x=\infty} \exp(-a^2 x^2) \operatorname{erf}^2(x) dx. \tag{A3.30}$$

Now

$$\int_{x=0}^{x=\infty} \exp(-a^2 x^2) \operatorname{erf}^2(x) dx$$

$$= \int_{x=0}^{x=\infty} \exp(-a^2 x^2) \left(\frac{2}{\sqrt{\pi}} \int_{u=0}^{u=x} \exp(-u^2) du\right)^2 dx$$

$$= \frac{4}{\pi} \int_{x=0}^{x=\infty} \int_{u=0}^{u=x} \int_{v=0}^{v=x} \exp(-(u^2 + v^2 + a^2 x^2)) du\, dv\, dx. \tag{A3.31}$$

Letting $w = ax$ and performing a change of variable gives

$$f(a) = \frac{8}{a\pi} \int_{w=0}^{w=\infty} \int_{u=0}^{u=w/a} \int_{v=0}^{v=w/a} \exp(-(u^2 + v^2 + w^2)) du\, dv\, dw. \tag{A3.32}$$

We now convert this triple integral to polar coordinates with $r \in [0, \infty)$, $\theta \in [0, 2\pi)$, and $\phi \in [0, \pi]$. This gives $u = r \cos\theta \sin\phi$, $v = r \sin\theta \sin\phi$, and $w = r \cos\phi$, with the volume element being $du\, dv\, dw = r^2 \sin\phi\, d\phi\, d\theta\, dr$. With this conversion we get $u^2 + v^2 + w^2 = r^2$. Thus

$$f(a) = \frac{8}{a\pi} \int_r \int_\theta \int_\phi r^2 \exp(-r^2) \sin\phi\, d\phi\, d\theta\, dr. \tag{A3.33}$$

The limits of integration for θ and ϕ can be determined as follows. Since u and v are integrated between zero and w/a, this imposes the following inequalities

$$0 \leq \cos\theta \sin\phi \leq \frac{\cos\phi}{a},$$

$$0 \leq \sin\theta \sin\phi \leq \frac{\cos\phi}{a}. \tag{A3.34}$$

Since ϕ is defined on $[0, \pi]$, $\sin\phi$ is always nonnegative. Therefore both $\cos\theta \geq 0$ and $\sin\theta \geq 0$, which implies that $\theta \in [0, \pi/2]$. Also, since $a \geq 0$, $\cos\phi \geq 0$ and therefore $\phi \in [0, \pi/2]$.

Thus, Inequalities (A3.34) can be written as

$$0 \leq \tan \phi \leq \frac{1}{a \cos \theta},$$

$$0 \leq \tan \phi \leq \frac{1}{a \sin \theta}. \tag{A3.35}$$

This can be restated as

$$\phi \leq \phi_1 = \arctan \left(\frac{1}{a \cos \theta} \right), \quad \theta \in \left[0, \frac{\pi}{4} \right],$$

$$\phi \leq \phi_2 = \arctan \left(\frac{1}{a \sin \theta} \right), \quad \theta \in \left[\frac{\pi}{4}, \frac{\pi}{2} \right]. \tag{A3.36}$$

Furthermore, since in the original integral of Eq. (A3.32), w is integrated between zero and ∞, and $\cos \phi \geq 0$, r must be integrated over $[0, \infty)$. Therefore

$$f(a) = \frac{8}{a\pi} \int_\theta \int_\phi \sin \phi \int_{r=0}^\infty r^2 \exp(-r^2) dr d\phi d\theta. \tag{A3.37}$$

The inner integral is simply $\int_{r=0}^{r=\infty} r^2 \exp(-r^2) dr = \sqrt{\pi}/4$. Thus

$$
\begin{aligned}
f(a) &= \frac{2}{a\sqrt{\pi}} \int_\theta \int_\phi \sin \phi \, d\phi d\theta \\
&= \frac{2}{a\sqrt{\pi}} \left(\int_{\theta=0}^{\theta=\pi/4} \int_{\phi=0}^{\phi=\phi_1} \sin \phi \, d\phi d\theta + \int_{\theta=\pi/4}^{\theta=\pi/2} \int_{\phi=0}^{\phi=\phi_2} \sin \phi \, d\phi d\theta \right) \\
&= \frac{2}{a\sqrt{\pi}} \left(\int_{\theta=0}^{\theta=\pi/4} [-\cos \phi]_0^{\phi_1} \, d\theta + \int_{\theta=\pi/4}^{\theta=\pi/2} [-\cos \phi]_0^{\phi_2} \, d\theta \right) \\
&= \frac{2}{a\sqrt{\pi}} \left(\int_{\theta=0}^{\theta=\pi/4} 1 - \cos \left(\arctan \left(\frac{1}{a \cos \theta} \right) \right) d\theta \right. \\
&\qquad \left. + \int_{\theta=\pi/4}^{\theta=\pi/2} 1 - \cos \left(\arctan \left(\frac{1}{a \sin \theta} \right) \right) d\theta \right) \\
&= \frac{2}{a\sqrt{\pi}} \left(\frac{\pi}{2} - 2 \int_{\theta=0}^{\theta=\pi/4} \cos \left(\arctan \left(\frac{1}{a \cos \theta} \right) \right) d\theta \right) \\
&= \frac{2}{a\sqrt{\pi}} \left(\frac{\pi}{2} - 2 \int_{\theta=0}^{\theta=\pi/4} \frac{a \cos \theta}{\sqrt{1 + a^2 \cos^2 \theta}} d\theta \right) \\
&= \frac{2}{a\sqrt{\pi}} \left(\frac{\pi}{2} - 2 \left[\arcsin \left(\frac{a \sin \theta}{\sqrt{1 + a^2}} \right) \right]_0^{\pi/4} \right) \\
&= \frac{2}{a\sqrt{\pi}} \left(\frac{\pi}{2} - 2 \arcsin \left(\frac{a}{\sqrt{2}\sqrt{1 + a^2}} \right) \right) \\
&= \frac{2}{a\sqrt{\pi}} \left(\frac{\pi}{2} - 2 \arcsin \left(\frac{1}{\sqrt{2}} \sqrt{1 - \frac{1}{1 + a^2}} \right) \right). \tag{A3.38}
\end{aligned}
$$

Given the identity $A = 2 \arcsin \left(\frac{1}{\sqrt{2}} \sqrt{1 - \cos A} \right)$, it can be seen that Eq. (A3.38) can be written as

$$
f(a) = \frac{2}{a\sqrt{\pi}} \left(\frac{\pi}{2} - \arccos \left(\frac{1}{1+a^2} \right) \right)
$$

$$
= \frac{2}{a\sqrt{\pi}} \arcsin \left(\frac{1}{1+a^2} \right). \tag{A3.39}
$$

This completes the proof.

A3.5 Minimum mean square error distortion decoding

This section proves the results stated in Section 6.5 of Chapter 6, regarding the minimum possible MSE distortion.

MMSE reconstruction points and distortion

The reconstruction points that provide the minimum mean square error (MMSE) distortion for a quantization encoding, y, consisting of possible states, $n = 0, \ldots, N$, are

$$
\hat{x}_n = \mathrm{E}_x[x|n] = \int_x x P_{x|y}(x|n) dx = \frac{1}{P_y(n)} \int_x x P_{y|x}(n|x) f_x(x) dx. \tag{A3.40}
$$

A proof of this follows.

If $\hat{x}$, with possible values given by $\{\hat{x}_0, \ldots, \hat{x}_N\}$, is a decoding of a quantized version of a signal, x, the mean square error distortion is

$$
\mathrm{MSE} = \mathrm{E}[(x - \hat{x})^2]
$$

$$
= \int_{x=-\infty}^{\infty} \sum_{n=0}^{N} (x - \hat{x}_n)^2 f_{xy}(x, n) dx
$$

$$
= \mathrm{E}[x^2] + \sum_{n=0}^{N} (\hat{x}_n^2 - 2x\hat{x}_n) P_y(n) P_{x|y}(x|n) dx
$$

$$
= \mathrm{E}[x^2]
$$

$$
+ \sum_{n=0}^{N} P_y(n) \left(\hat{x}_n^2 \int_{x=-\infty}^{\infty} P_{x|y}(x|n) dx - 2\hat{x}_n \int_{x=-\infty}^{\infty} x P_{x|y}(x|n) dx \right)
$$

$$
= \mathrm{E}[x^2] + \sum_{n=0}^{N} P_y(n) \left(\hat{x}_n^2 - 2\hat{x}_n \mathrm{E}[x|n] \right). \tag{A3.41}
$$

Notice that since we wish to find the set of $\hat{x}_n$ that minimizes the MSE distortion, and since $P_y(n)$ is always positive, we can simply differentiate the term inside the summation in Eq. (A3.41) with respect to $\hat{x}_n$ and set to zero. Also, the second derivative with respect to $\hat{x}_n$ is equal to 2, which is always greater than zero. This gives $\hat{x}_n = E[x|n]$ as a minimum and we have completed the proof.

Note also that substituting for $\hat{x}_n = E[x|n]$ in Eq. (A3.41) gives the minimum MSE distortion as

$$\text{MMSE} = E[x^2] - E[\hat{x}^2].$$ (A3.42)

MMSE decoded output is uncorrelated with the error

We begin by showing that $E[x\hat{x}] = E[\hat{x}^2]$

$$E[x\hat{x}] = \sum_n \int_x x\hat{x} f_{xy}(x, n)dx$$

$$= \sum_n \int_x x\hat{x} P_y(n) P_{x|y}(x|n)dx$$

$$= \sum_n P_y(n)\hat{x} \int_x x P_{x|y}(x|n)dx$$

$$= \sum_n P_y(n)\hat{x}\hat{x}$$

$$= E[\hat{x}^2].$$ (A3.43)

Thus, if the optimal decoding, $\hat{x}$, is used, the mean square error is

$$E[(\hat{x} - x)^2] = E_n[\hat{x}^2] - 2E[x\hat{x}] + E_x[x^2]$$
$$= E_x[x^2] - E_n[\hat{x}^2],$$ (A3.44)

just as in Eq. (A3.42).

We also have

$$E[\hat{x} - x] = E[\hat{x}] - 0$$

$$= \sum_n P_y(n)E[x|n]$$

$$= \sum_n P_y(n) \int_x x P_{x|y}(x|n)dx$$

$$= \int_x x \sum_n P_y(n) P_{x|y}(x|n)dx$$

$$= \int_x x f_x(x) dx$$

$$= E[x]$$

$$= 0. \tag{A3.45}$$

Thus, the mean of the decoded output is zero, the mean error is zero and therefore the MMSE distortion is also the minimum error variance.

Furthermore, with the decoding, $\hat{x}$, the encoded output, y, is uncorrelated with the error, $\epsilon = x - \hat{x}$. The proof of this is

$$E[\epsilon y] = E[xy] - E[\hat{x}y]$$

$$= \sum_n \int_x x n f_{xy}(x, n) dx - E[\hat{x}y]$$

$$= \sum_n \int_x x n P_y(n) P_{x|y}(x|n) dx - E[\hat{x}y]$$

$$= \sum_n n P_y(n) \int_x x P_{x|y}(x|n) dx - E[\hat{x}y]$$

$$= \sum_n n P_y(n) E[x|n] - E[\hat{x}y]$$

$$= \sum_n n \hat{x}_n P_y(n) - E[\hat{x}y]$$

$$= E[\hat{x}y] - E[\hat{x}y]$$

$$= 0. \tag{A3.46}$$

Relationship of MMSE to backwards conditional variance

We also consider the mean square value of x given output state n. This gives an idea of how variable x is, given that output state. This is

$$E_x[x^2|n] = \int_x x^2 P_{x|y}(x|n) dx = \frac{1}{P_y(n)} \int_x x^2 P_{y|x}(n|x) f_x(x) dx. \tag{A3.47}$$

Consider the variance of x given output state n. We call this the backwards conditional variance (BCV), which we label as

$$\mathrm{BCV}(n) = \mathrm{var}_x[x|n] = E_x[x^2|n] - E_x[x|n]^2 = E_x[x^2|n] - \hat{x}_n^2. \tag{A3.48}$$

Note that the BCV is a function of n. If we take the expected value of the BCV over all N, we get

$$
\begin{aligned}
E_n[\mathrm{BCV}(n)] &= E_n\left[E_x[x^2|n] - \hat{x}_n^2\right] \\
&= \sum_n P_y(n) \int_x x^2 P_{x|y}(x|n)dx - E[\hat{x}_n^2] \\
&= \int_x x^2 f_x(x) \left(\sum_n P_{y|x}(n|x)\right) dx - E[\hat{x}_n^2] \\
&= \int_x x^2 f_x(x) 1 dx - E[\hat{x}_n^2] \\
&= E[x^2] - E[\hat{x}_n^2].
\end{aligned}
\tag{A3.49}
$$

The rhs of Eq. (A3.49) is precisely the MMSE that is obtained using the estimator $\hat{x} = E[x|n]$. So we make note of this by writing explicitly

$$
\mathrm{MMSE} = E_n[\mathrm{BCV}(n)] = E_x[x^2] - E_n[\hat{x}_n^2].
\tag{A3.50}
$$

A3.6 Fisher information

This section provides two alternative derivations for the Fisher information in the SSR model, an expression which is used in Section 5.4 of Chapter 5 and Section 6.7 of Chapter 6.

First derivation

Consider the individual output signal of each threshold device in the SSR model, $y_i \in \{0, 1\}$, to be an estimator for the input, x. There are two output states, zero and one, with conditional probability functions $P_{y_i|x}(0|x) = 1 - P_{1|x}$ and $P_{y_i|x}(1|x) = P_{1|x}$ respectively. Hence, from Eq. (6.104), the score for each state is

$$
V(0) = \frac{1}{(1 - P_{1|x})} \frac{d(1 - P_{1|x})}{dx},
\tag{A3.51}
$$

and

$$
V(1) = \frac{1}{P_{1|x}} \frac{dP_{1|x}}{dx}.
\tag{A3.52}
$$

Therefore, upon substituting into Eq. (6.114) the Fisher information for each comparator is

$$
J_i(x) = (1 - P_{1|x}) \frac{1}{(1 - P_{1|x})^2} \left(\frac{d(1 - P_{1|x})}{dx}\right)^2 + P_{1|x} \frac{1}{P_{1|x}^2} \left(\frac{dP_{1|x}}{dx}\right)^2,
\tag{A3.53}
$$

which simplifies to

$$J_i(x) = \left(\frac{d P_{1|x}}{dx}\right)^2 \frac{1}{P_{1|x}(1 - P_{1|x})}. \tag{A3.54}$$

The Fisher information for N *iid* samples is N times the individual information (Cover and Thomas 1991). Therefore since in the SSR array of N threshold devices, the set of N random variables y_i are all conditionally *iid* given x, the overall Fisher information is N times the formula given in Eq. (A3.54). Thus

$$J(x) = \left(\frac{d P_{1|x}}{dx}\right)^2 \frac{N}{P_{1|x}(1 - P_{1|x})}. \tag{A3.55}$$

This is in agreement with the formula derived for the Fisher information in Hoch *et al.* (2003a) and Hoch *et al.* (2003b).

Second derivation

We now give a derivation of the Fisher information in the SSR model that does not require using the fact that the overall Fisher information is the sum of the Fisher information in each individual threshold device. Recall that the transition probabilities, $P_{y|x}(n|x)$, are given by the binomial formula in terms of $P_{1|x}$ as

$$P_{y|x}(n|x) = \binom{N}{n}(P_{1|x})^n (1 - P_{1|x})^{N-n}, \quad n = 0, \ldots, N. \tag{A3.56}$$

Differentiation of $P_{y|x}(n|x)$ with respect to x gives

$$\frac{d P_{y|x}(n|x)}{dx} = \frac{(N P_{1|x} - n)\frac{d P_{1|x}}{dx}}{P_{1|x}(1 - P_{1|x})} P_{y|x}(n|x). \tag{A3.57}$$

Substituting Eq. (A3.57) into Eq. (6.116) gives

$$J(x) = \frac{\frac{d P_{1|x}}{dx}^2}{P_{1|x}^2(1 - P_{1|x})^2} \sum_{n=0}^{N} P_{y|x}(n|x)(N P_{1|x} - n)^2. \tag{A3.58}$$

Noting that $\sum_{n=0}^{N} n P_{y|x}(n|x) = N P_{1|x}$ and $\sum_{n=0}^{N} n^2 P_{y|x}(n|x) = N P_{1|x}(1 - P_{1|x}) + N^2 P_{1|x}^2$, Eq. (A3.58) simplifies to

$$J(x) = \left(\frac{d P_{1|x}}{dx}\right)^2 \frac{N}{P_{1|x}(1 - P_{1|x})}. \tag{A3.59}$$

This result is in agreement with Eq. (A3.55).

A3.7 Proof of the information and Cramer–Rao bounds

This section gives proofs of the information bound, and the Cramer–Rao bound. These bounds are used in Section 6.7 of Chapter 6.

The Cramer–Rao bound can be easily proved using the Cauchy–Schwarz inequality, which states that

$$(\text{cov}[U, y|x])^2 \le \text{var}[U|x]\text{var}[y|x]. \tag{A3.60}$$

Here, we set U to be the *score function* – the gradient of the log-likelihood function – which is

$$U(x, y) = \frac{d}{dx} \ln P_{y|x}(n|x). \tag{A3.61}$$

To prove the Cramer–Rao bound, note that the covariance of U and y can be simplified using the fact that $\text{E}[U|x] = 0$ as

$$
\begin{aligned}
\text{cov}[U, y|x] &= \text{E}\left[(U - \text{E}[U|x])(y - \text{E}[y|x])|x\right] \\
&= \text{E}[U(y - \text{E}[y|x]|x) \\
&= \text{E}[Uy|x] - \text{E}[y|x]\text{E}[U|x] \\
&= \text{E}[Uy|x] \\
&= \text{E}\left[y\frac{d}{dx}\ln P_{y|x}(n|x)|x\right] \\
&= \text{E}\left[y\frac{1}{P_{y|x}(n|x)}\frac{d}{dx}P_{y|x}(n|x)|x\right] \\
&= \sum_{n=0}^{N} n\frac{d}{dx}P_{y|x}(n|x) \\
&= \frac{d}{dx}\sum_{n=0}^{N} nP_{y|x}(n|x) \\
&= \frac{d}{dx}\text{E}[y|x].
\end{aligned}
\tag{A3.62}
$$

Noting that $J(x) = \text{var}[U|x]$

$$\left(\frac{d}{dx}\text{E}[y|x]\right)^2 \le J(x)\text{var}[y|x]. \tag{A3.63}$$

Therefore

$$\text{var}[y|x] \ge \frac{\left(\frac{d}{dx}\text{E}[y|x]\right)^2}{J(x)}. \tag{A3.64}$$

This proves the information bound for a biased estimator. Replacing y with a decoding $\hat{y}$ in the above derivation gives

$$\mathrm{var}[\hat{y}|x] \geq \frac{\left(\frac{d}{dx}E[\hat{y}|x]\right)^2}{J(x)}. \tag{A3.65}$$

For an unbiased estimator,

$$\frac{d}{dx}E[y|x] = \frac{d}{dx}x = 1$$

and

$$\mathrm{var}[y|x] \geq \frac{1}{J(x)}, \tag{A3.66}$$

which proves the Cramer–Rao bound for an unbiased estimator.

References

Abarbanel, H. D. I. and Rabinovich, M. I. (2001). Neurodynamics: nonlinear dynamics and neurobiology, *Current Opinion in Neurobiology*, **11**, 423–430.

Abbott, D. (2001). Overview: unsolved problems of noise and fluctuations, *Chaos*, **11**, 526–538.

Abramowitz, M. and Stegun, I. A. (1972). *Handbook of Mathematical Functions*, Washington DC: National Bureau of Standards.

Akyildiz, I. F., Su, W., Sankarasubramaniam, Y., and Cayirci, E. (2002). A survey on sensor networks, *IEEE Communications Magazine*, **40**, 102–114.

Allingham, D., Stocks, N. G., and Morse, R. P. (2003). The use of suprathreshold stochastic resonance in cochlear implant coding, in S. M. Bezrukov, H. Frauenfelder and F. Moss (eds.), *Proc. SPIE Fluctuations and Noise in Biological and Biomedical Systems*, Vol. 5110, 92–101.

Allingham, D., Stocks, N. G., Morse, R. P., and Meyer, G. F. (2004). Noise enhanced information transmission in a model of multichannel cochlear implantation, in D. Abbott, S. M. Bezrukov, A. Der, and A. Sánchez (eds.), *Proc. SPIE Fluctuations and Noise in Biological, Biophysical, and Biomedical Systems II*, Vol. 5467, 139–148.

Allison, A. and Abbott, D. (2000). Some benefits of random variables in switched control systems, *Microelectronics Journal*, **31**, 515–522.

Amblard, P. O. and Zozor, S. (1999). Cyclostationarity and stochastic resonance in threshold devices, *Physical Review E*, **59**, 5009–5020.

Amblard, P. O., Zozor, S., and Michel, O. J. J. (2006). Networks of the pooling type and optimal quantization, in *Proc. 2006 IEEE International Conference on Acoustics, Speech, and Signal Processing*, Vol. 3, 716–719.

Anderson, V. C. (1960). Digital array phasing, *The Journal of the Acoustical Society of America*, **32**, 867–870.

Anderson, V. C. (1980). Nonstationary and nonuniform oceanic background in a high-gain acoustic array, *The Journal of the Acoustical Society of America*, **67**, 1170–1179.

Andò, B. (2002). Stochastic resonance and dithering: a matter of classification!, *IEEE Instrumentation and Measurement Magazine*, **5**, 60–63.

Andò, B. and Graziani, S. (2000). *Stochastic Resonance: Theory and Applications*, Kluwer Academic Publishers.

Andò, B. and Graziani, S. (2001). Adding noise to improve measurement, *IEEE Instrumentation and Measurement Magazine*, **4**, 24–31.

Andrews, G. E. (1976). *The Theory of Partitions*, London: Addison-Wesley Publishing Company.

Anishchenko, V. S., Astakov, V. V., Neiman, A. B., Vadivasova, T. E., and Schimansky-Geier, L. (2002). *Nonlinear Dynamics of Chaotic and Stochastic Systems*, New York: Springer.

Anishchenko, V. S., Neiman, A. B., and Safanova, M. A. (1993). Stochastic resonance in chaotic systems, *Journal of Statistical Physics*, **70**, 183–196.

Anishchenko, V. S., Neiman, A. B., Moss, F., and Schimansky-Geier, L. (1999). Stochastic resonance: noise enhanced order, *Uspekhi Fizicheskikh Nauk*, **169**, 7–38.

Anishchenko, V. S., Safonova, M. A., and Chua, L. O. (1994). Stochastic resonance in Chua's circuit driven by amplitude or frequency-modulated signals, *International Journal of Bifurcation and Chaos*, **4**, 441–446.

Apostolico, F., Gammaitoni, L., Marchesoni, F., and Santucci, S. (1997). Resonant trapping: a failure mechanism in switch transitions, *Physical Review E*, **55**, 36–39.

Arimoto, S. (1972). An algorithm for computing the capacity of arbitrary discrete memoryless channels, *IEEE Transactions on Information Theory*, **18**, 14–20.

Astumian, R. D., Adair, R. K., and Weaver, J. C. (1997). Stochastic resonance at the single-cell level, *Nature*, **388**, 632–633.

Astumian, R. D. and Moss, F. (1998). Overview: The constructive role of noise in fluctuation driven transport and stochastic resonance, *Chaos*, **8**, 533–538.

Ayanoğlu, E. (1990). On optimal quantization of noisy sources, *IEEE Transactions on Information Theory*, **36**, 1450–1452.

Badzey, R. L. and Mohanty, P. (2005). Coherent signal amplification in bistable nanomechanical oscillators by stochastic resonance, *Nature*, **437**, 995–998.

Balasubramanian, V., Kimber, D., and Berry II, M. J. (2001). Metabolically efficient information processing, *Neural Computation*, **13**, 799–815.

Barron, A. R. and Cover, T. M. (1991). Minimum complexity density estimation, *IEEE Transactions on Information Theory*, **37**, 1034–1054.

Bartussek, R., Hänggi, P., and Jung, P. (1994). Stochastic resonance in optical bistable systems, *Physical Review E*, **49**, 3930–3939.

Basak, G. K. (2001). Stabilization of dynamical systems by adding a colored noise, *IEEE Transactions on Automatic Control*, **46**, 1107–1111.

Beasley, D., Bull, D. R., and Martin, R. R. (1993). An overview of genetic algorithms: Part I, fundamentals, *University Computing*, **15**, 58–69.

Behnam, S. E. and Zeng, F. G. (2003). Noise improves suprathreshold discrimnation in cochlear-implant listeners, *Hearing Research*, **186**, 91–93.

Bench, J., Kowal, A., and Bamford, J. (1979). The BKB (Bamford–Kowal–Bench) sentence lists for partially hearing children, *British Journal of Audiology*, **13**, 108–112.

Bennett, M., Wiesenfeld, K., and Jaramillo, F. (2004). Stochastic resonance in hair cell mechanoelectrical transduction, *Fluctuation and Noise Letters*, **4**, L1–L10.

Benzi, R., Parisi, G., Sutera, A., and Vulpiani, A. (1982). Stochastic resonance in climatic change, *Tellus*, **34**, 10–16.

Benzi, R., Parisi, G., Sutera, A., and Vulpiani, A. (1983). A theory of stochastic resonance in climatic change, *SIAM Journal on Applied Mathematics*, **43**, 565–578.

Benzi, R., Sutera, A., and Vulpiani, A. (1981). The mechanism of stochastic resonance, *Journal of Physics A: Mathematical and General*, **14**, L453–L457.

Benzi, R., Sutera, A., and Vulpiani, A. (1985). Stochastic resonance in the Landau-Ginzburg equation, *Journal of Physics A: Mathematical and General*, **18**, 2239–2245.

Berger, A. (ed.) (1980). *Climatic Variations and Variability: Facts and Theories*, NATO Advanced Study Institutes Series, D. Riedel Publishing Company.

Berger, T. (1971). *Rate Distortion Theory: A Mathematical Basis for Data Compression*, New Jersey: Prentice-Hall Inc.

Berger, T. and Gibson, J. D. (1998). Lossy source coding, *IEEE Transactions on Information Theory*, **44**, 2693–2723.

Berger, T., Zhang, Z., and Viswanathan, H. (1996). The CEO problem, *IEEE Transactions on Information Theory*, **42**, 887–902.

Berndt, H. (1968). Correlation function estimation by a polarity method using stochastic reference signals, *IEEE Transactions on Information Theory*, **14**, 796–801.

Berndt, H. and Jentschel, H. J. (2001). Differentially randomized quantization in sigma-delta analog-to-digital converters, *Proc. 8th IEEE International Conference on Electronics, Circuits and Systems*, Vol. 2, 1057–1060.

Berndt, H. and Jentschel, H. J. (2002). Stochastic quantization transfer functions for high resolution signal estimation, *Proc. 14th International Conference on Digital Signal Processing*, Vol. 2, 881–884.

Bershad, N. J. and Feintuch, P. L. (1974). Sonar array detection of Gaussian signals in Gaussian noise of unknown power, *IEEE Transactions on Aerospace and Electronic Systems*, **10**, 94–99.

Bethge, M. (2003). *Codes and goals of neuronal representation*, PhD thesis, Universität Bremen.

Bethge, M., Rotermund, D., and Pawelzik, K. (2002). Optimal short-term population coding: when Fisher information fails, *Neural Computation*, **14**, 2317–2351.

Bezrukov, S. M. (1998). Stochastic resonance as an inherent property of rate-modulated random series of events, *Physics Letters A*, **248**, 29–36.

Bezrukov, S. M. and Voydanoy, I. (1995). Noise-induced enhancement of signal transduction across voltage-dependent ion channels, *Nature*, **378**, 362–364.

Bezrukov, S. M. and Voydanoy, I. (1997a). Signal transduction across alamethicin ion channels in the presence of noise, *Biophysical Journal*, **73**, 2456–2464.

Bezrukov, S. M. and Voydanoy, I. (1997b). Stochastic resonance in non-dynamical systems without response thresholds, *Nature*, **385**, 319–321.

Bialek, W., DeWeese, M., Rieke, F., and Warland, D. (1993). Bits and brains: information flow in the nervous system, *Physica A*, **200**, 581–593.

Blahut, R. E. (1972). Computation of channel capacity and rate-distortion functions, *IEEE Transactions on Information Theory*, **18**, 460–473.

Blarer, A. and Doebeli, M. (1999). Resonance effects and outbreaks in ecological time series, *Ecology Letters*, **2**, 167–177.

Blum, R. S. (1995). Quantization in multisensor random signal detection, *IEEE Transactions on Information Theory*, **41**, 204–215.

Borst, A. and Theunissen, F. E. (1999). Information theory and neural coding, *Nature Neuroscience*, **2**, 947–957.

Bowen, G. and Mancini, S. (2004). Noise enhancing the classical information capacity of a quantum channel, *Physics Letters A*, **321**, 1–5.

Brewster, J. F., Graham, M. R., and Mutch, W. A. C. (2005). Convexity, Jensen's inequality and benefits of noisy mechanical ventilation, *Journal of the Royal Society Interface*, **2**, 393–396.

Broomhead, D. S., Luchinskaya, E. A., McClintock, P. V. E., and Mullin, T. (eds.) (2000). *Stochastic and Chaotic Dynamics in the Lakes (STOCHAOS), Ambleside, UK*, Vol. 502, American Institute of Physics Conference Proceedings.

Browning, E. B. (1850). *Sonnets from the Portuguese*.

Bruce, I. C., White, M. W., Irlicht, L. S., O'Leary, S. J., Dynes, S., Javel, E., and Clark, G. M. (1999). A stochastic model of the electrically stimulated auditory nerve: Single-pulse response, *IEEE Transactions on Biomedical Engineering*, **46**, 617–629.

Brunel, N. and Nadal, J. (1998). Mutual information, Fisher information and population coding, *Neural Computation*, **10**, 1731–1757.

Bucklew, J. A. (1981). Companding and random quantization in several dimensions, *IEEE Transactions on Information Theory*, **IT-27**, 207–211.

Bulsara, A., Chillemi, S., Kiss, L. B., McClintock, P. V. E., Mannella, R., Marchesoni, F., Nicolis, K., and Wiesenfeld, K. (eds.) (1994). Proc. International Workshop on Fluctuations in Physics and Biology: Stochastic Resonance, Signal Processing, and Related Phenomena, Elba, Italy, Published in *Il Nuovo Cimento*, **17**, 661–981 (1995).

Bulsara, A., Hänggi, P., Marchesoni, F., Moss, F., and Shlesinger, M. (1993). Proceedings of the NATO advanced research workshop – stochastic resonance in physics and biology – preface, *Journal of Statistical Physics*, **70**, 1–2.

Bulsara, A., Jacobs, E. W., and Zhou, T. (1991). Stochastic resonance in a single neuron model – theory and analog simulation, *Journal of Theoretical Biology*, **152**, 531–555.

Bulsara, A. R. (2005). No-nuisance noise, *Nature*, **437**, 962–963.

Bulsara, A. R. and Gammaitoni, L. (1996). Tuning in to noise, *Physics Today*, **49**, 39–45.

Bulsara, A. R. and Inchiosa, M. E. (1996). Noise-controlled resonance behavior in nonlinear dynamical systems with broken symmetry, *Physical Review Letters*, **77**, 2162–2165.

Bulsara, A. R. and Moss, F. E. (1991). Single neuron dynamics: noise-enhanced signal processing, *IEEE International Joint Conference on Neural Networks*, Vol. 1, 420–425.

Bulsara, A. R. and Schmera, G. (1993). Stochastic resonance in globally coupled nonlinear oscillators, *Physical Review E*, **47**, 3734–3737.

Bulsara, A. R. and Zador, A. (1996). Threshold detection of wideband signals: a noise induced maximum in the mutual information, *Physical Review E*, **54**, R2185–R2188.

Carbone, P. and Petri, D. (2000). Performance of stochastic and deterministic dithered quantizers, *IEEE Transactions on Instrumentation and Measurement*, **49**, 337–340.

Casado-Pascual, J., Denk, C., Gomez-Ordonez, J., Marillo, M., and Hänggi, P. (2003). Gain in stochastic resonance: precise numerics versus linear response theory beyond the two-mode approximation, *Physical Review E*, **67**, Art. No. 036109.

Castanie, F. (1979). Signal processing by random reference quantizing, *Signal Processing*, **1**, 27–43.

Castanie, F. (1984). Linear mean transfer random quantization, *Signal Processing*, **7**, 99–117.

Castanie, F., Hoffman, J. C., and Lacaze, B. (1974). On the performance of a random reference correlator, *IEEE Transactions on Information Theory*, **IT-20**, 266–269.

Chang, C. and Davisson, L. D. (1990). Two iterative algorithms for finding minimax solutions, *IEEE Transactions on Information Theory*, **36**, 126–140.

Chang, C. I. and Davisson, L. D. (1988). On calculating the capacity of an infinite-input finite (infinite)–output channel, *IEEE Transactions on Information Theory*, **34**, 1004–1010.

Chapeau-Blondeau, F. (1996). Stochastic resonance in the Heaviside nonlinearity with white noise and arbitrary periodic signal, *Physical Review E*, **53**, 5469–5472.

Chapeau-Blondeau, F. (1997a). Input–output gains for signal in noise in stochastic resonance, *Physics Letters A*, **232**, 41–48.

Chapeau-Blondeau, F. (1997b). Noise-enhanced capacity via stochastic resonance in an asymmetric binary channel, *Physical Review E*, **55**, 2016–2019.

Chapeau-Blondeau, F. (1999). Periodic and aperiodic stochastic resonance with output signal-to-noise ratio exceeding that at the input, *International Journal of Bifurcation and Chaos*, **9**, 267–272.

Chapeau-Blondeau, F. and Godivier, X. (1996). Stochastic resonance in nonlinear transmission of spike signals: an exact model and an application to the neuron, *International Journal of Bifurcation and Chaos*, **6**, 2069–2076.

Chapeau-Blondeau, F. and Godivier, X. (1997). Theory of stochastic resonance in signal transmission by static nonlinear systems, *Physical Review E*, **55**, 1478–1495.

Chapeau-Blondeau, F. and Rojas-Varela, J. (2001). Estimation and Fisher information enhancement via noise addition with nonlinear sensors, *Proceedings 2nd International Symposium on Physics in Signal and Image Processing*, Marseille, France, 47–50.

Chapeau-Blondeau, F. and Rousseau, D. (2004). Noise-enhanced performance for an optimal Bayesian estimator, *IEEE Transactions on Signal Processing*, **52**, 1327–1334.

Chatterjee, M. and Oba, S. I. (2005). Noise improves modulation detection by cochlear implant listeners at moderate carrier levels, *Journal of the Acoustical Society of America*, **118**, 993–1002.

Chatterjee, M. and Robert, M. E. (2001). Noise enhances modulation sensitivity in cochlear implant listeners: stochastic resonance in a prosthetic sensory system? *Journal of the Association for Research in Otolaryngology*, **2**, 159–171.

Chen, H., Varshney, P. K., Kay, S. M., and Michels, J. H. (2007). Theory of the stochastic resonance effect in signal detection: Part I: fixed detectors, *IEEE Transactions on Signal Processing*, **55**, 3172–3184.

Chialvo, D. R. and Apkarian, A. V. (1993). Modulated noisy biological dynamics – three examples, *Journal of Statistical Physics*, **70**, 375–391.

Chialvo, D. R., Longtin, A., and Muller-Gerking, J. (1997). Stochastic resonance in models of neuronal ensembles, *Physical Review E*, **55**, 1798–1808.

Choi, M. H., Fox, R. F., and Jung, P. (1998). Quantifying stochastic resonance in bistable systems: response vs residence-time distribution functions, *Physical Review E*, **57**, 6335–6344.

Chong, C. and Kumar, S. P. (2003). Sensor networks: evolution, opportunities and challenges, *Proceedings of the IEEE*, **91**, 1247–1256.

Clark, G. (2000). *Sounds from Silence*, Allen Unwin.

Clark, G. M. (1986). The University of Melbourne/Cochlear Corporation (Nucleus) program, *Otolaryngologic Clinics of North America*, **19**, 329–354.

Clarke, B. S. and Barron, A. R. (1990). Information-theoretic asymptotics of Bayes methods, *IEEE Transactions on Information Theory*, **36**, 453–471.

Cogdell, J. R. (1996). *Foundations of Electrical Engineering*, second edn, Prentice Hall International.

Collins, J. J. (1999). Fishing for function in noise, *Nature*, **402**, 241–242.

Collins, J. J., Chow, C. C., and Imhoff, T. T. (1995a). Aperiodic stochastic resonance in excitable systems, *Physical Review E*, **52**, R3321–R3324.

Collins, J. J., Chow, C. C., and Imhoff, T. T. (1995b). Stochastic resonance without tuning, *Nature*, **376**, 236–238.

Collins, J. J., Chow, C. C., and Imhoff, T. T. (1995c). Tuning stochastic resonance, *Nature*, **378**, 341–342.

Collins, J. J., Chow, C. C., Capela, A. C., and Imhoff, T. T. (1996a). Aperiodic stochastic resonance, *Physical Review E*, **54**, 5575–5584.

Collins, J. J., Imhoff, T. T., and Grigg, P. (1996b). Noise-enhanced information transmission in rat SA1 cutaneous mechanoreceptors via aperiodic stochastic resonance, *Journal of Neurophysiology*, **76**, 642–645.

Collins, J. J., Priplata, A. A., Gravelle, D. C., Niemi, J., Harry, J., and Lipsitz, L. A. (2003). Noise-enhanced human sensorimotor control, *IEEE Engineering in Medicine and Biology Magazine*, **22**, 76–83.

Coullet, P. H., Elphick, C., and Tirapegui, E. (1985). Normal form of a Hopf bifurcation with noise, *Physics Letters*, **111A**, 277–282.

Courant, R. and Robbins, H. (1996). *What Is Mathematics?: An Elementary Approach to Ideas and Methods*, second edn, Oxford: Oxford University Press.

Cover, T. M. and Thomas, J. A. (1991). *Elements of Information Theory*, New York: John Wiley and Sons.

Cox, R. M., Matesich, J. S., and Moore, J. N. (1988). Distribution of short-term RMS levels in conversational speech, *Journal of the Acoustical Society of America*, **84**, 1100–1104.

Crawford, A. C. and Fettiplace, R. (1986). The mechanical properties of ciliary bundles of turtle cochlear hair-cells, *Journal of Physiology- London*, **364**, 359–379.

Damgaard, P. O. and Huffel, H. (1988). *Stochastic Quantization*, Amsterdam: World Scientific.

Damper, R. I. (1995). *Introduction to Discrete-Time Signals and Systems*, London: Chapman and Hall.

Das, A. (2006). Enhanced signal coding using stochastic resonance with application to the auditory system, Ph.D. thesis, School of Engineering, The University of Warwick.

Davisson, L. D. and Leon-Garcia, A. (1980). A source matching approach to finding minimax codes, *IEEE Transactions on Information Theory*, **IT-26**, 166–174.

Debnath, G., Zhou, T., and Moss, F. (1989). Remarks on stochastic resonance, *Physical Review A*, **39**, 4323–4326.

Debye, P. J. W. (1929). *Polar Molecules*, New York: Chemical Catalog Co.

Delbridge, A., Bernard, J. R. L., Blair, D., Butler, S., Peters, P., and Yallop, C. (1997). *The Macquarie Dictionary*, third edn, The Macquarie Library Pty Ltd.

Denk, W. and Webb, W. W. (1992). Forward and reverse transduction at the limit of sensitivity studied by correlating electrical and mechanical fluctuations in frog saccular hair-cells, *Hearing Research*, **60**, 89–102.

Denk, W., Webb, W. W., and Hudspeth, A. J. (1989). Mechanical properties of sensory hair bundles are reflected in their brownian motion measured with a laser differential interferometer, *Proceedings of the National Academy of Sciences of the USA*, **86**, 5371–5375.

DeWeese, M. (1996). Optimization principles for the neural code, *Network: Computation in Neural Systems*, **7**, 325–331.

DeWeese, M. and Bialek, W. (1995). Information flow in sensory neurons, *Il Nuovo Cimento*, **17**, 733–741.

Dimitrov, A. G. and Miller, J. P. (2001). Neural coding and decoding: communication channels and quantization, *Network: Computation in Neural Systems*, **12**, 441–472.

Dimitrov, A. G., Miller, J. P., Gedeon, T., Aldworth, Z., and Parker, A. E. (2003). Analysis of neural coding through quantization with an information-based distortion measure, *Network: Computation in Neural Systems*, **14**, 151–176.

Doering, C. R. (1995). Randomly rattled ratchets, *Il Nuovo Cimento*, **17**, 685–697.

Donaldson, G. S., Kreft, H. A., and Litvak, L. (2005). Place–pitch discrimination of single- versus dual-electrode stimuli by cochlear implant users, *Journal of the Acoustical Society of America*, **118**, 623–626.

Dorman, M. F. and Wilson, B. S. (2004). The design and function of cochlear implants, *American Scientist*, **92**, 436–445.

Douglass, J. K., Wilkens, L., Pantazelou, E., and Moss, F. (1993). Noise enhancement of information transfer in crayfish mechanoreceptors by stochastic resonance, *Nature*, **365**, 337–339.

Draper, S. C. and Wornell, G. W. (2004). Side information aware coding strategies for sensor networks, *IEEE Journal on Selected Areas in Communications*, **22**, 966–976.

Drozhdin, K. (2001). Stochastic resonance in ferroelectric TGS crystals, Ph.D. thesis, Mathematisch-Naturwissenschaftlich-Technischen Fakultät der Martin-Luther-Universität Halle-Wittenberg.

Duan, F. B., Rousseau, D., and Chapeau-Blondeau, F. (2004). Residual aperiodic stochastic resonance in a bistable dynamic system transmitting a suprathreshold binary signal, *Physical Review E*, **69**, Art. No. 011109.

Duan, F., Chapeau-Blondeau, F., and Abbott, D. (2006). Noise-enhanced SNR gain in parallel array of bistable oscillators, *Electronics Letters*, **42**, 1008–1009.

Dunay, R., Kollár, I., and Widrow, B. (1998). Dithering for floating-point number representation, 1st International On-Line Workshop on Dithering in Measurement, 1–31 March 9/1–9/12.

Dunlop, J. and Smith, D. G. (1984). *Telecommunications Engineering*, New York: Van Nostrand Reinhold.

Dykman, M. I. and McClintock, P. V. E. (1998). What can stochastic resonance do?, *Nature*, **391**, 344.

Dykman, M. I., Luchinsky, D. G., Mannella, R., McClintock, P. V. E., Short, H. E., Stein, N. D., and Stocks, N. G. (1994). Noise-induced linearisation, *Physics Letters A*, **193**, 61–66.

Dykman, M. I., Luchinsky, D. G., Mannella, R., McClintock, P. V. E., Stein, N. D., and Stocks, N. G. (1995). Stochastic resonance in perspective, *Il Nuovo Cimento*, **17**, 661–683.

Dykman, M. I., Luchinsky, D. G., McClintock, P. V. E., Stein, N. D., and Stocks, N. G. (1992). Stochastic resonance for periodically modulated noise intensity, *Physical Review A*, **46**, R1713–R1716.

Dykman, M. I., Mannella, R., McClintock, P. V. E., and Stocks, N. G. (1990a). Comment on stochastic resonance in bistable systems, *Physical Review Letters*, **65**, 2606.

Dykman, M. I., McClintock, P. V. E., Mannella, R., and Stocks, N. (1990b). Stochastic resonance in the linear and nonlinear responses of a bistable system to a periodic field, *Journal of Experimental and Theoretical Physics Letters*, **52**, 141–144.

Eckmann, J.-P. and Thomas, L. E. (1982). Remarks on stochastic resonances, *Journal of Physics A: Mathematical and General*, **15**, L261–L266.

Eichwald, C. and Walleczek, J. (1997). Aperiodic stochastic resonance with chaotic input signals in excitable systems, *Physical Review E*, **55**, R6315–R6318.

Eliasmith, C. and Anderson, C. H. (2003). *Neural Engineering: Computation, Representation, and Dynamics in Neurobiological Systems*, Cambridge, MA: MIT Press.

Evans, E. F. (1975). Cochlear nerve and cochlear nucleus, in W. D. Keidel and W. D. Neff (eds.), *Handbook of Sensory Physiology*, New York: Springer, 1–108.

Evans, E. F. (1978). Place and time coding of frequency in the peripheral auditory system: some physiological pros and cons, *Audiology*, **17**, 369–420.

Evans, M., Hastings, N., and Peacock, B. (2000). *Statistical Distributions*, third edn, New York: John Wiley & Sons.

Fakir, R. (1998a). Nonstationary stochastic resonance, *Physical Review E*, **57**, 6996–7001.

Fakir, R. (1998b). Nonstationary stochastic resonance in a single neuronlike system, *Physical Review E*, **58**, 5175–5178.

Fauve, S. and Heslot, F. (1983). Stochastic resonance in a bistable system, *Physics Letters A*, **97A**, 5–7.

Finley, C. C., Wilson, B. S., and White, M. W. (1990). Models of neural responsiveness to electrical stimulation, in J. J. Miller and F. A. Spelman (eds.), *Cochlear Implants: Models of the Electrically Stimulated Ear*, New York: Springer.

Fitelson, M. (1970). Asymptotic expressions for the mean and variance of the output of clipped DIMUS arrays in a field of uniform uncorrelated Gaussian noise plus signal, *The Journal of the Acoustical Society of America*, **48**, 27–31.

Fox, R. F. (1989). Stochastic resonance in a double well, *Physical Review A*, **39**, 4148–4153.

Freund, J. A., Schimansky-Geier, L., Beisner, B., Neiman, A., Russell, D. F., Yakusheva, T., and Moss, F. (2002). Behavioral stochastic resonance: how the noise from a daphnia swarm enhances individual prey capture by juvenile paddlefish, *Journal of Theoretical Biology*, **214**, 71–83.

Frisch, U., Froeschle, C., Scheidecker, J. P., and Sulem, P. L. (1973). Stochastic resonance in one-dimensional random media, *Physical Review A – General Physics*, **8**, 1416–1421.

Fry, T. (1928). *Probability and its Engineering Uses*, New York: D. Van Nostrand Co.

Gailey, P. C., Neiman, A., Collins, J. J., and Moss, F. (1997). Stochastic resonance in ensembles of nondynamical elements: The role of internal noise, *Physical Review Letters*, **79**, 4701–4704.

Galdi, V., Pierro, V., and Pinto, I. M. (1998). Evaluation of stochastic-resonance-based detectors of weak harmonic signals in additive white Gaussian noise, *Physical Review E*, **57**, 6470–6479.

Gammaitoni, L. (1995a). Stochastic resonance and the dithering effect in threshold physical systems, *Physical Review E*, **52**, 4691–4698.

Gammaitoni, L. (1995b). Stochastic resonance in multi-threshold systems, *Physics Letters A*, **208**, 315–322.

Gammaitoni, L., Hänggi, P., Jung, P., and Marchesoni, F. (1998). Stochastic resonance, *Reviews of Modern Physics*, **70**, 223–287.

Gammaitoni, L., Löcher, M., Bulsara, A., Hänggi, P., Neff, J., Wiesenfeld, K., Ditto, W., and Inchiosa, M. E. (1999). Controlling stochastic resonance, *Physical Review Letters*, **82**, 4574–4577.

Gammaitoni, L., Marchesoni, F., and Santucci, S. (1995a). Stochastic resonance as a bona fide resonance, *Physical Review Letters*, **74**, 1052–1055.

Gammaitoni, L., Marchesoni, F., Menichella-Saetta, E., and Santucci, S. (1989a). Stochastic resonance in bistable systems, *Physical Review Letters*, **62**, 349–352.

Gammaitoni, L., Marchesoni, F., Menichella-Saetta, E., and Santucci, S. (1990). Reply on comment on stochastic resonance in bistable systems, *Physical Review Letters*, **65**, 2607.

Gammaitoni, L., Marchesoni, F., Menichella-Saetta, E., and Santucci, S. (1995b). Stochastic resonance in the strong forcing limit, *Physical Review E*, **51**, R3799–R3802.

Gammaitoni, L., Martinelli, M., Pardi, L., and Santucci, S. (1991). Observation of stochastic resonance in bistable electron-paramagnetic-resonance systems, *Physical Review Letters*, **67**, 1799–1802.

Gammaitoni, L., Menichella-Saetta, E., Marchesoni, F., and Presilla, C. (1989b). Periodically time-modulated bistable systems: stochastic resonance, *Physical Review A*, **40**, 2114–2119.

Gang, H., Ditzinger, T., Ning, C. Z., and Haken, H. (1993). Stochastic resonance without external periodic force, *Physical Review Letters*, **71**, 807–810.

Gaudet, V. C. and Rapley, A. C. (2003). Iterative decoding using stochastic computation, *Electronics Letters*, **39**, 299–301.

Gebeshuber, I. C. (2000). The influence of stochastic behavior on the human threshold of hearing, *Chaos Solitons and Fractals*, **11**, 1855–1868.

Gershenfeld, N. (1999). *The Nature of Mathematical Modeling*, Cambridge: Cambridge University Press.

Gersho, A. and Gray, R. M. (1992). *Vector Quantization and Signal Compression*, Dordrecht: Kluwer Academic Publishers.

Gerstner, W. and Kistler, W. M. (2002). *Spiking Neuron Models*, Cambridge: Cambridge University Press.

Giacomelli, G., Marin, F., and Rabbiosi, I. (1999). Stochastic and bona fide resonance: an experimental investigation, *Physical Review Letters*, **82**, 675–678.

Gingl, Z. (2002). Special issue on stochastic resonance I: fundamental and special aspects – introduction, *Fluctuation and Noise Letters*, **2**, L125–L126.

Gingl, Z., Kiss, L. B., and Moss, F. (1995a). Non-dynamical stochastic resonance: theory and experiments with white and arbitrarily coloured noise, *Europhysics Letters*, **29**, 191–196.

Gingl, Z., Kiss, L. B., and Moss, F. (1995b). Non-dynamical stochastic resonance: theory and experiments with white and various coloured noises, *Il Nuovo Cimento*, **17**, 795–802.

Gingl, Z., Makra, P., and Vajtai, R. (2001). High signal-to-noise ratio gain by stochastic resonance in a double well, *Fluctuation and Noise Letters*, **1**, L181–L188.

Gingl, Z., Vajtai, R., and Kiss, L. B. (2000). Signal-to-noise ratio gain by stochastic resonance in a bistable system, *Chaos, Solitons and Fractals*, **11**, 1929–1932.

Godivier, X. and Chapeau-Blondeau, F. (1997). Noise-assisted signal transmission by a nonlinear electronic comparator: experiment and theory, *Signal Processing*, **56**, 293–303.

Godivier, X. and Chapeau-Blondeau, F. (1998). Stochastic resonance in the information capacity of a nonlinear dynamic system, *International Journal of Bifurcation and Chaos*, **8**, 581–589.

Godivier, X., Rojas-Varela, J., and Chapeau-Blondeau, F. (1997). Noise-assisted signal transmission via stochastic resonance in a diode nonlinearity, *Electronics Letters*, **33**, 1666–1668.

Goychuk, I. (2001). Information transfer with rate-modulated Poisson processes: a simple model for nonstationary stochastic resonance, *Physical Review E*, **64**, Art. No. 021909.

Goychuk, I. and Hänggi, P. (1999). Quantum stochastic resonance in parallel, *New Journal of Physics*, **1**, 14.1–14.14.

Gray, R. M. and Neuhoff, D. L. (1998). Quantization, *IEEE Transactions on Information Theory*, **44**, 2325–2383.

Gray, R. M. and Stockham, T. G. (1993). Dithered quantizers, *IEEE Transactions on Information Theory*, **39**, 805–812.

Greenwood, P. E., Miller, U. U., Ward, L. M., and Wefelmeyer, W. (2003). Statistical analysis of stochastic resonance in a thresholded detector, *Austrian Journal of Statistics*, **32**, 49–70.

Greenwood, P. E., Ward, L. M., and Wefelmeyer, W. (1999). Statistical analysis of stochastic resonance in a simple setting, *Physical Review E*, **60**, 4687–4695.

Greenwood, P. E., Ward, L. M., Russell, D. F., Neiman, A., and Moss, F. (2000). Stochastic resonance enhances the electrosensory information available to paddlefish for prey capture, *Physical Review Letters*, **84**, 4773–4776.

Grifoni, M. and Hänggi, P. (1996). Coherent and incoherent quantum stochastic resonance, *Physical Review Letters*, **76**, 1611–1614.

Grunwald, P. D., Myung, I. J., and Pitt, M. A. (2005). *Advances in Minimum Description Length: Theory and Applications (Neural Information Processing)*, Cambridge, MA: MIT Press.

György, A. and Linder, T. (2000). Optimal entropy-constrained scalar quantization of a uniform source, *IEEE Transactions on Information Theory*, **46**, 2702–2711.

Hänggi, P. (2002). Stochastic resonance in biology: how noise can enhance detection of weak signals and help improve biological information processing, *Chemphyschem*, **3**, 285–290.

Hänggi, P., Inchiosa, M. E., Fogliatti, D., and Bulsara, A. R. (2000). Nonlinear stochastic resonance: the saga of anomalous output-input gain, *Physical Review E*, **62**, 6155–6163.

Hänggi, P., Jung, P., Zerbe, C., and Moss, F. (1993). Can colored noise improve stochastic resonance, *Journal of Statistical Physics*, **70**, 25–47.

Harmer, G. P. (2001). Stochastic processing for enhancement of artificial insect vision, Ph.D. thesis, Department of Electrical and Electronic Engineering, Adelaide University, Australia.

Harmer, G. P. and Abbott, D. (1999). Losing strategies can win by Parrondo's paradox, *Nature*, **402**, 864.

Harmer, G. P. and Abbott, D. (2001). Motion detection and stochastic resonance in noisy environments, *Microelectronics Journal*, **32**, 959–967.

Harmer, G. P., Davis, B. R., and Abbott, D. (2002). A review of stochastic resonance: circuits and measurement, *IEEE Transactions on Instrumentation and Measurement*, **51**, 299–309.

Harrington, J. V. (1955). An analysis of the detection of repeated signals in noise by binary integration, *IRE Transactions on Information Theory*, **1**, 1–9.

Harry, J. D., Niemi, J. B., Priplata, A. A., and Collins, J. J. (2005). Balancing act (noise based sensory enhancement technology), *IEEE Spectrum*, **42**, 36–41.

Hartmann, W. M. (1996). Pitch, periodicity, and auditory organization, *Journal of the Acoustical Society of America*, **100**, 3491–3502.

Heneghan, C., Chow, C. C., Collins, J. J., Imhoff, T. T., Lowen, S. B., and Teich, M. C. (1996). Information measures quantifying aperiodic stochastic resonance, *Physical Review E*, **54**, R2228–R2231.

Henry, K. R. and Lewis, E. R. (2001). Cochlear nerve acoustic envelope response detection is improved by the addition of random-phased tonal stimuli, *Hearing Research*, **155**, 91–102.

Hibbs, A. D., Singsaas, A. L., Jacobs, E. W., Bulsara, A. R., Bekkedahl, J. J., and Moss, F. (1995). Stochastic resonance in a superconducting loop with a Josephson-junction, *Journal of Applied Physics*, **77**, 2582–2590.

Hinojosa, R. and Lindsay, J. R. (1980). Profound deafness: associated sensory and neural degeneration, *Archives of Otolaryngology – Head and Neck Surgery*, **106**, 193–209.

Hoch, T., Wenning, G. and Obermayer, K. (2003a). Adaptation using local information for maximizing the global cost, *Neurocomputing*, **52–54**, 541–546.

Hoch, T., Wenning, G., and Obermayer, K. (2003b). Optimal noise-aided signal transmission through populations of neurons, *Physical Review E*, **68**, Art. No. 011911.

Hogg, R. V., McKean, J. W., and Craig, A. T. (2005). *Introduction to Mathematical Statistics*, sixth edn, London: Pearson Prentice Hall.

Hohn, N. (2001). Stochastic resonance in a neuron model with applications to the auditory pathway, Ph.D. thesis, Department of Otaryngology, The University of Melbourne, Australia.

Hohn, N. and Burkitt, A. N. (2001). Shot noise in the leaky integrate-and-fire neuron, *Physical Review E*, **63**, Art. No. 031902.

Hong, R. S. and Rubinstein, J. T. (2003). High-rate conditioning pulse trains in cochlear implants: dynamic range measures with sinusoidal stimuli, *Journal of the Acoustical Society of America*, **114**, 3327–3342.

Hong, R. S. and Rubinstein, J. T. (2006). Conditioning pulse trains in cochlear implants: effects on loudness growth, *Otology and Neurotology*, **27**, 50–56.

Horsthemke, W. and Lefever, R. (1980). Voltage-noise-induced transitions in electrically excitable membranes, *Biophysical Journal*, **35**, 415–432.

Hu, G., Gong, D. C., Wen, X. D., Yang, C. Y., Qing, G. R., and Li, R. (1992). Stochastic resonance in a nonlinear-system driven by an aperiodic force, *Physical Review A*, **46**, 3250–3254.

Hu, G., Nicolis, G., and Nicolis, C. (1990). Periodically forced Fokker–Planck equation and stochastic resonance, *Physical Review A*, **42**, 2030–2041.

Hudspeth, A. J. (1989). How the ears works work, *Nature*, **341**, 397–404.

Iannelli, J. M., Yariv, A., Chen, T. R., and Zhuang, Y. H. (1994). Stochastic resonance in a semiconductor distributed feedback laser, *Applied Physics Letters*, **65**, 1983–1985.

Imkeller, P. and Pavlyukevich, I. (2001). Stochastic resonance in two-state Markow chains, *Archiv der Mathematik*, **77**, 107–115.

Inchiosa, M. E. and Bulsara, A. R. (1995). Nonlinear dynamic elements with noisy sinusoidal forcing: enhancing response via nonlinear coupling, *Physical Review E*, **52**, 327–339.

Inchiosa, M. E. and Bulsara, A. R. (1996). Signal detection statistics of stochastic resonators, *Physical Review E*, **53**, R2021–2024.

Inchiosa, M. E. and Bulsara, A. R. (1998). DC signal detection via dynamical asymmetry in a nonlinear device, *Physical Review E*, **58**, 115–127.

Inchiosa, M. E., Robinson, J. W. C., and Bulsara, A. R. (2000). Information-theoretic stochastic resonance in noise-floor limited systems: the case for adding noise, *Physical Review Letters*, **85**, 3369–3372.

Iyengar, S. S. and Brooks, R. R. (2004). Special issue introduction – the road map for distributed sensor networks in the context of computing and communication, *Journal of Parallel and Distributed Computing*, **64**, 785–787.

Jaramillo, F. and Wiesenfeld, K. (1998). Mechanoelectrical transduction assisted by Brownian motion: a role for noise in the auditory system, *Nature Neuroscience*, **1**, 384–388.

Jaramillo, F. and Wiesenfeld, K. (2000). Physiological noise level enhances mechanoelectrical transduction in hair cells, *Chaos Solitons and Fractals*, **11**, 1869–1874.

Jayant, N. S. and Noll, P. (1984). *Digital Coding of Waveforms: Principles and Applications to Speech and Video*, Prentice Hall.

Jaynes, E. T. (2003). *Probability Theory: The Logic of Science*, Cambridge: Cambridge University Press.

Johnson, D. H. (1980). The relationship between spike rate and synchrony in the responses of auditory nerve fibers to single tones, *Journal of the Acoustical Society of America*, **68**, 1115–1122.

Johnson, D. H. (2004). Neural population structures and consequences for neural coding, *Journal of Computational Neuroscience*, **16**, 69–80.

Jung, P. (1993). Periodically driven stochastic systems, *Physics Reports – Review Section of Physics Letters*, **234**, 175–295.

Jung, P. (1994). Threshold devices: fractal noise and neural talk, *Physical Review E*, **50**, 2513–2522.

Jung, P. (1995). Stochastic resonance and optimal design of threshold detectors, *Physics Letters A*, **207**, 93–104.

Jung, P. and Hänggi, P. (1991). Amplification of small signals via stochastic resonance, *Physical Review A*, **44**, 8032–8042.

Jung, P. and Mayer-Kress, G. (1995). Stochastic resonance in threshold devices, *Il Nuovo Cimento*, **17**, 827–834.

Jung, P. and Mayerkress, G. (1995). Spatiotemporal stochastic resonance in excitable media, *Physical Review Letters*, **74**, 2130–2133.

Jung, P., Behn, U., Pantazelou, E., and Moss, F. (1992). Collective response in globally coupled bistable systems, *Physical Review A*, **46**, R1709–R1712.

Kadtke, J. B. and Bulsara, A. (eds.) (1997). *Applied Nonlinear Dynamics and Stochastic Systems Near the Millenium, San Diego, USA*, Vol. 411, American Institute of Physics Conference Proceedings.

Kalmykov, Y. P., Coffey, W. T., and Titov, S. V. (2004). Bimodal approximation for anomalous diffusion in a potential, *Physical Review E*, **69**, Art. No. 021105.

Kandel, E. R., Schwartz, J. H., and Jessell, T. M. (1991). *Principles of Neural Science*, third edn, New York: Elsevier.

Kanefsky, M. (1966). Detection of weak signals with polarity coincidence arrays, *IEEE Transactions on Information Theory*, **IT-12**, 260–268.

Kang, K. and Sompolinsky, H. (2001). Mutual information of population codes and distance measures in probability space, *Physical Review Letters*, **86**, 4958–4961.

Kay, S. (2000). Can detectability be improved by adding noise? *IEEE Signal Processing Letters*, **7**, 8–10.

Khovanov, I. A. and McClintock, P. V. E. (2003). Comment on 'signal-to-noise ratio gain in neuronal systems', *Physical Review E*, **67**, Art. No. 043901.

Kiang, N. Y. S. (1966). *Discharge Patterns of Single Fibers in the Cat's Auditory Nerve*, MIT Press.

Kiang, N. Y. S., Moxon, E. C., and Levine, R. A. (1970). Auditory nerve activity in cats with normal and abnormal cochleas, in G. E. W. Wolstenholme and J. Knight (eds.), *Sensorineural Hearing Loss*, Churchill, 241–273.

Kikkert, C. J. (1995). Improving ADC performance by adding dither, *Proc. 13th Australian Microelectronics Conference, Microelectronics: Technology Today for the Future (MICRO '95)*, IREE Soc, Milsons Point, NSW, Australia, 97–102.

Kirkpatrick, S., Gelatt, C. D., and Vecchi, M. P. (1983). Optimisation by simulated annealing, *Science*, **220**, 671–680.

Kish, L. B., Harmer, G. P., and Abbott, D. (2001). Information transfer rate of neurons: stochastic resonance of Shannon's information channel capacity, *Fluctuation and Noise Letters*, **1**, L13–L19.

Kiss, L. B. (1996). Possible breakthrough: significant improvement of signal to noise ratio by stochastic resonance, in R. Katz (ed.), *Chaotic, Fractal and Nonlinear Signal Processing*, Vol. 375, American Institute of Physics, 382–396.

Kiss, L. B., Gingl, Z., Marton, Z., Kertesz, J., Moss, F. and Schmera, G. (1993). 1/F noise in systems showing stochastic resonance, *Journal of Statistical Physics*, **70**, 451–462.

Kosko, B. (2006). *Noise*, Viking.

Kosko, B. and Mitaim, S. (2001). Robust stochastic resonance: signal detection and adaptation in impulsive noise, *Physical Review E*, **64**, Art. No. 051110.

Kosko, B. and Mitaim, S. (2003). Stochastic resonance in noisy threshold neurons, *Neural Networks*, **16**, 755–761.

Kosko, B. and Mitaim, S. (2004). Robust stochastic resonance for simple threshold neurons, *Physical Review E*, **70**, Art. No. 031911.

Kreyszig, E. (1988). *Advanced Engineering Mathematics*, 6th edn, New York: John Wiley & Sons.

Lanzara, E., Mantegna, R. N., Spagnolo, B., and Zangara, R. (1997). Experimental study of a nonlinear system in the presence of noise: the stochastic resonance, *American Journal of Physics*, **65**, 341–349.

Lathi, B. P. (1998). *Modern Digital and Analog Communication Systems*, 3rd edn, Oxford: Oxford University Press.

Laughlin, S. B. (2001). Energy as a constraint on the coding and processing of sensory information, *Current Optinion in Neurobiology*, **11**, 475–480.

Lecar, H. and Nossal, R. (1971a). Theory of threshold fluctuations in nerves. I: Relationships between electrical noise and fluctuations in axon firing, *Biophysical Journal*, **11**, 1048–1067.

Lecar, H. and Nossal, R. (1971b). Theory of threshold fluctuations in nerves. II: Analysis of various sources of membrane noise, *Biophysical Journal*, **11**, 1068–1084.

Lee, I., Liu, X., Zhou, C., and Kosko, B. (2006). Noise-enhanced detection of subthreshold signals with carbon nanotubes, *IEEE Transactions on Nanotechnology*, **5**, 613–627.

Lee, I. Y., Liu, X. L., Kosko, B., and Zhou, C. W. (2003). Nanosignal processing: Stochastic resonance in carbon nanotubes that detect subthreshold signals, *Nano Letters*, **3**, 1683–1686.

Lefevre, G. R., Kowalski, S. E., Girling, L. G., Thiessen, D. B., and Mutch, W. A. C. (1996). Improved arterial oxygenation after oleic acid lung injury in the pig using a computer-controlled mechanical ventilator, *American Journal of Respiratory and Critical Care Medicine*, **154**, 1567–1572.

Lehmann, E. L. and Casella, G. (1998). *Theory of Point Estimation*, New York: Springer.

Leonard, D. S. and Reichl, L. E. (1994). Stochastic resonance in a chemical-reaction, *Physical Review E*, **49**, 1734–1737.

Levin, J. E. and Miller, J. P. (1996). Broadband neural encoding in the cricket cercal sensory system enhanced by stochastic resonance, *Nature*, **380**, 165–168.

Levy, W. B. and Baxter, R. A. (2002). Energy-efficient neuronal computation via quantal synaptic failures, *The Journal of Neuroscience*, **22**, 4746–4755.

Lewis, E. R., Henry, K. R., and Yamada, W. M. (2000). Essential roles of noise in neural coding and in studies of neural coding, *Biosystems*, **58**, 109–115.

Liberman, M. C. and Dodds, L. W. (1984). Single-neuron labeling and chronic cochlear pathology. II: Steriocilia damage and alterations of spontaneous discharge rates, *Hearing Research*, **16**, 43–53.

Lim, M. and Saloma, C. (2001). Noise-enhanced measurement of weak doublet spectra with a Fourier-transform spectrometer and a 1-bit analog-to-digital converter, *Applied Optics*, **40**, 1767–1775.

Linder, T. and Zamir, R. (1999). High-resolution source coding for non-difference distortion measures: the rate-distortion function, *IEEE Transactions on Information Theory*, **45**, 533–547.

Lindner, J. F., Bennett, M., and Wiesenfeld, K. (2005). Stochastic resonance in the mechanoelectrical transduction of hair cells, *Physical Review E*, **72**, Art. No. 051110.

Lindner, J. F., Meadows, B. K., Ditto, W. L., Inchiosa, M. E., and Bulsara, A. R. (1995). Array enhanced stochastic resonance and spatiotemporal synchronization, *Physical Review Letters*, **75**, 3–6.

Lindner, J. F., Meadows, B. K., Ditto, W. L., Inchiosa, M. E., and Bulsara, A. R. (1996). Scaling laws for spatiotemporal synchronization and array enhanced stochastic resonance, *Physical Review E*, **53**, 2081–2086.

Liu, F., Yu, Y., and Wang, W. (2001). Signal-to-noise ratio gain in neuronal systems, *Physical Review E*, **63**, Art. No. 051912.

Lloyd, S. P. (1982). Least squares quantization in PCM, *IEEE Transactions on Information Theory*, **IT-28**, 129–137.

Löcher, M., Inchiosa, M. E., Neff, J., Bulsara, A., Wiesenfeld, K., Gammaitoni, L., Hänggi, P., and Ditto, W. (2000). Theory of controlling stochastic resonance, *Physical Review E*, **62**, 317–327.

Löcher, M., Johnson, G. A., and Hunt, E. R. (1996). Spatiotemporal stochastic resonance in a system of coupled diode resonators, *Physical Review Letters*, **77**, 4698–4701.

Loeb, G. E. (1985). The functional replacement of the ear, *Scientific American*, **252**, 86–92.

Loerincz, K., Gingl, Z., and Kiss, L. B. (1996). A stochastic resonator is able to greatly improve signal-to-noise ratio, *Physics Letters A*, **224**, 63–67.

Lofstedt, R. and Coppersmith, S. N. (1994). Quantum stochastic resonance, *Physical Review Letters*, **72**, 1947–1950.

Loizou, P. C. (1998). Mimicking the human ear, *IEEE Signal Processing Magazine*, **15**, 101–130.

Loizou, P. C. (1999). Signal processing techniques for cochlear implants – a review of progress in deriving electrical stimuli for the speech signal, *IEEE Engineering in Medicine and Biology Magazine*, **18**, 34–46.

Longtin, A. (1993). Stochastic resonance in neuron models, *Journal of Statistical Physics*, **70**, 309–327.

Longtin, A., Bulsara, A., and Moss, F. (1991). Time-interval sequences in bistable systems and the noise-induced transmission of information by sensory neurons, *Physical Review Letters*, **67**, 656–659.

Longtin, A., Bulsara, A., Pierson, D., and Moss, F. (1994). Bistability and the dynamics of periodically forced sensory neurons, *Biological Cybernetics*, **70**, 569–578.

Luchinsky, D. G., Mannella, R., McClintock, P. V. E., and Stocks, N. G. (1999a). Stochastic resonance in electrical circuits – I: Conventional stochastic resonance, *IEEE Transactions on Circuits and Systems-II: Analog and Digital Signal Processing*, **46**, 1205–1214.

Luchinsky, D. G., Mannella, R., McClintock, P. V. E., and Stocks, N. G. (1999b). Stochastic resonance in electrical circuits – II: Nonconventional stochastic resonance, *IEEE Transactions on Circuits and Systems-II: Analog and Digital Signal Processing*, **46**, 1215–1224.

Luchinsky, D. G., McClintock, P. V. E., and Dykman, M. I. (1998). Analogue studies of nonlinear systems, *Reports on Progress in Physics*, **61**, 889–997.

Maddox, J. (1991). Towards the brain-computer's code? *Nature*, **352**, 469.

Makra, P., Gingl, Z., and Kish, L. B. (2002). Signal-to-noise ratio gain in non-dynamical and dynamical bistable stochastic resonators, *Fluctuation and Noise Letters*, **2**, L147–L155.

Mao, X. M., Sun, K., and Ouyang, Q. (2002). Stochastic resonance in a financial model, *Chinese Physics*, **11**, 1106–1110.

Marchesoni, F., Gammaitoni, L., and Bulsara, A. R. (1996). Spatiotemporal stochastic resonance in a phi(4) model of kink antikink nucleation, *Physical Review Letters*, **76**, 2609–2612.

Marchesoni, F., Gammaitoni, L., Apostolico, F., and Santucci, S. (2000). Numerical verification of bona fide stochastic resonance, *Physical Review E*, **62**, 146–149.

Martinez, K., Hart, J. K., and Ong, R. (2004). Sensor network applications, *Computer*, **37**, 50–56.

Martorell, F., McDonnell, M. D., Abbott, D., and Rubio, A. (2004). Generalized noise resonance: using noise for signal enhancement, in D. Abbott, S. M. Bezrukov, A. Der and A. Sánchez (eds.), *Proc. SPIE Fluctuations and Noise in Biological, Biophysical, and Biomedical Systems II*, Vol. 5467, 163–174.

Martorell, F., McDonnell, M. D., Rubio, A., and Abbott, D. (2005). Using noise to break the noise barrier in circuits, in S. F. Al-Sarawi (ed.), *Proc. SPIE Smart Structures, Devices, and Systems II*, Vol. 5649, 53–66.

Matsumoto, K. and Tsuda, I. (1985). Information theoretical approach to noisy dynamics, *Journal of Physics A: Mathematical and General*, **18**, 3561–3566.

Matsuoka, A. J., Abbas, P. J., Rubinstein, J. T., and Miller, C. A. (2000a). The neuronal response to electrical constant-amplitude pulse train stimulation: additive Gaussian noise, *Hearing Research*, **149**, 129–137.

Matsuoka, A. J., Abbas, P. J., Rubinstein, J. T., and Miller, C. A. (2000b). The neuronal response to electrical constant-amplitude pulse train stimulation: evoked compound action potential recordings, *Hearing Research*, **149**, 115–128.

Max, J. (1960). Quantizing for minimum distortion, *IRE Transactions on Information Theory*, **IT-6**, 7–12.

McCulloch, W. S. and Pitts, W. (1943). A logical calculus of ideas immanent in nervous activity, *Bulletin of Mathematical Biophysics*, **5**, 115–133.

McDonnell, M. D., Abbott, D., and Pearce, C. (2003a). Neural mechanisms for analog to digital conversion, in D. V. Nicolau (ed.), *Proc. SPIE BioMEMS and Nanoengineering*, Vol. 5275, 278–286.

McDonnell, M. D., Abbott, D., and Pearce, C. E. M. (2002a). An analysis of noise enhanced information transmission in an array of comparators, *Microelectronics Journal*, **33**, 1079–1089.

McDonnell, M. D., Abbott, D., and Pearce, C. E. M. (2002b). A characterization of suprathreshold stochastic resonance in an array of comparators by correlation coefficient, *Fluctuation and Noise Letters*, **2**, L205–L220.

McDonnell, M. D., Amblard, P. O., Stocks, N. G., Zozor, S., and Abbott, D. (2006a). High resolution optimal quantization for stochastic pooling networks, in A. Bender (ed.), *Proc. SPIE Complexity and Nonlinear Dynamics*, Vol. 6417, 641706.

McDonnell, M. D. and Abbott, D. (2002). Open questions for suprathreshold stochastic resonance in sensory neural models for motion detection using artificial insect vision, in S. M. Bezrukov (ed.), *UPoN 2002: Third International Conference on*

Unsolved Problems of Noise and Fluctuations in Physics, Biology, and High Technology, Vol. 665, American Institute of Physics, 51–58.

McDonnell, M. D. and Abbott, D. (2004a). Optimal quantization in neural coding, *Proc. IEEE International Symposium on Information Theory*, Chicago, 496.

McDonnell, M. D. and Abbott, D. (2004b). Signal reconstruction via noise through a system of parallel threshold nonlinearities, *Proc. 2004 IEEE International Conference on Acoustics, Speech, and Signal Processing*, Vol. 2, Montreal, Canada, 809–812.

McDonnell, M. D., Pearce, C. E. M., and Abbott, D. (2001). Neural information transfer in a noisy environment, in N. W. Bergmann (ed.), *Proc. SPIE Electronics and Structures for MEMS II*, Vol. 4591, 59–69.

McDonnell, M. D., Sethuraman, S., Kish, L. B., and Abbott, D. (2004a). Cross-spectral measurement of neural signal transfer, in Z. Gingl, J. M. Sancho, L. Schimansky-Geier, and J. Kertesz (eds.), *Proc. SPIE Noise in Complex Systems and Stochastic Dynamics II*, Vol. 5471, 550–559.

McDonnell, M. D., Stocks, N. G., and Abbott, D. (2007). Optimal stimulus and noise distributions for information transmission via suprathreshold stochastic resonance, *Physical Review E*, **75**, Art. No. 061105.

McDonnell, M. D., Stocks, N. G., Pearce, C. E. M., and Abbott, D. (2002c). Maximising information transfer through nonlinear noisy devices, in D. V. Nicolau (ed.), *Proc. SPIE Biomedical Applications of Micro and Nanoengineering*, Vol. 4937, 254–263.

McDonnell, M. D., Stocks, N. G., Pearce, C. E. M., and Abbott, D. (2003b). The data processing inequality and stochastic resonance, in L. Schimansky-Geier, D. Abbott, A. Neiman, and C. Van den Broeck (eds.), *Proc. SPIE Noise in Complex Systems and Stochastic Dynamics*, Vol. 5114, 249–260.

McDonnell, M. D., Stocks, N. G., Pearce, C. E. M., and Abbott, D. (2003c). Stochastic resonance and the data processing inequality, *Electronics Letters*, **39**, 1287–1288.

McDonnell, M. D., Stocks, N. G., Pearce, C. E. M., and Abbott, D. (2004b). Optimal quantization for energy-efficient information transfer in a population of neuron-like devices, in Z. Gingl, J. M. Sancho, L. Schimansky-Geier, and J. Kertesz (eds.), *Proc. SPIE Noise in Complex Systems and Stochastic Dynamics II*, Vol. 5471, 222–232.

McDonnell, M. D., Stocks, N. G., Pearce, C. E. M., and Abbott, D. (2005a). Analog to digital conversion using suprathreshold stochastic resonance, in S. F. Al-Sarawi (ed.), *Proc. SPIE Smart Structures, Devices, and Systems II*, Vol. 5649, 75–84.

McDonnell, M. D., Stocks, N. G., Pearce, C. E. M., and Abbott, D. (2005b). How to use noise to reduce complexity in quantization, in A. Bender (ed.), *Proc. SPIE Complex Systems*, Vol. 6039, 115–126.

McDonnell, M. D., Stocks, N. G., Pearce, C. E. M., and Abbott, D. (2005c). Optimal quantization and suprathreshold stochastic resonance, in N. G. Stocks, D. Abbott and R. P. Morse (eds.), *Proc. SPIE Noise in Biological, Biophysical, and Biomedical Systems III*, Vol. 5841, 164–173.

McDonnell, M. D., Stocks, N. G., Pearce, C. E. M., and Abbott, D. (2005d). Quantization in the presence of large amplitude threshold noise, *Fluctuation and Noise Letters*, **5**, L457–L468.

McDonnell, M. D., Stocks, N. G., Pearce, C. E. M., and Abbott, D. (2006b). Optimal information transmission in nonlinear arrays through suprathreshold stochastic resonance, *Physics Letters A*, **352**, 183–189.

McNamara, B. and Wiesenfeld, K. (1989). Theory of stochastic resonance, *Physical Review A*, **39**, 4854–4869.

McNamara, B., Wiesenfeld, K., and Roy, R. (1988). Observation of stochastic resonance in a ring laser, *Physical Review Letters*, **60**, 2626–2629.

Meddis, R. (1986). Simulation of mechanical to neural transduction in the auditory receptor, *Journal of Acoustical Society of America*, **79**, 702–711.

Meddis, R., O'Mard, L. P., and Lopez-Paveda, E. A. (2001). A computational algorithm for computing nonlinear auditory frequency selectivity, *Journal of Acoustical Society of America*, **109**, 2852–2861.

Melnikov, V. I. (1993). Schmitt trigger: a solvable model of stochastic resonance, *Physical Review E*, **48**, 2481–2489.

Merzenich, M. M., Schindler, D. N., and White, M. W. (1974). Feasibility of multichannel scala tympani stimulation, *Laryngoscope*, **84**, 1887–1893.

Michelson, R. P. (1971). Electrical stimulation of the human cochlea, *Archives of Otolaryngology*, **93**, 317–323.

Mickan, S., Abbott, D., Munch, J., Zhang, X.-C., and van Doorn, T. (2000). Analysis of system trade-offs for terahertz imaging, *Microelectronics Journal*, **31**, 503–514.

Mickan, S. P. and Zhang, X. C. (2003). T-ray sensing and imaging, *International Journal of High Speed Electronics and Systems*, **13**, 606–676.

Mingesz, R., Makra, P., and Gingl, Z. (2005). Cross-spectral analysis of signal improvement by stochastic resonance in bistable systems, in L. B. Kish, K. Lindenberg, and Z. Gingl (eds.), *Proc. SPIE Noise in Complex Systems and Stochastic Dynamics III*, Vol. 5845, 283–292.

Mitaim, S. and Kosko, B. (1998). Adaptive stochastic resonance, *Proceedings of the IEEE*, **86**, 2152–2183.

Mitaim, S. and Kosko, B. (2004). Adaptive stochastic resonance in noisy neurons based on mutual information, *IEEE Transactions on Neural Networks*, **15**, 1526–1540.

Moore, B. C. J. (2003). Coding of sounds in the auditory system and its relevance to signal processing and coding in cochlear implants, *Otology and Neurotology*, **24**, 243–254.

Morse, R. P., Allingham, D., and Stocks, N. G. (2002). An information-theoretic approach to cochlear implant coding, in S. M. Bezrukov (ed.), *UPoN 2002: Third International Conference on Unsolved Problems of Noise and Fluctuations in Physics, Biology, and High Technology*, Vol. 665, American Institute of Physics, 125–132.

Morse, R. P. and Evans, E. F. (1996). Enhancement of vowel coding for cochlear implants by addition of noise, *Nature Medicine*, **2**, 928–932.

Morse, R. P. and Evans, E. F. (1999a). Additive noise can enhance temporal coding in a computational model of analogue cochlear implant stimulation, *Hearing Research*, **133**, 107–119.

Morse, R. P. and Evans, E. F. (1999b). Preferential and non-preferential transmission of formant information by an analogue cochlear implant using noise: the role of the nerve threshold, *Hearing Research*, **133**, 120–132.

Morse, R. P. and Meyer, G. F. (2000). The practical use of noise to improve speech coding by analogue cochlear implants, *Chaos, Solitons and Fractals*, **11**, 1885–1894.

Morse, R. P. and Roper, P. (2000). Enhanced coding in a cochlear-implant model using additive noise: aperiodic stochastic resonance with tuning, *Physical Review E*, **61**, 5683–5692.

Morse, R. P. and Stocks, N. G. (2005). Enhanced cochlear implant coding using multiplicative noise, in N. G. Stocks, D. Abbott, and R. P. Morse (eds.), *Proc. SPIE Fluctuations and Noise in Biological, Biophysical and Biomedical Systems III*, Vol. 5841, 23–30.

Morse, R. P., Morse, P. F., Nunn, T. B., Archer, K. A. M., and Boyle, P. (2007). The effect of Gaussian noise on the threshold, dynamic range, and loudness of analogue cochlear implant stimuli, *Journal of the Association for Research in Otolaryngology*, **8**, 42–53.

Moss, F. and Milton, J. G. (2003). Balancing the unbalanced, *Nature*, **425**, 911–912.

Moss, F. and Pei, X. (1995). Neurons in parallel, *Nature*, **376**, 211–212.

Moss, F. and Wiesenfeld, K. (1995). The benefits of background noise, *Scientific American*, **273**, 50–53.

Moss, F., Bulsara, A., and Shlesinger, M. F. (eds.) (1992). Proc. NATO Advanced Research Workshop on Stochastic Resonance in Physics and Biology, San Diego, USA, Published in *Journal of Statistical Physics*, **70**, 1–512 (1993).

Moss, F., Douglass, J. K., Wilkens, L., Pierson, D., and Pantazelou, E. (1993). Stochastic resonance in an electronic FitzHugh-Nagumo model, in J. R. Buchler and H. E. Kandrup (eds.), *Stochastic Processes in Astrophysics*, Vol. 706, Annals of the New York Academy of Sciences, 26–41.

Moss, F., Pierson, D., and Ogorman, D. (1994). Stochastic resonance – tutorial and update, *International Journal of Bifurcation and Chaos*, **4**, 1383–1397.

Moss, F., Ward, L. M., and Sannita, W. G. (2004). Stochastic resonance and sensory information processing: a tutorial and review of application, *Clinical Neurophysiology*, **115**, 267–281.

Nadal, J. and Parga, N. (1994). Nonlinear neurons in the low noise limit: a factorial code maximizes information-transfer, *Network: Computation in Neural Systems*, **5**, 565–581.

Nadol, Jr., J. B. (1981). Histopathology of human aminoglycoside ototoxicity, in S. A. Lerner, G. J. Matz, and J. E. Hawkins (eds.), *Aminoglycoside Ototoxicity*, Little and Brown and Co., 409–434.

Nadol Jr., J. B. (1988). Comparative anatomy of the cochlea and auditory nerve in mammals, *Hearing Research*, **34**, 253–266.

Narayanan, R. M. and Kumru, C. (2005). Implementation of fully polarimetric random noise radar, *IEEE Antennas and Wireless Propagation Letters*, **4**, 125–128.

Neiman, A. (1994). Synchronizationlike phenomena in coupled stochastic bistable systems, *Physical Review E*, **49**, 3484–3487.

Neiman, A. and Schimansky-Geier, L. (1994). Stochastic resonance in bistable systems driven by harmonic noise, *Physical Review Letters*, **72**, 2988–2991.

Neiman, A., Schimansky-Geier, L., and Moss, F. (1997). Linear response theory applied to stochastic resonance in models of ensembles of oscillators, *Physical Review E*, **56**, R9–R12.

Neiman, A., Shulgin, D., Anishchenko, V., Ebeling, W., Schimansky-Geier, L., and Freund, J. (1996). Dynamical entropies applied to stochastic resonance, *Physical Review Letters*, **76**, 4299–4302.

Neiman, A., Silchenko, A., Anishchenko, V., and Schimansky-Geier, L. (1998). Stochastic resonance: noise-enhanced phase coherence, *Physical Review E*, **58**, 7118–7125. Part A.

Nguyen, T. (2007). Robust data-optimized stochastic analog-to-digital converters, *IEEE Transactions on Signal Processing*, **55**, 2735–2740.

Nicolis, C. (1981). Solar variability and stochastic effects on climate, *Solar Physics*, **74**, 473–478.

Nicolis, C. (1982). Stochastic aspects of climatic transitions – response to a periodic forcing, *Tellus*, **34**, 1–9.

Nikitin, A. and Stocks, N. G. (2004). The application of Gaussian channel theory to the estimation of information transmission rates in neural systems, in D. Abbott, S. M. Bezrukov, A. Der, and A. Sánchez (eds.), *Proc. SPIE Fluctuations and Noise in Biological, Biophysical, and Biomedical Systems II*, Vol. 5467, Maspalomas, Spain, 202–211.

Nikitin, A., Stocks, N. G., and Morse, R. P. (2005). Enhanced information transmission mediated by multiplicative noise, in N. G. Stocks, D. Abbott, and R. P. Morse (eds.), *Proc. SPIE Fluctuations and Noise in Biological, Biophysical and Biomedical Systems III*, Vol. 5841, 31–39.

Nikitin, A., Stocks, N. G., and Morse, R. P. (2007). Enhanced information transmission with signal-dependent noise in an array of nonlinear elements, *Physical Review E*, **75**, Art. No. 021121.

Nocedal, J. and Wright, S. J. (1999). *Numerical Optimization*, New York: Springer-Verlag.

Noest, A. J. (1995). Tuning stochastic resonance, *Nature*, **378**, 341.

Øksendal, B. (1998). *Stochastic Differential Equations: An Introduction with Applications*, fifth edn, Berlin: Springer.

Oliaei, O. (2003). Stochastic resonance in sigma-delta modulators, *Electronics Letters*, **39**, 173–174.

Paradiso, J. A. and Starner, T. (2005). Energy scavenging for mobile and wireless electronics, *IEEE Pervasive Computing*, **4**, 18–27.

Patel, A. and Kosko, B. (2005). Stochastic resonance in noisy spiking retinal and sensory neuron models, *Neural Networks*, **18**, 467–478.

Paulsson, J. and Ehrenberg, M. (2000). Random signal fluctuations can reduce random fluctuations in regulated components of chemical regulatory networks, *Physical Review Letters*, **84**, 5447–5450.

Paulsson, J., Berg, O. G., and Ehrenberg, M. (2000). Stochastic focusing: fluctuation-enhanced sensitivity of intracellular regulation, *Proceedings of the National Academy of Sciences of the USA*, **97**, 7148–7153.

Pei, X., Wilkens, L., and Moss, F. (1996). Noise mediated spike timing precision from aperiodic stimuli in an array of Hodgkin–Huxley-type neurons, *Physical Review Letters*, **77**, 4679–4682.

Petracchi, D. (2000). What is the role of stochastic resonance?, *Chaos, Solitons and Fractals*, **11**, 1827–1834.

Petracchi, D., Gebeshuber, I. C., DeFelice, L. J., and Holden, A. V. (2000). Introduction: stochastic resonance in biological systems, *Chaos, Solitons and Fractals*, **11**, 1819–1822.

Pinsker, M. S. (1964). *Information and Information Stability of Random Variables and Processes*, San Francisco, CA: Holden-Day.

Plaskota, L. (1996a). How to benefit from noise, *Journal of Complexity*, **12**, 175–184.

Plaskota, L. (1996b). Worst case complexity of problems with random information noise, *Journal of Complexity*, **12**, 416–439.

Pohlmann, K. C. (2005). *Principles of Digital Audio*, fifth edn, McGraw Hill.

Pradhan, S. S. and Ramchandran, K. (2005). Generalized coset codes for distributed binning, *IEEE Transactions on Information Theory*, **51**, 3457–3474.

Pradhan, S. S., Kusuma, J., and Ramchandran, K. (2002). Distributed compression in a dense microsensor network, *IEEE Signal Processing Magazine*, **19**, 51–60.

Press, W. H., Teukolsky, S. A., Vetterling, W. T., and Flannery, B. P. (1992). *Numerical Recipes in C: The Art of Scientific Computing*, second edn, Cambridge: Cambridge University Press.

Price, R. (1958). A useful theorem for nonlinear devices having Gaussian inputs, *IRE Transactions on Information Theory*, **IT-4**, 69–72.

Priplata, A. A., Niemi, J. B., Harry, J. D., Lipsitz, L. A., and Collins, J. J. (2003). Vibrating insoles and balance control in elderly people, *Lancet*, **362**, 1123–1124.

Priplata, A., Niemi, J., Salen, M., Harry, J., Lipsitz, L. A., and Collins, J. J. (2004). Noise-enhanced human balance control, *Physical Review Letters*, **89**, Art. No. 238101.

Proakis, J. G. and Manolakis, D. G. (1996). *Digital Signal Processing: Principles, Algorithms, and Applications*, 3rd edn, Prentice Hall.

Proakis, J. G. and Salehi, M. (1994). *Communication Systems Engineering*, Prentice Hall.

Prudnikov, A. P., Brychkov, Y. A., and Marichev, O. I. (1986). *Integrals and Series, Vol. 2: Special Functions*, Gordon and Breach Science Publishers.

Rattay, F., Gebeshuber, I. C. and Gitter, A. H. (1998). The mammalian auditory hair cell: a simple electric circuit model, *Journal of Acoustical Society of America*, **103**, 1558–1565.

Remley, W. R. (1966). Some effects of clipping in array processing, *The Journal of the Acoustical Society of America*, **39**, 702–707.

Rényi, A. (1970). *Probability Theory*, Amsterdam: North-Holland Publishing Company.

Rice, S. O. (1944). Mathematical analysis of random noise, *The Bell System Technical Journal*, **23**, 282–332.

Rice, S. O. (1945). Mathematical analysis of random noise, *The Bell System Technical Journal*, **24**, 46–156.

Rice, S. O. (1948). Statistical properties of a sine wave plus random noise, *The Bell System Technical Journal*, **27**, 109–157.

Rieke, F., Warland, D., de Ruyter van Steveninck, R., and Bialek, W. (1997). *Spikes: Exploring the Neural Code*, Cambridge, MA: MIT Press.

Rissanen, J. J. (1996). Fisher information and stochastic complexity, *IEEE Transactions on Information Theory*, **42**, 40–47.

Roberts, L. G. (1962). Picture coding using pseudo-random noise, *IRE Transactions on Information Theory*, **IT-8**, 145–154.

Robinson, J. W. C., Asraf, D. E., Bulsara, A. R., and Inchiosa, M. E. (1998). Information-theoretic distance measures and a generalization of stochastic resonance, *Physical Review Letters*, **81**, 2850–2853.

Robinson, J. W. C., Rung, J., Bulsara, A. R., and Inchiosa, M. E. (2001). General measures for signal-noise separation in nonlinear dynamical systems, *Physical Review E*, **63**, Art. No. 011107.

Robles, L. and Roggero, M. A. (2001). Mechanics of the mammalian cochlear, *Physiological Reviews*, **81**, 1305–1352.

Rose, K. (1994). A mapping approach to rate-distortion computation and analysis, *IEEE Transactions on Information Theory*, **40**, 1939–1952.

Rose, K. (1998). Deterministic annealing for clustering, compression, classification, regression and related optimization problems, *Proceedings of the IEEE*, **86**, 2210–2239.

Rose, K., Gurewitz, E., and Fox, G. C. (1990). Statistical mechanics and phase transitions in clustering, *Physical Review Letters*, **65**, 945–948.

Rose, K., Gurewitz, E., and Fox, G. C. (1992). Vector quantization by deterministic annealing, *IEEE Transactions on Information Theory*, **38**, 1249–1257.

Rosen, S. (1996). Cochlear implants, in D. Stephens (ed.), *Scott-Brown's Otolarangology*, Vol. 2, Butterworth Heinemann.

Rousseau, D. and Chapeau-Blondeau, F. (2004). Suprathreshold stochastic resonance and signal-to-noise ratio improvement in arrays of comparators, *Physics Letters A*, **321**, 280–290.

Rousseau, D. and Chapeau-Blondeau, F. (2005). Constructive role of noise in signal detection from parallel arrays of quantizers, *Signal Processing*, **85**, 571–580.

Rousseau, D., Duan, F., and Chapeau-Blondeau, F. (2003). Suprathreshold stochastic resonance and noise-enhanced Fisher information in arrays of threshold devices, *Physical Review E*, **68**, Art. No. 031107.

Rozenfeld, R. and Schimansky-Geier, L. (2000). Array-enhanced stochastic resonance in finite systems, *Chaos, Solitons and Fractals*, **11**, 1937–1944.

Rubinstein, J. T. and Hong, R. (2003). Signal coding in cochlear implants: exploiting stochastic effects of electrical stimulation, *Annals of Otology Rhinology and Laryngology*, **112**, 14–19.

Rubinstein, J. T., Wilson, B. S., Finley, C. C., and Abbas, P. J. (1999). Pseudospontaneous activity: stochastic independence of auditory nerve fibers with electrical stimulation, *Hearing Research*, **127**, 108–118.

Rudnick, P. (1960). Small signal detection in the DIMUS array, *The Journal of the Acoustical Society of America*, **32**, 871–877.

Russell, D. F., Wilkens, L. A., and Moss, F. (1999). Use of behavioural stochastic resonance by paddle fish for feeding, *Nature*, **402**, 291–294.

Sato, A., Ueda, M., and Munakata, T. (2004). Signal estimation and threshold optimization using an array of bithreshold elements, *Physical Review E*, **70**, Art. No. 021106.

Schreiber, S., Machens, C. K., Herz, A. V., and Laughlin, S. B. (2002). Energy-efficient coding with discrete stochastic events, *Neural Computation*, **14**, 1323–1346.

Schuchman, L. (1964). Dither signals and their effect on quantization noise, *IEEE Transactions on Communications*, **COM-12**, 162–165.

Sellick, P. M., Patuzzi, R., and Johnstone, B. M. (1982). Measurement of basilar-membrane motion in the guinea-pig using the Mossbauer technique, *Journal of the Acoustical Society of America*, **72**, 131–141.

Shannon, C. E. (1948). A mathematical theory of communication, *The Bell System Technical Journal*, **27**, 379–423, 623–656.

Shatokhin, V., Wellens, T., and Buchleitner, A. (2004). The noise makes the signal: What a small fry should know about stochastic resonance, *Journal of Modern Optics*, **51**, 851–860.

Shepherd, R. K., Hatsushika, S., and Clark, G. M. (1993). Electrical stimulation of the auditory nerve: the effect of electrode position on neural excitation, *Hearing Research*, **66**, 108–120.

Shiryaev, A. N. (1996). *Probability*, second edn, New York: Springer.

Simmons, F. B. and Glattke, T. J. (1972). Comparison of electrical and acoustical stimulation of the cat ear, *The Annals of Otology, Rhinology, and Laryngology*, **81**, 731–737.

Slaney, M. (1993). An efficient implementation of the patterson-holdsworth filter bank, *Technical Report Apple Technical Report 35*, Apple.

Spiegel, M. R. and Liu, J. (1999). *Mathematical Handbook of Formulas and Tables*, McGraw-Hill.

Stein, R. B., French, A. S., and Holden, A. V. (1972). The frequency response, coherence, and information capacity of two neuronal models, *Biophysical Journal*, **12**, 295–322.

Stemmler, M. (1996). A single spike suffices: the simplest form of stochastic resonance in model neurons, *Network: Computation in Neural Systems*, **7**, 687–716.

Stocks, N. G. (1995). A theoretical study of the nonlinear response of a periodically driven bistable system, *Nuovo Cimento D*, **17**, 925–940.

Stocks, N. G. (2000a). Optimising information transmission in model neuronal ensembles: the role of internal noise, in J. A. Freund and T. Poschel (eds.), *Future Directions for Intelligent Systems and Information Sciences*, Springer-Verlag, 150–159.

Stocks, N. G. (2000b). Suprathreshold stochastic resonance, in D. S. Broomhead, E. A. Luchinskaya, P. V. E. McClintock, and T. Mullin (eds.), *Stochastic and Chaotic Dynamics in the Lakes (STOCHAOS)*, Vol. 502, American Institute of Physics, 415–421.

Stocks, N. G. (2000c). Suprathreshold stochastic resonance in multilevel threshold systems, *Physical Review Letters*, **84**, 2310–2313.

Stocks, N. G. (2001a). Information transmission in parallel threshold arrays: suprathreshold stochastic resonance, *Physical Review E*, **63**, Art. No. 041114.

Stocks, N. G. (2001b). Information transmission in parallel threshold networks: suprathreshold stochastic resonance and coding efficieny, in G. Bosman (ed.), *Noise in Physical Systems and 1/f Fluctuations, ICNF 2001*, 594–597.

Stocks, N. G. (2001c). Suprathreshold stochastic resonance: an exact result for uniformly distributed signal and noise, *Physics Letters A*, **279**, 308–312.

Stocks, N. G., Allingham, D., and Morse, R. P. (2002). The application of suprathreshold stochastic resonance to cochlear implant coding, *Fluctuation and Noise Letters*, **2**, L169–L181.

Stocks, N. G. and Mannella, R. (2000). Suprathreshold stochastic resonance in a neuronal network model: a possible strategy for sensory coding, in N. Kosabov (ed.), *Stochastic Processes in Physics, Chemistry and Biology*, Physica-Verlag, 236–247.

Stocks, N. G. and Mannella, R. (2001). Generic noise enhanced coding in neuronal arrays, *Physical Review E*, **64**, Art. No. 030902(R).

Stocks, N. G., Stein, N. D., and McClintock, P. V. E. (1993). Stochastic resonance in monostable systems, *Journal of Physics A-Mathematical and General*, **26**, L385–L390.

Suki, B., Alencar, A. M., Sujeer, M. K., Lutchen, K. R., Collins, J. J., Andrade Jr., J. S., Ingenito, E. P., Zapperi, S., and Stanley, H. E. (1998). Life-support system benefits from noise, *Nature*, **393**, 127–128.

Sulcs, S., Oppy, G., and Gilbert, B. C. (2000). Effective linearization by noise addition in threshold detection and implications for stochastic optics, *Journal of Physics A: Mathematical and General*, **33**, 3997–4007.

Sumner, C. J., Lopez-Poveda, E. A., O'Mard, L. P., and Meddis, R. (2002). A revised model of the inner-hair cell and auditory-nerve complex, *Journal of Acoustical Society of America*, **111**, 2178–2188.

Tessone, C. J., Mirasso, C. R., Toral, R., and Gunton, J. D. (2006). Diversity-induced resonance, *Physical Review Letters*, **97**, Art. No. 194101.

Tishby, N., Pereira, F. C., and Bialek, W. (1999). The information bottleneck method, in B. Hajek and R. S. Sreenivas (eds.), *Proc. 37th Annual Allerton Conference on Communication, Control and Computing*, University of Illinois, 368–377.

Toral, R., Mirasso, C. R., Hernandez-Garcia, E., and Piro, O. (1999). Synchronization of chaotic systems by common random forcing, in D. Abbott and L. B. Kish (eds.), *UPoN 1999: Second International Conference on Unsolved Problems of Noise and Fluctuations*, Vol. 511, American Institute of Physics, 255–260.

Tsitsiklis, J. N. (1993). Extremal properties of likelihood-ratio quantizers, *IEEE Transactions on Communications*, **41**, 550–558.

Tuckwell, H. C. (1988). *Introduction to Theoretical Neurobiology*, Vol. 1, Cambridge University Press.

Tuteur, F. B. and Presley, J. A. (1981). Spectral estimation of space-time signals with a DIMUS array, *The Journal of the Acoustical Society of America*, **70**, 80–89.

Urick, R. J. (1967). *Principles of Underwater Sound for Engineers*, New York: McGraw Hill Book Company.

van den Honert, C. and Stypulkowski, P. H. (1987). Temporal response patterns of single auditory-nerve fibers elicited by periodic electrical simuli, *Hearing Research*, **29**, 207–222.

Van Vleck, J. H. and Middleton, D. (1966). The spectrum of clipped noise, *Proceedings of the IEEE*, **54**, 2–19.

Vaudelle, F., Gazengel, J., Rivoire, G., Godivier, X., and Chapeau-Blondeau, F. (1998). Stochastic resonance and noise-enhanced transmission of spatial signals in optics: the case of scattering, *Journal of the Optical Society of America*, **15**, 2674–2680.

Vemuri, G. and Roy, R. (1989). Stochastic resonance in a bistable ring laser, *Physical Review A*, **39**, 4668–4674.

Vilar, J. M. and Rubí, J. M. (1996). Divergent signal-to-noise ratio and stochastic resonance in monostable systems, *Physical Review Letters*, **77**, 2863–2866.

Vilar, J. M. G. and Rubi, J. M. (1997). Spatiotemporal stochastic resonance in the swift-hohenberg equation, *Physical Review Letters*, **78**, 2886–2889.

von Baeyer, H. C. (2003). *Information: The New Language of Science*, London: Orion Book Ltd.

von Neumann, J. and Morgenstern, O. (1944). *Theory of Games and Economic Behaviour*, Princeton, NJ: Princeton University Press.

Wallace, R., Wallace, D., and Andrews, H. (1997). AIDS, tuberculosis, violent crime, and low birthweight in eight US metropolitan areas: public policy, stochastic resonance, and the regional diffusion of inner-city markers, *Environment and Planning A*, **29**, 525–555.

Wang, H. S. C. (1972). Quantizer functions and their use in the analyses of digital beamformer performance, *The Journal of the Acoustical Society of America*, **53**, 929–945.

Wang, H. S. C. (1976). Postfiltering in digital beamformers, *IEEE Transactions on Aerospace and Electronic Systems*, **AES-12**, 718–727.

Wang, Y. and Wu, L. (2005). Stochastic resonance and noise-enhanced Fisher information, *Fluctuation and Noise Letters*, **5**, L435–L442.

Wannamaker, R. A. (1997). The theory of dithered quantization, Ph.D. thesis, The University of Waterloo, Canada.

Wannamaker, R. A., Lipshitz, S. P., and Vanderkooy, J. (2000a). Stochastic resonance as dithering, *Physical Review E*, **61**, 233–236.

Wannamaker, R. A., Lipshitz, S. P., Vanderkooy, J., and Wright, J. N. (2000b). A theory of non-subtractive dither, *IEEE Transactions on Signal Processing*, **48**, 499–516.

Ward, L. M., Nieman, A., and Moss, F. (2002). Stochastic resonance in psychophysics and in animal behavior, *Biological Cybernetics*, **87**, 91–101.

Warren, D. and Willett, P. (1999). Optimum quantization for detector fusion: some proofs, examples, and pathology, *Journal of the Franklin Insitute – Engineering and Applied Mathematics*, **336**, 323–359.

Wellens, T., Shatokhin, V., and Buchleitner, A. (2004). Stochastic resonance, *Reports on Progress in Physics*, **67**, 45–105.

Wenning, G. (2004). Aspects of noisy neural information processing, Ph.D. thesis, Technische Universität Berlin.

Widrow, B., Kollár, I., and Liu, M. (1996). Statistical theory of quantization, *IEEE Transactions on Instrumentation and Measurement*, **45**, 353–361.

Wiesenfeld, K. (1991). Amplification by globally coupled arrays – coherence and symmetry, *Physical Review A*, **44**, 3543–3551.

Wiesenfeld, K. (1993a). An introduction to stochastic resonance, in J. R. Buchler and H. E. Kandrup (eds.), *Stochastic Processes in Astrophysics*, Vol. 706 of the Annals of the New York Academy of Sciences, The New York Academy of Sciences, 13–25.

Wiesenfeld, K. (1993b). Signals from noise: stochastic resonance pays off, *Physics World*, **6**, 23–24.

Wiesenfeld, K. and Jaramillo, F. (1998). Minireview of stochastic resonance, *Chaos*, **8**, 539–548.

Wiesenfeld, K. and Moss, F. (1995). Stochastic resonance and the benefits of noise: from ice ages to crayfish and SQUIDS, *Nature*, **373**, 33–36.

Wiesenfeld, K., Pierson, D., Pantazelou, E., Dames, C., and Moss, F. (1994). Stochastic resonance on a circle, *Physical Review Letters*, **72**, 2125–2129.

Wilke, S. D. and Eurich, C. W. (2001). Representational accuracy of stochastic neural populations, *Neural Computation*, **14**, 155–189.

Wilson, B., Finley, C., Zerbi, M., and Lawson, D. T. (1994). Speech processors for auditory prostheses, Technical Report 9th Quarterly NIH Report N01-DC-2-2401, Center for Auditory Prosthesis Research, Research Triangle Institute.

Wilson, B. S. (1997). The future of cochlear implants, *British Journal of Audiology*, **31**, 205–225.

Wolff, S. S., Thomas, J. B., and Williams, T. R. (1962). The polarity-coincidence correlator: a nonparametric detection device, *IRE Transactions on Information Theory*, **8**, 5–9.

Xiong, Z., Liveris, A. D., and Cheng, S. (2004). Distributed source coding for sensor networks, *IEEE Signal Processing Magazine*, **21**, 80–94.

Xu, Y. and Collins, L. M. (2003). Predicting the threshold of single-pulse electrical stimuli using a stochastic auditory nerve model: the effects of noise, *IEEE Transactions on Biomedical Engineering*, **50**, 825–835.

Xu, Y. and Collins, L. M. (2004). Predicting the threshold of pulse train electrical stimuli using a stochastic auditory nerve model: the effects of stimulus noise, *IEEE Transactions on Biomedical Engineering*, **51**, 590–603.

Xu, Y. and Collins, L. M. (2005). Predicting dynamic range and intensity discrimination for electrical pulse-train stimuli using a stochastic auditory nerve model: the effects of stimulus noise, *IEEE Transactions on Biomedical Engineering*, **52**, 1040–1049.

Yates, R. D. and Goodman, D. J. (2005). *Probability and Stochastic Processes: A Friendly Introduction for Electrical and Computer Engineers*, second edn, John Wiley & Sons.

Yu, X. and Lewis, E. R. (1989). Studies with spike initiators: linearization by noise allows continuous signal modulation in neural networks, *IEEE Transactions on Biomedical Engineering*, **36**, 36–43.

Zador, P. L. (1982). Asymptotic quantization error of continuous signals and the quantization dimension, *IEEE Transactions on Information Theory*, **IT-28**, 139–149.

Zames, G. and Shneydor, N. A. (1976). Dither in nonlinear systems, *IEEE Transactions on Automatic Control*, **AC-21**, 660–667.

Zamir, R. and Feder, M. (1992). On universal quantization by randomized uniform/lattice quantizers, *IEEE Transactions on Information Theory*, **38**, 428–436.

Zeng, F. G., Fu, Q. J., and Morse, R. (2000). Human hearing enhanced by noise, *Brain Research*, **869**, 251–255.

Zhou, T. and Moss, F. (1990). Analog simulations of stochastic resonance, *Physical Review A*, **41**, 4255–4264.

Zozor, S., Amblard, P. O., and Duchêne, C. (2007). On pooling networks and fluctuation in suboptimal detection framework, *Fluctuation and Noise Letters*, **7**, L39–L60.

Zozor, S. and Amblard, P. O. (1999). Stochastic resonance in discrete time nonlinear AR(1) models, *IEEE Transactions on Signal Processing*, **47**, 108–122.

Zozor, S. and Amblard, P. O. (2001). Erratum: Stochastic resonance in discrete time nonlinear AR(1) models, *IEEE Transactions on Signal Processing*, **49**, 1107–1109.

Zozor, S. and Amblard, P. O. (2002). On the use of stochastic resonance in sine detection, *Signal Processing*, **82**, 353–367.

Zozor, S. and Amblard, P. O. (2005). Noise-aided processing: revisiting dithering in a sigma-delta quantizer, *IEEE Transactions on Signal Processing*, **53**, 3202–3210.

Abbreviations

ADC	analogue-to-digital conversion, page 47
AESR	array enhanced stochastic resonance, page 28
AIB	average information bound, page 221
ASR	aperiodic stochastic resonance, page 17
BCPD	backward conditional probability distribution, page 196
BCV	backwards conditional variance, page 387
BF	best frequencies, page 338
BFGS	Broyden–Fletcher–Goldfarb–Shanno, page 258
BKB	Bamford–Kowal–Bench, page 351
BM	basilar membrane, page 330
CDF	cumulative distribution function, page 67
dB	decibels, page 65
DIMUS	DIgital MUltibeam Steering, page 44
DPI	data processing inequality, page 37
DRNL	Dual Resonance NonLinear, page 334
FWHM	full width at half maximum, page 77
ICDF	inverse cumulative distribution function, page 70
IHC	inner hair cell, page 330
iid	independent and identically distributed, page 66
LCD	level crossing detector, page 27
LHS	left-hand side, page 72
MAP	maximum a posteriori, page 196
MASE	mean asymptotic square error, page 242
MDL	minimum description length, page 166
MGF	moment generating function, page 254
ML	maximum likelihood, page 198
MMSE	minimum mean square error, page 200
MSE	mean square error, page 167

NAS	no analytical solution, page 80
PDF	probability density function, page 48
PMF	probability mass function, page 67
PSD	power spectral density, page 10
RHS	right-hand side, page 71
rms	root mean square, page 53
SAS	simultaneous analogue strategy, page 347
SNR	signal-to-noise ratio, page 3
SQNR	signal-to-quantization-noise ratio, page 51
SR	stochastic resonance, page 1
SSR	suprathreshold stochastic resonance, page 2

Index

Biographies

Mark D. McDonnell was born in Adelaide, Australia. He received a B.Sc. in Mathematical and Computer Sciences (1997), a BE (Hons) in Electrical and Electronic Engineering (1998), and a B.Sc. (Hons) in Applied Mathematics (2001) all from the University of Adelaide, Australia. He received his Ph.D. in Electrical and Electronic Engineering (2006), with Special Commendation, under Derek Abbott and Charles Pearce, also from the University of Adelaide. He was awarded the Postgraduate Alumni University Medal for his thesis. Several times during the course of his Ph.D., he was also a visiting scholar at the University of Warwick, UK, under Nigel Stocks, funded by a D. R. Stranks Fellowship, a Doreen MacCarthy Bursary, and an Australian Research Council (ARC) Academy of Science (AAS) Young Researchers award. In 2003, McDonnell was awarded a Santa Fe Institute Complex Systems Fellowship. In 2007, he won a Fresh Science award, the Gertrude Rohan Prize, and an Australian Research Council (ARC) Postdoctoral Fellowship that he took up at the Institute for Telecommunications Research at the University of South Australia. His current research interests are in the field of nonlinear signal processing, with applications in computational neuroscience and the reliable communication and coding of noisy signals. He is a member of the IEEE and SPIE. Mark lives with his wife, Juliet, a high school English teacher, and two adored cats.

Nigel G. Stocks was born in Bradford, West Yorkshire, UK. He received a B.Sc. in Applied Physics and Electronics (1987) and a Ph.D. in stochastic nonlinear dynamics (1991), under Peter V. E. McClintock, at Lancaster University, UK. His early research work was undertaken in the Lancaster Nonlinear Group and focused on the development of the theory of nonequilibrium dynamical systems and, in particular, on stochastic resonance. Stocks moved to the University of Warwick in 1993 where he joined the Fluid Dynamics Research Centre and undertook studies on transition to turbulence. In 1996 he was awarded a TMR EU Fellowship and worked with Riccardo Mannella at Pisa University before returning to Warwick as a University of Warwick Research Fellow. He was promoted to Senior Lecturer in 2002, Reader in 2005, and full Professor in 2007. Stocks' research interests are in the general area of stochastic nonlinear systems and biomimetics. In particular, his research has focused on neural coding mechanisms for cochlear implants and the development of biomimetic signal processing techniques. Stocks is the discoverer of a form of stochastic resonance – termed suprathreshold stochastic resonance – that promises to improve signal coding in a wide range of potential applications.

Charles E. M. Pearce was born in Wellington, New Zealand, in 1940. He obtained his B.Sc. in 1961 from the University of New Zealand, with a double major in Physics and Mathematical Physics and a further double major in Pure and Applied Mathematics. In 1962 he gained an M.Sc. with honours in Mathematics from Victoria University of Wellington. In 1966 he was awarded a Ph.D. in Mathematical Statistics from the Australian National University, where he was supervised by the late Patrick A. P. Moran. From 1966 to 1968 he lectured at the University of Sheffield. In 1968 he joined the University of Adelaide, where he now holds the Thomas Elder Chair of Mathematics. His main research interests lie in probabilistic and statistical modelling and analysis, convex analysis, and optimization. He has nearly 300 publications, mostly in peer-reviewed academic journals. This number includes about 30 book chapters. In 2001 he was awarded the ANZIAM Medal by the Australian Mathematical Society. This is the premier award of the Society for contributions to Applied and Industrial Mathematics in Australia and New Zealand. In 2007 he was awarded the Ren Potts Medal of the Australian Society for Operations Research for outstanding contributions to Operations Research in Australia. His work is recognized internationally in frequent invitations to speak at major conferences. For the past 16 years he has been Editor-in-Chief of the *ANZIAM Journal*, Australia's principal journal of Applied Mathematics. He is editor of a dozen other journals, including *Nonlinear Functional Analysis and Applications*, the *Journal of Management and Industrial Optimization*, and the *International Journal of Innovative Computing and Control*. He is an Executive Member of the Council of the Australian Mathematical Society and has three times been National President of the Australian Society for Operations Research.

Derek Abbott was born in South Kensington, London, UK, in 1960. He received a B.Sc. (Hons) in Physics (1982) from Loughborough University of Technology (LUT), UK. He received a Ph.D. in Electrical and Electronic Engineering (1995), with commendation, from The University of Adelaide, under Kamran Eshraghian and Bruce R. Davis. From 1978 to 1986 he worked at the GEC Hirst Research Centre, Wembley, UK, and in 1986–1987 he was with Austek Microsystems, Adelaide, Australia. He joined the University of Adelaide in 1987, where he is presently a full professor and Director of the Centre for Biomedical Engineering (CBME). His work has received scientific reportage in *New Scientist*, *The Sciences*, *Scientific American*, *Nature*, *The New York Times*, and *Sciences et Avenir* and he holds over 300 publications/patents. Abbott has been an invited speaker at over 80 institutions around the world, including Princeton, NJ; MIT, MA; Santa Fe Institute, NM; Los Alamos National Laboratories, NM; Cambridge, UK; and EPFL, Lausanne, Switzerland. He won the GEC Bursary (1977), the Stephen Cole the Elder Prize (1998), the E. R. H. Tiekink Memorial Award (2002), a Tall Poppy Award for Science (2004), and the SA Great Award in Science and Technology for outstanding contributions to South Australia (2004). He has served as an editor and/or guest editor for a number of journals including *Chaos* (AIP), *Smart Structures and Materials* (IOP), *Journal of Optics B* (IOP), *IEEE Journal of Solid-State Electronics*, *Microelectronics Journal* (Elsevier), *Proceedings of the IEEE*, and *Fluctuation Noise Letters* (World Scientific). He is a chartered physicist and Fellow of the Institute of Physics (IOP), with honourary life membership. He is also a chartered engineer and Fellow of the Institution of Electrical and Electronic Engineers (IEEE). Abbott is co-founder of two international conference series: *Microelectronics in the New Millennium* and *Fluctuations and Noise*. He is a co-editor of the book *Quantum Aspects of Life*, published by Imperial College Press (ICP).

Scientific Genealogy of the Authors

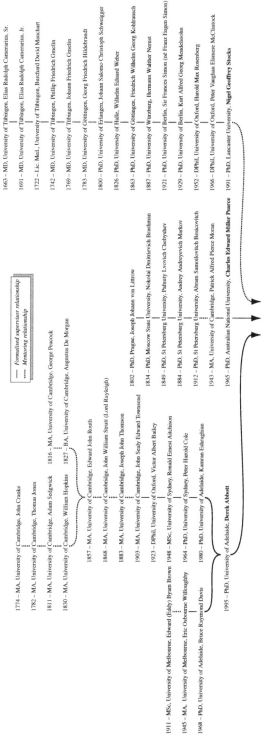